Atomic and Molecular Spectra and Lasers

SECOND EDITION

Atomic and Molecular Spectra and Lasers

SECOND EDITION

AK Saxena MSc, MTech, PhD

Associate Professor
Department of Physics
APS University
Rewa (MP)

CBS

CBS Publishers & Distributors Pvt Ltd

New Delhi • Bengaluru • Chennai • Kochi • Mumbai • Pune
Hyderabad • Kolkata • Nagpur • Patna • Vijayawada

Atomic and Molecular Spectra and Lasers

SECOND EDITION

ISBN: 978-81-239-2509-7

Copyright © Author and Publisher

Second Edition: 2015
First Edition: 2009

Published by Satish Kumar Jain for
CBS Publishers & Distributors Pvt Ltd
4819/XI Prahlad Street, 24 Ansari Road, Daryaganj, New Delhi 110 002, India.
Ph: 23289259, 23266861, 23266867   Website: www.cbspd.com
Fax: 011-23243014   e-mail: delhi@cbspd.com; cbspubs@airtelmail.in.
Corporate Office: 204 FIE, Industrial Area, Patparganj, Delhi 110 092
Ph: 4934 4934   Fax: 4934 4935   e-mail: publishing@cbspd.com; publicity@cbspd.com

Branches

- **Bengaluru:** Seema House 2975, 17th Cross, K.R. Road,
  Banasankari 2nd Stage, Bengaluru 560 070, Karnataka
  Ph: +91-80-26771678/79   Fax: +91-80-26771680   e-mail: bangalore@cbspd.com
- **Chennai:** 20, West Park Road, Shenoy Nagar, Chennai 600 030, Tamil Nadu
  Ph: +91-44-26260666, 26208620   Fax: +91-44-42032115   e-mail: chennai@cbspd.com
- **Kochi:** 36/14 Kalluvilakam, Lissie Hospital Road, Kochi 682 018, Kerala
  Ph: +91-484-4059061-65   Fax: +91-484-4059065   e-mail: kochi@cbspd.com
- **Mumbai:** 83-C, Dr E Moses Road, Worli, Mumbai-400018, Maharashtra
  Ph: +91-22-24902340/41   Fax: +91-22-24902342   e-mail: mumbai@cbspd.com
- **Pune:** Bhuruk Prestige, Sr. No. 52/12/2+1+3/2 Narhe, Haveli
  (Near Katraj-Dehu Road Bypass), Pune 411 041, Maharashtra
  Ph: +91-20-64704058/59, 32392277   Fax: +91-20-24300160   e-mail: pune@cbspd.com

Representatives

- **Hyderabad** 0-9885175004
- **Nagpur** 0-9021734563
- **Vijayawada** 0-9000660880
- **Kolkata** 0-9831437309, 0-9051152362
- **Patna** 0-9334159340

Printed at: Swastik Packagings, 506 F.I.E. Patparganj, Delhi - 110092

Preface to the Second Edition

Thanks are due to the readers for the interest and applause offered to this book. This second edition has been revised by inclusion of more topics such as matter waves, Schrödinger equation for hydrogem atom and spherically symmetric potential, the periodic table and periodicity in the properties of elements, the vector atom model and space quantization, spectral terms of equivalent electrons, selection rules in atoms in central field approximation, thermal distribution of quantum states in vibrational bands and interpretation of vibrational spectra, symmetry classification of electronic states, symmetry properties of rotational levels, coupling of nuclear rotational and electronic interactions, Hund's coupling cases and uncoupling phenomena, Raman spectra and structure determination using IR and Raman spectroscopy, stimulated Raman effect, magnetic resonance imaging, resonance fluorescence, the CO_2 laser and neodymium–glass laser, lastly, two new chapters on X-ray spectra and selection rules in spectroscopy, and an Appendix on dirac notations.

The author wishes to thank Mr YN Arjuna (Vice President) and the CBS Publishers & Distributors for bringing out the book promptly and in the least possible time.

Ajay Kumar Saxena

Preface to the First Edition

This book is meant for honours and postgraduate students of physics and chemistry of the Indian universities. It has been written, keeping in view the difficulties faced by the students in understanding the concepts in Atomic and Molecular Spectroscopy and lasers. The emphasis has been laid on the conceptual development of the subject. For convenience of the students, the subject matter has been divided into three parts, viz. atomic spectra, molecular spectra and lasers.

Chapter 1 is introductory, covering basic aspects of spectroscopy. Chapter 2 incorporates Bohr's theory, and explanation of hydrogen spectra on the basis of Bohr's theory, Sommerfeld's extension of Bohr's theory and relativistic correction. Chapter 3 considers quantum mechanical origin of spectral lines and various line width broadening mechanisms. Angular momentum of the atom plays an important role in atomic spectra. This has been discussed in Chapter 4, which includes vector model of the atom, spectroscopic notations, spin orbit interaction and its effect on hydrogen fine structure. This chapter also encompasses Lamb shift in relation to H_α line. Chapter 5 is on spectra of He atom and its quantum mechanical explanation, and Alkali spectra and explanation of their fine structure on the basis of spin orbit interaction. Chapter 6 is on normal and anomalous Zeeman effects, Paschen-Back and Stark effects. Chapter 7 is on effect of nuclear spin on hyperfine structure. Chapter 8 considers basic aspects of molecular spectroscopy, Born-Oppenheimer approximation and Heitler and London theory (basic concept), basic aspects of microwave and infrared spectroscopy. Chapter 9 encompasses pure rotational spectra (quantum mechanical treatment), isotope effect and non-rigid rotator, symmetric and asymmetric top molecules. Chapter 10 is on vibration of molecules, normal modes and vibrations rotational spectra. Chapter 11 discusses molecular electronic spectra. Franck-Condon principle and P, Q, R branches and Fortrat parabola. Chapter 12 is on classical and quantum theory of Raman effect, Raman spectrography, rotational and vibration-rotation Raman spectra. Chapter 13 is on symmetry elements. This consists of kinds of symmetry elements, classification of groups, matrix representation of symmetry operations. Multiplication tables for the point groups, the great orthogonality theorem and character tables. Chapters 14 and 15 consider physical aspects of nuclear magnetic resonance and electron spin resonance and basic instrumentation. Chapter 16 discusses basic principles of Fourier infrared spectroscopy (Fourier transform spectroscopy). Chapter 17 describes photo acoustic effect and photo acoustic spectroscopy. Chapter 18 considers physical principles of Mossbauer spectroscopy. Chapter 19 is on Lasers. This has been discussed a bit in detail starting from basic aspects and working principles, Fabry Perot and rectangular cavity modes, laser rate equations have been derived followed by various kinds of lasers and applications of lasers in industry and medical fields. At the end there are three appendices, namely about Auger process, ammonia maser and mode locking in lasers. A basic knowledge of under-graduate physics is a prerequisite for understanding the subject.

I am thankful to the authors and publishers of various books who have directly or indirectly helped me in preparing the manuscript.

I am grateful to Prof ON Srivastava (BHU, Varanasi), Prof SP Agrawal and Prof SK Nigam (ex-Vice Chancellors, APS University), Prof DP Tiwari (Head, Department of Physics, APS University), and Dr AP Mishra and Dr SL Agrawal (Department of Physics, APS University) for providing me the moral encouragement.

I am also thankful to my wife Alka and sons Ankur and Akshat for maintaining endurance and providing morale-boost during the completion of this task.

Last, I wish to express my thanks to Mr YN Arjuna and CBS Publishers & Distributors for bringing out the book in time.

Ajay Kumar Saxena

Contents

1

Atomic and Molecular Spectra: An Introduction

1.1 EMISSION AND ABSORPTION SPECTRA

There are two basic mechanisms that can excite an atom to a higher level than the ground state, thereby enabling the atom to radiate:

(*i*) Excitation by collision

(*ii*) Absorption of radiation.

The process (*i*) is illustrated in Fig. 1.1.

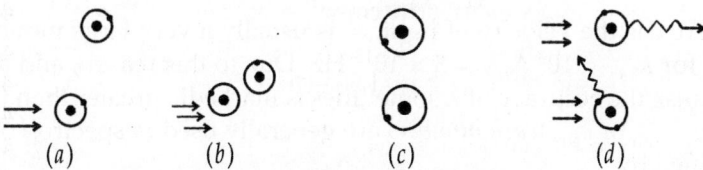

Fig. 1.1: Excitation by collision

In Figure 1.1 (*a*), both atoms are in ground state. In Figure 1.1 (*b*), during collision, some kinetic energy gets transformed to excitation energy. An excited atom retains its excited state for about 10^{-8} s [Fig. 1.1 (*c*)]. Then subsequently, just after collision, atoms return to ground state by emitting photons [Fig. 1.1 (*d*)]. The collision can be effected by producing an electric discharge in a rarefied gas when accelerated electrons collide with one another due to their kinetic energy.

1

In the second mechanism, an atom absorbs a photon of right energy to raise the atom to a higher level. This energy must equal the energy difference of two permitted orbits so that the emitted photon has an energy equal to the difference of the values of the two energy levels. This is depicted in Fig. 1.2.

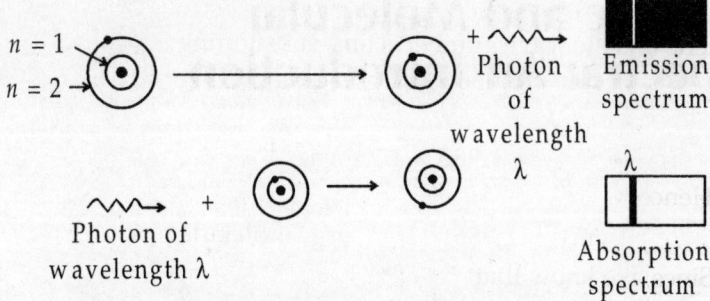

Fig. 1.2: The origin of emission and absorption spectra

1.2 WAVE NUMBER

For the purpose of investigating regularities in spectra and their connection with atomic structure, it is convenient to use (instead of the wavelength) the frequency of a given line. The frequency is given by

$$\nu = \frac{c_{air}}{\lambda_{air}} = \frac{c_{vac.}}{\lambda_{vac.}} \qquad ..(1)$$

where c is the velocity of light, ν is usually a very large number, e.g. for $\lambda_{vac.} = 10^3$ Å, $\nu = 3 \times 10^{15}$ Hz. Due to this reason, and also because the accuracy of λ sometimes is markedly greater than that of c_{medium} or c_{air}, *wave numbers* are generally used in spectroscopy, defined by

$$\overline{\nu} = \frac{\nu}{c_{vac.}} = \frac{1}{\lambda_{vac.}} = \frac{1}{n_{air}\lambda_{air}} \qquad ..(2)$$

where n_{air} is the refractive index for the wavelength considered. That is, the wave number is simply the reciprocal of wavelength in vacuum (The number of waves in 1 cm in vacuum). Its dimensions are cm^{-1}.

Since the frequency ν and the energy E of a light quantum are related by the relation $E = h\nu$ where h is Planck's constant ($h = 6.626 \times 10^{-27}$ erg-sec $= 6.626 \times 10^{-34}$ J-s), the frequency ν or the wave number $\overline{\nu}$ serves as a measure of the energy.

Units of Wave Number

Wave number of a given monochromatic light is expressed in cm^{-1}. Suppose a single atom or a molecule emits a light of wave number $1 \ cm^{-1}$ then emitted light quantum will have an energy

$$E = h\nu$$
$$= hc\bar{\nu}$$

where $\bar{\nu}$ is the wave number. Thus, substituting the values, we get

$$E = (6.626 \times 10^{-27} \ \text{erg-s}) \times (2.998 \times 10^{10} \ \text{cm s}^{-1})$$
$$(1 \ cm^{-1})$$
$$= 1.986 \times 10^{-16} \ \text{erg}$$

Hence

$$1 \ cm^{-1} \equiv 1.986 \times 10^{-16} \ \text{erg/molecule}$$

Since, we know that

$$1 \ eV = 1.602 \times 10^{-12} \ \text{erg/molecule}$$

$$\therefore \quad 1 \ cm^{-1} = \frac{1.986 \times 10^{-16}}{1.602 \times 10^{-12}} \ eV$$

or $\quad 1 \ cm^{-1} = 1.239 \times 10^{-4} \ eV$ ..(3)

Conversely,

$$1 \ eV = \frac{1.602 \times 10^{-12}}{1.986 \times 10^{-16}} \ cm^{-1}$$

or $\quad 1 \ eV = 8066 \ cm^{-1}$ ..(4)

1,3 SPECTRA AND THEIR CLASSIFICATION

The spectrum is defined as the ordered arrangement of electromagnetic radiation according to wavelength, wave number or frequency. Spectroscopy is that branch of Physics that is concerned with the theory and interpretation of spectra. There are two principal types of spectra: The emission spectra and the absorption spectra.

Emission Spectra	*Absorption Spectra*
(*i*) Continuous emission spectra	(*i*) Continuous absorption spectra
(*ii*) Line emission spectra	(*ii*) Line absorption spectra
(*iii*) Band emission spectra	(*iii*) Band absorption spectra

1.3.1 Emission Spectra

Emission spectra are obtained when the light coming directly from a source is examined with a spectroscope. Absorption spectra are obtained when the light from a source showing a continuous emission spectrum is passed through an absorbing material and then into the spectroscope.

(a) **Continuous Spectra:** The most common sources of continuous emission spectra are solids when incandescent at high temperature, e.g. in the tungsten lamp, the filament is heated to about 2100° C by the dissipation of electric energy due to its resistance. Similarly, in the carbon arc in air, the temperature of the positive pole is ~4000° C and that of the negative pole ~3000° C and it constitutes the brightest thermal source of light available in the laboratory. The heating results from the bombardment of the positive pole by electrons drawn from the gaseous part of the arc. Relatively, lesser light comes from the gas itself. In the continuous spectrum, the distribution in different wave-lengths of the energy is given by Kirchhoff's law, and depends on the ability of the surface to absorb light of different wavelengths. In general, the reflectance spectrum of a solid gives a clue to its emission spectrum.

(b) **Line Spectra:** When the slit placed before the prism or grating spectroscope is illuminated with the light from a mercury arc, several lines of different colour are seen in the eyepiece. Each of these lines is an image of the slit (formed by the telescope lens) by light of a particular wavelength. The different wavelengths are diffracted through different angles by the prism or the grating, hence, the line images are separated. It is to be evinced that to obtain line spectra, a slit has to be used.

The most intense sources of line spectra are metallic arcs and sparks, although vacuum tubes containing hydrogen or one of the rare gases are very suitable. All common sources of line emission or line absorption spectra are gases. Further, only the individual atoms give true line spectra, i.e. when a molecular compound is used (in the source) in discharge tube, the lines observed are due to the elements and not due to the molecules.

(i) **Important Features of Line Spectra:**

1. The line spectrum is the characteristic of the atom of an element. No matter in what form does the atom exist, i.e. in whatever form any element or a compound, its atomic spectrum will be a characteristic of the atom concerned.

2. The line emitted by a specific atom occupies the same position when emitted by different substances having that atom as the constituent.

3. In atomic spectra, either dark lines on the bright background or bright lines on the dark background are observed.

4. The different lines of the line spectra show multiple characters and possess fine or hyperfine structure.

5. Another characteristic of the line spectra is that the lines can be grouped in series. The separation between the succeeding lines in a series decreases gradually towards the shorter wavelength side, i.e. towards the violet end of the spectrum.

(ii) **Series of Spectral Lines:** In the spectra of some elements, lines are observed which all together form a series in which the spacing and intensities (of the lines) change in a regular manner. For example, in the Balmer series of hydrogen, the spacing of the lines decreases gradually, as they proceed into the ultraviolet toward shorter wavelengths and their intensity also falls off gradually.

(iii) **Intensity of Spectral Lines:** In general, the intensity of a spectral line depends upon the method of excitation. The DC arc is generally considered a thermal source and in this type of source the population of atoms in any excited state tends to follow Maxwell-Boltzmann statistics. Accordingly, the excited states close to the ground state are more heavily populated. Allowed transitions from these states to lower states predominate in the spectrum, and hence the intensity of these lines is greater than those involving higher energy states. These lines are called *arc lines*. On the other hand, in AC spark excitation, due to the high energy of the free electrons in the source, the atoms are predominantly excited to the higher energy states.

Allowed transitions from these states produce greater intensities for lines involving the high energy states. These are generally called *spark lines*.

(*iv*) **Ultimate Lines:** Ultimate lines are generally those involving transitions from the energy states immediately above the ground state and are normally the most intense lines in the spectrum of the given element. As the number of atoms of a given element decreases in a matrix, the intensity of these ultimate lines decreases, and they are the last lines of an element to disappear in the spectrum. These lines are used in the qualitative analysis.

(*c*) **Band Spectra:** Band spectra can be obtained in the laboratory by carbon arc cored with a metallic salt, along with the vacuum tube and the flame. Band spectra can be observed by using a spectroscope of small dispersion. Various evidences reveal that band spectra arise from molecules and not by atoms. It is found that although the line spectrum of calcium is independent of which salt we put in the arc, we obtain different band spectra by using CaF_2, $CaCl_2$.

(*i*) *Important Features of Band Spectra*

1. The bands are observed in a dark region and they are arranged in a regular sequence forming a group of bands.

2. A regular arrangement of group of bands forms a band system. In a complete molecular spectrum, several such bands are present. In certain spectra, the band systems overlap each other.

3. Each band contains several lines close together. The lines are closer on one end and successively wider on the other end. The intensity in each band falls off from a definite limit (termed band head). The intensity at other end is indistinct, hence they are also called fluted spectra.

4. The lines in a band are the characteristics of the molecules and depend upon the mode of oscillations.

5. The band spectra are characteristics of a particular molecule hence its appearance changes from molecule to molecule (as mentioned above that different band spectra are observed for CaF_2 and $CaCl_2$).

1.3.2 Absorption Spectra

When light from a source of continuous spectrum is passed through an absorbing layer, certain radiations (wavelengths) are found missing which give rise to dark lines or bands in a bright background in the emergent light. These missing radiations thus constitute the absorption spectrum. One of the most familiar examples of an absorption spectrum is of the sun. The bright continuous spectrum of the photosphere is crossed by dark lines, known as *Fraunhoffer lines* which are present due to absorption of certain radiations by the presence of vapours of some elements in the cooler surroundings of the photosphere. In general, the absorption lines of the spectrum of a gas/vapour are in coincidence with the bright emission lines of the same substance.

Besides the solar spectrum, other examples of absorption spectra are as follows:

(*i*) By putting the sodium vapour in the path of white light, two famous dark sodium lines (D_1 and D_2) are obtained with a bright background.

(*ii*) If aqueous solution of $CoCl_2$ or aqueous solution of $KMnO_4$ is placed in the path of continuous (white) light, we get the band spectrum of absorption type.

A source of a continuous spectrum in the ultraviolet region is sometimes desired for the study of absorption spectra in this region. Hot solids are unsuitable for this purpose, because they emit very small amount of ultraviolet light even at the highest temperature available. For this purpose, a vacuum tube discharge through H_2 gas (at 5 to 10 mm Hg pressure) is very convenient.

1.4 OBSERVATION OF SPECTRA

The separation of light into its spectral components can be accomplished by refraction or diffraction. Both of these depend on wavelength but in opposite ways: the greater the wavelength, the greater is the diffraction of light, but with greater wavelength, smaller is the refraction of light. For diffraction, gratings are used, while for refraction, we use prisms. Both methods may be employed except in the region below 1250 Å, a grating is necessary. The prism method has the advantage of greater light intensity whereas the grating offers greater resolving power.

On the basis of instrumentation, we may divide spectroscopy investigations into following categories:

Microwave Spectrometer: Employs Klystron source, waveguide, crystal detector (for molecular rotation spectra).

Infrared Spectrometer: Employs hot ceramic sources, rock salt prism or grating, thermocouple detector (molecular vibration spectra).

Visible and Ultraviolet Spectrometer: Employs tungsten lamp or hydrogen discharge tube, glass or quartz prism or grating, photomultiplier detector (electronic spectra).

Spectra in the far infrared: Can be investigated only with thermopiles or bolometers; however below 13000 Å, photographic plates are generally used. By using a photographic plate, a large region of the spectrum may be obtained at one time.

1.5 SPECTRAL ANALYSIS

Since each chemical element gives rise to a characteristic line spectrum by suitable excitation (flame/arc/spark/electric discharge), therefore the appearance of a line spectrum can be used as an analytical test for the presence of an element and thus extraordinarily small amounts of an element in a sample can be detected. This method of analysis is called spectral analysis. A knowledge of the structure of the spectrum is of some importance to the spectra analyst, particularly in the choice of suitable lines for tests for chemical analysis of unknown substances.

1.6 LIGHT SOURCES FOR PRODUCING SPECTRA

There are many ways for production of light for spectroscopic investigations. The important ones are temperature radiation and all kinds of luminescence* fluorescence, electroluminescence, etc. In temperature radiation of gases, the atoms/molecules are excited to light emission by collision amongst themselves, the necessary energy as emission is derived from the kinetic energy of the colliding particles. (A high temperature is necessitated, e.g. in flames or by electric furnaces).

---

*The phenomenon of emission of optical radiation.

Luminescence includes all forms of light emission in which kinetic energy (high temperature) is not essential for the mechanism of excitation. Luminescence includes all forms of light emission not only by heat energy but other ways also.

Electroluminescence includes luminescence from all kinds of electrical discharges—such as sparks, arcs, etc. or different kinds operating on direct or alter-nating current of low or high frequency—the kinetic energy (due to electrical excitation) of electrons or ions accelerated in an electric field is given up to the atoms or molecules of the gas present.

Chemiluminescence results when energy set free in a chemical reaction is converted to high energy, e.g. $2Na + Cl_2 \rightarrow 2NaCl +$ energy.

1.7 LUMINESCENCE

Some bodies generally emit light when heated. Some materials emit light after they have been irradiated with visible or ultraviolet light, with X-rays, γ-rays, electrons or other particles or when placed in an electric field, etc. The emitted light may be in the visible part of the spectrum although the temperature of the emitting body is low (room temperature and below). Such cold emission of light is termed *luminescence* and the bodies exhibiting it are termed *luminophors*. The luminescence excited by light is termed *photoluminescence*. In contrast to thermal radiation, luminescence is a non-equilibrium process. An important feature of luminescence is its long duration in comparison with the period of optic oscillation equal to 10^{-13} to 10^{-15} s. The emission of light in the process of luminescence continues at least 10^{-10} s after the excitation has ceased. In some instances, the emission of light may continue for seconds, minutes, hours and even months after the excitation has ceased. In accordance with the duration of light emission, photoluminescence is divided into *phosphorescence* and *fluorescence*.

1.8 FLUORESCENCE AND PHOSPHORESCENCE

The energy gained by an atom or a molecule by the absorption of a photon does not remain in that molecule but is lost by any of several mechanisms, e.g. conversion of a part into heat, lowering the net energy of the molecule to the lowest vibrational and rotational level within the same electronic (singlet) level. The rest of the energy is then released as radiation and the molecule then returns to the ground state. This is known as *fluorescence*. The emitted radiation has less energy per photon than the exciting radiation and hence a longer wavelength, e.g. a number of organic compounds (and some inorganic compounds) fluoresce in the visible region when irradiated with ultraviolet light.

In some molecules, it is possible for a non-radiative transition to occur from an excited singlet state to the corresponding triplet state from which the remaining energy is radiated when the molecule comes to the ground state. The probability of transition from the triplet state to the ground state is however very low. Hence radiation may last for some time after exciting radiation is cut off. This persistent radiation is called *phosphorescence*.

Fluorescence occurs more rapidly than phosphorescence. Fluorescence is completed generally after about 10^{-5} seconds (or less) from the time of excitation. Phosphorescence takes place over periods longer than 10^{-5} seconds and may continue longer than few minutes or hours after irradiation has ceased.

Resonance fluorescence is a process in which the emitted radiation is identical in frequency to the radiation employed for excitation. Phosphorescence is characterised by existence of a metastable (excited) electronic state having an average lifetime of greater than about 10^{-5} seconds. It is to be noted that direct excitation to the triplet state is forbidden as it involves a change in multiplicity (which has a very low probability of occurrence).

Questions

1.1 Give a brief qualitative explanation for origin of absorption and emission spectra.

1.2 Define each of the following terms: absorption, emission, fluorescence, phosphorescence, luminescence, electroluminescence, chemiluminescence.

1.3 The wavelength of sodium D-line is 589 nanometres, calculate (*a*) the frequency in Hertz and (*b*) the wave number in cm^{-1}.

[**Ans.** (*a*) 5.09×10^{14} Hz (*b*) 1.7×10^4 cm^{-1}]

1.4 Calculate $\nu, \bar{\nu}$ and energy E for typical ultraviolet radiation of wavelength 1000 Å. (Express E in ergs).

[**Ans.** $\nu = 1.5 \times 10^{15}$ Hz, $\bar{\nu} = 5 \times 10^4$ cm^{-1}, $E = 9.93 \times 10^{-12}$ ergs]

Theory of the Atom and Origin of Spectra

2.1 HYDROGEN SPECTRUM

A characteristic feature of line spectra is that the spectral lines could be grouped in one or more series in which the separation and intensity of lines decrease gradually towards shorter wavelength side. For example, in hydrogen, the line spectrum revealed a series of lines with separation and intensity decreasing regularly towards the shorter wavelengths and the series converges to a limit at about 3646 Å. The line with the longest wavelength (6563 Å) is designated H_α, the next ($\lambda = 4861$ Å) is designated H_β and so on (see Fig. 2.1).

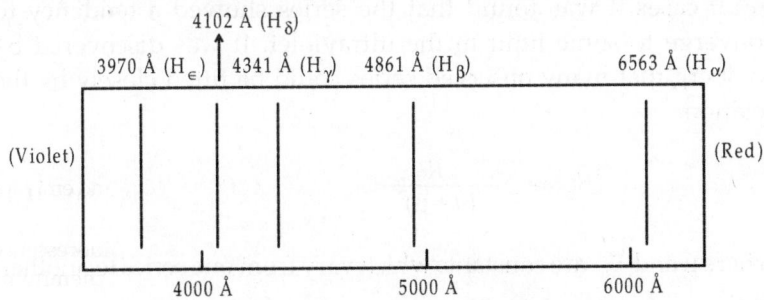

Fig. 2.1: The hydrogen spectrum (Balmer lines)

Attempts were made to explain such line spectra in analogy of overtones in acoustics to find harmonic relations in lines found in the spectrum of a given element. In this regard, though, initially, attempts were in vain, however, in 1880, Liveing and Dewar emphasized that there existed a certain sort of series relation between successive pairs of lines of the same type. In 1883, Hartley discovered that if frequencies instead of wavelengths are used, it was observed that the difference in frequency between the components of a multiplet, e.g. doublet or a triplet, in a particular spectrum, is the same for all similar multiplets. This law enabled to isolate from the large number of lines in any given spectrum those line-groups which were related to one another.

In 1885, the wavelengths of the line then known lines in the spectrum of hydrogen atom could be expressed by the following formula given by Balmer.

$$\lambda = b \frac{n^2}{n^2 - 2^2} = 3646 \left(\frac{n^2}{n^2 - 4} \right) \mathring{A} \qquad ..(1)$$

where $n = 3, 4, 5, ..$, etc. for respectively the first (beginning at the red), second, third, .., etc. lines in the spectrum. Yet, this could not provide a clue to the explanation for mechanism of atomic radiation.

In this respect Balmer proposed that his formula might be a special case of a more general formula applicable to other series of lines in other elements. Rydberg made an attempt to discover such formula using Hartley's law of constant wave number separation as applied to comparatively large mass wavelength data of alkalis. In all cases it was found that the series showed a tendency to converge to some limit in the ultraviolet. It was discovered by Rydberg that many observed series could be fitted closely by the relation.

$$\overline{\nu}_n = \overline{\nu}_\infty - \frac{R}{(n + \mu)^2} \qquad ..(2)$$

where μ and $\overline{\nu}_\infty$ are constants which vary from one series to another.

The constant $\overline{\nu}_\infty$ represents the high frequency limit to which the lines in the series ultimately converge. The Balmer formula is a special case of the above Rydberg formula, since we can write

$$\overline{\nu} - \frac{1}{\lambda} - \frac{1}{b} - \frac{4}{bn^2} = \frac{1}{b} - \frac{R}{n^2} = \overline{\nu}_\infty - \frac{R}{n^2} \qquad ..(3)$$

with $R = 4/b$ and $\overline{\nu}_\infty = 1/b = R/4$.

or $$\bar{v} = \frac{R}{4} - \frac{R}{n^2} = R\left[\frac{1}{2^2} - \frac{1}{n^2}\right] \quad \text{(Balmer series)}$$

$$= \frac{5R}{36}, \frac{3R}{16}, \frac{21R}{100} \qquad\qquad ..(4)$$

with $n = 3, 4, 5, \ldots$

This is of the type of eqn (2) with $\mu = 0$ and $R/4$ represents the *convergence limit* corresponding to $n = \infty$. This corresponds to state of ionization, i.e. complete removal of electron from the nucleus. Thus spectral lines obtained were correlated empirically by Rydberg. The constant R in eqn (2) (now called the *Rydberg constant*) was found to have the same value for a large group of series for each substance and very nearly the same for all substances. From spectroscopic observations, its value for the hydrogen atom is

$$R = 109677.6 \text{ cm}^{-1} = 1.09678 \times 10^7 \text{ m}^{-1}$$

Its slight variation from one atom to another is now known to be due to difference in the atomic weights.

Each line series (such as Balmer series, or Lyman series) shows a continuous emission (or absorption) to high wave numbers of the convergence limits. The convergence limit corresponds to atomic electron absorbing sufficient energy to escape from the nucleus with zero velocity, however, more energy could be absorbed by the electron and hence escape with higher velocities. Now, since the electron has escaped, its energy is not quantized, and any energy higher than convergence limit can be absorbed, so spectrum in the region beyond the convergence limit is continuous.

The Balmer series contain only those spectral lines of the hydrogen atom which occur in the visible and near ultraviolet parts of the spectrum. The Lyman series occur in the ultraviolet region and correspond to the formula

$$\bar{v} = \frac{1}{\lambda} = R\left(\frac{1}{1^2} - \frac{1}{n^2}\right); n = 2, 3, 4\ldots\text{(Lyman series)} \qquad ..(5)$$

In the infrared, three spectral series have been observed which correspond to the formulas

$$\bar{v} = \frac{1}{\lambda} = R\left(\frac{1}{3^2} - \frac{1}{n^2}\right); n = 4, 5, 6\ldots\text{(Paschen series)} \qquad ..(6)$$

$$\bar{v} = \frac{1}{\lambda} = R\left(\frac{1}{4^2} - \frac{1}{n^2}\right); n = 5, 6, 7\ldots\text{(Brackett series)} \qquad ..(7)$$

$$\bar{\nu} = \frac{1}{\lambda} = R\left(\frac{1}{5^2} - \frac{1}{n^2}\right); n = 6, 7, 8, \ldots \text{(Pfund series)} \qquad ..(8)$$

Similar types of regularities are found in the spectra of more complex atoms.

2.2 RITZ COMBINATION PRINCIPLE

The structure of formulae for H-series persuaded Rydberg to think that combination of terms giving two spectral lines of a series might correspond to a spectral line of another series. This idea of Rydberg was generalized by Ritz (in 1908) and is known as Ritz combination principle. It may be stated as follows:

If $\bar{\nu}_1$ and $\bar{\nu}_2$ are the wave numbers of two lines in the spectrum of certain atoms, then we are also likely to get lines with wave numbers $\bar{\nu}_1 + \bar{\nu}_2$ or $\bar{\nu}_1 - \bar{\nu}_2$.

For example, the H_α and H_β lines of Balmer series of hydrogen spectrum, viz.

$$\bar{\nu}_\alpha = R\left(\frac{1}{2^2} - \frac{1}{3^2}\right) \text{ and } \bar{\nu}_\beta = R\left(\frac{1}{2^2} - \frac{1}{4^2}\right) \qquad ..(9)$$

may be combined as

$$\bar{\nu}_\beta - \bar{\nu}_\alpha = R\left(\frac{1}{3^2} - \frac{1}{4^2}\right) \qquad ...(10)$$

which represent first line of the Paschen series. The second line of this series is obtained by combining H_γ and H_α lines of the Balmer series. Similarly, the lines of other series may also be obtained.

2.3 BOHR'S THEORY

Planck's quantum theory involves two essential features:

1. An oscillator can exist only in one of a number of a discrete quantum states, and to each of these states, there corres-ponds a definite allowed value of its energy.

2. No radiation is emitted while the oscillator remains in one of its quantum states, but it is capable of jumping from one quantum state to another one of lower energy, the energy lost in doing so being emitted in the form of a quantum of radiation.

   Niels Bohr in 1913 discovered how to apply similar ideas to a hydrogen atom of the Rutherford type and was successful in giving a theoretical formulation for H-spectrum agreeing with the observations. Bohr suggested the following postulates to explain the electron motion in an atom:

(*i*) An electron in an atom moves in a circular orbit about the nucleus under the influence of Coulomb's force of the nucleus which is balanced by the Newtonian centrifugal force.

(*ii*) An electron cannot revolve round the nucleus in all possible orbits. It can revolve only in a few widely separated permitted orbits. While moving in these permitted orbits round the nucleus, the electron does not radiate energy. These orbits are called stationary orbits.

(*iii*) The permitted orbits are those for which the angular momentum of the electron is an integral multiple of $h/(2\pi)$, where h is Planck's constant.

(*iv*) When an electron jumps from an outer (initial) orbit of higher energy to an inner (final) orbit of lower energy, the energy difference of the two orbits is radiated in the form of quantum of radiation whose frequency ν is given by

$$h\nu = (E_i - E_f) \qquad \qquad \text{... (11)}$$

This is called *Bohr's frequency condition* and is concerned with the origin of spectral lines.

2.3.1 Bohr's Model of a One Electron Atom

Let us consider a one electron atom, e.g. a neutral hydrogen atom ($Z = 1$), a singly ionized He atom ($Z = 2$) or a doubly ionized Li atom ($Z = 3$). In either case, an electron of charge $-e$ and mass m revolves round a nucleus of charge $+Ze$ and mass M, in a circular orbit. For the moment, if we assume nucleus to be infinitely heavy compared to the electron then nucleus may be considered to be at rest. Under equilibrium, Coulomb force equals the centripetal force, so

$$\frac{Ze^2}{4\pi\varepsilon_0 r_n^2} = \frac{mv^2}{r_n} \qquad \qquad ..(12)$$

where v is the orbital speed of electron and r_n is the radius of the orbit. Since the force acting on the electron acts in radial direction, and 'a' is fixed, therefore, Bohr's quantisation condition requires for angular momentum

$$|\mathbf{L}| = mvr_n = \frac{nh}{2\pi} \qquad ..(13)$$

where $n = 1, 2, 3, \ldots$and is called the principal quantum number. Substituting for v from eqn. (13) in eqn. (12), we get

$$\frac{Ze^2}{4\pi\varepsilon_0} = mv^2 r_n = mr_n \frac{n^2 h^2}{4\pi^2 m^2 r_n^2} = \frac{n^2 h^2}{4\pi^2 m r_n}$$

$$\Rightarrow \qquad r_n = \frac{n^2 \hbar^2 (4\pi\varepsilon_0)}{mZe^2} \qquad \left(\text{where } \hbar = \frac{h}{2\pi} \right)$$
$$..(14)$$

and
$$v = \frac{nh}{2\pi m r_n}$$

$$= \frac{nh}{2\pi m} \times \frac{4\pi^2 m Z e^2}{n^2 h^2 (4\pi\varepsilon_0)}$$

or
$$v = \frac{Ze^2}{n\hbar(4\pi\varepsilon_0)} \qquad ..(15)$$

Thus electron orbits are restricted to those having radii given by eqn. (14) and the radii of successive allowed orbits are proportional to n^2. For hydrogen $Z = 1$ and the radius of the smallest allowed orbits, known as the Bohr radius is given by

$$a_0 = r_1 = (4\pi\varepsilon_0)\frac{\hbar^2}{me^2} \simeq 0.529 \times 10^{-10} \text{ m} \qquad ..(16)$$

The ratio of the speed of the electron in the first Bohr orbit of hydrogen (v_1) to the speed of light (c) is known as the fine structure constant and is given by

$$\alpha = \frac{v_1}{c} = \frac{1}{4\pi\varepsilon_0}\frac{e^2}{\hbar c} \simeq \frac{1}{137} \qquad ..(17)$$

Let us now calculate the total energy of an electron moving in one of the allowed orbits (The energy is partly kinetic and partly potential). The potential energy is defined to be zero when the electron is at infinite distance from the nucleus. Then the potential energy V at a finite distance r from the nucleus is equal to the work done in removing the electron from r to ∞ against the attractive electrostatic force $(-Ze^2/r^2)$, i.e.

$$V(r) = \frac{1}{(4\pi\varepsilon_0)}\int_r^\infty -\frac{Ze^2}{r^2}\,dr$$

$$= -\frac{Ze^2}{(4\pi\varepsilon_0)r} \qquad ..(18)$$

The kinetic energy of the electron is

$$K = \frac{1}{2}mv^2 = \frac{1}{2}\frac{Ze^2}{(4\pi\varepsilon_0)r} \qquad ..(19)$$

So, total energy

$$E = K + V = \frac{1}{(4\pi\varepsilon_0)}\left\{\frac{1}{2}\frac{Ze^2}{r} - \frac{Ze^2}{r}\right\}$$

$$= -\frac{1}{2}\frac{Ze^2}{r}\times\frac{1}{4\pi\varepsilon_0} \qquad ..(20)$$

Substituting for r, we get for the nth orbit

$$E_n = -\frac{2\pi^2 mZ^2 e^4}{n^2 h^2} \qquad \text{(CGS units)}$$

or
$$E_n = -\frac{1}{2}\frac{Ze^2 \times mZe^2}{n^2 h^2 (4\pi\varepsilon_0)^2} = -\left(\frac{1}{4\pi\varepsilon_0}\right)^2 \frac{me^4 Z^2}{2n^2 \hbar^2}$$
$$\text{(MKS units)} \ (n = 1, 2, 3, ..) \quad ..(21)$$

or
$$E_n = \frac{me^4 Z^2}{8\varepsilon_0^2 h^2 n^2}$$

We thus see that the quantisation of the orbital angular momentum of the electron leads to the quantisation of its total energy.

On substituting the values of m, h and e, we find that the energies of the various allowed orbits of the hydrogen atom are

$$E_1 = -13.6 \text{ eV}$$
$$E_2 = -3.40 \text{ eV} \ (1 \text{ eV} = 1.6 \times 10^{-19} \text{ J})$$
$$E_3 = -1.51 \text{ eV}$$
$$E_4 = -0.85 \text{ eV}$$

These are all negative, signifying that the electron is bound to the nucleus. The lowest energy level E_1 is called the *ground state* of the atom and the higher energy levels E_2, E_3, .., etc. are called the *excited states*.

The ground state energy E_1 of the hydrogen atom is a convenient energy unit to use in discussing various aspects of atomic and molecular spectra. This energy unit is called the *rydberg (ry)* and its numerical value is

$$1 \ ry = \frac{me^4}{8\varepsilon_0^2 h^2} = 2.17\times10^{-18} \text{ Joule}$$

$$= 13.6 \text{ eV}. \qquad ..(22)$$

It is the minimum amount of energy required to remove the electron from its ground state ($n = 1$) to infinity. Hence, it is the *binding energy of the hydrogen atom.*

2.3.2 The Energy Levels and Line Spectra of Hydrogen

The energy equation may also be written as

$$E_n = -\frac{R_\infty hcZ^2}{n^2} \qquad ..(23)$$

where $R_\infty \left(= \dfrac{me^4}{8\varepsilon_0^2(ch^3)} \right)$ is the *Rydberg constant* for an infinitely heavy nucleus.

The energy values of the various energy levels divided by $-hc$, i.e. ($-E_n/hc$) are called *term values* of the states and are denoted by T_n:

$$T_n = -\frac{E_n}{hc} = \frac{R_\infty}{n^2} Z^2 \text{ m}^{-1} \qquad ..(24)$$

where $\qquad R_\infty = \dfrac{me^4}{8\varepsilon_0^2 ch^3}$

$$= \frac{(9.1 \times 10^{-31} \text{ kg}) \times (1.6 \times 10^{-19} \text{ C})^4}{8 \times (8.85 \times 10^{-12} \text{ F/m})^2 \times 3 \times 10^8 \text{ m/s} \times (6.63 \times 10^{-34} \text{ J-s})}$$

$$= 1.097 \times 10^7 \text{ m}^{-1} \qquad ..(25)$$

This agrees well with the experimental value.

The presence of definite discrete energy levels in the hydrogen atom suggests a connection with the line spectra. When an electron in an excited state drops to a lower state, the lost energy is emitted as a single photon of light.

If the quantum number of the initial (higher) energy state is n_i and that of the final (lower) energy state is n_f then,

$$E_i - E_f = h\nu$$

where ν is the frequency of the emitted photon.

$$E_i = -\frac{me^4}{8\varepsilon_0^2 h^2} \left(\frac{1}{n_i^2} \right) \qquad ..(26)$$

$$E_f = -\frac{me^4}{8\varepsilon_0^2 h^2} \left(\frac{1}{n_f^2} \right) \qquad ..(27)$$

Hence $E_i - E_f = -\dfrac{me^4}{8\varepsilon_0^2 h^2}\left(\dfrac{1}{n_i^2} - \dfrac{1}{n_f^2}\right)$ ..(28)

The frequency ν of the photon released in this transition is

$$\nu = \frac{E_i - E_f}{h} = \frac{me^4}{8\varepsilon_0^2 h^3}\left(\frac{1}{n_f^2} - \frac{1}{n_i^2}\right)$$ ..(29)

In terms of wavelength λ , we have

$$\lambda = \frac{c}{\nu} \text{ or } \frac{1}{\lambda} = \frac{\nu}{c} = \frac{me^4}{8\varepsilon_0^2 ch^3}\left(\frac{1}{n_f^2} - \frac{1}{n_i^2}\right)$$

or $\bar{\nu} = \dfrac{1}{\lambda} = R_\infty\left(\dfrac{1}{n_f^2} - \dfrac{1}{n_i^2}\right)$ ..(30)

This equation suggests that the radiation emitted by excited hydrogen atom should contain certain wavelengths only. Further, these wavelengths fall into definite sequences that depend upon the quantum number n_f of the final energy level of the electron. Thus, according to the Bohr model, the formulae for the first five known series of the hydrogen spectrum are

$$n_f = 1, \bar{\nu} = \frac{1}{\lambda} = R_\infty\left(\frac{1}{1^2} - \frac{1}{n^2}\right); n = 2,3,4...$$

(The Lyman series) ..(31)

$$n_f = 2, \bar{\nu} = \frac{1}{\lambda} = R_\infty\left(\frac{1}{2^2} - \frac{1}{n^2}\right); n = 3,4,5...$$

(The Balmer series) ..(32)

$$n_f = 3, \bar{\nu} = \frac{1}{\lambda} = R_\infty\left(\frac{1}{3^2} - \frac{1}{n^2}\right); n = 4,5,6...$$

(The Paschen series) ..(33)

$$n_f = 4, \bar{\nu} = \frac{1}{\lambda} = R_\infty\left(\frac{1}{4^2} - \frac{1}{n^2}\right); n = 5,6,7...$$

(The Brackett series) ..(34)

$$n_f = 5, \bar{\nu} = \frac{1}{\lambda} = R_\infty\left(\frac{1}{5^2} - \frac{1}{n^2}\right); n = 6,7,8...$$

(The Pfund series) ..(35)

where $R_\infty = \dfrac{me^4}{8\varepsilon_0^2 ch^3} \equiv 1.097 \times 10^7 \text{ m}^{-1}$

These series are illustrated in terms of the energy level diagram in Fig. 2.2.

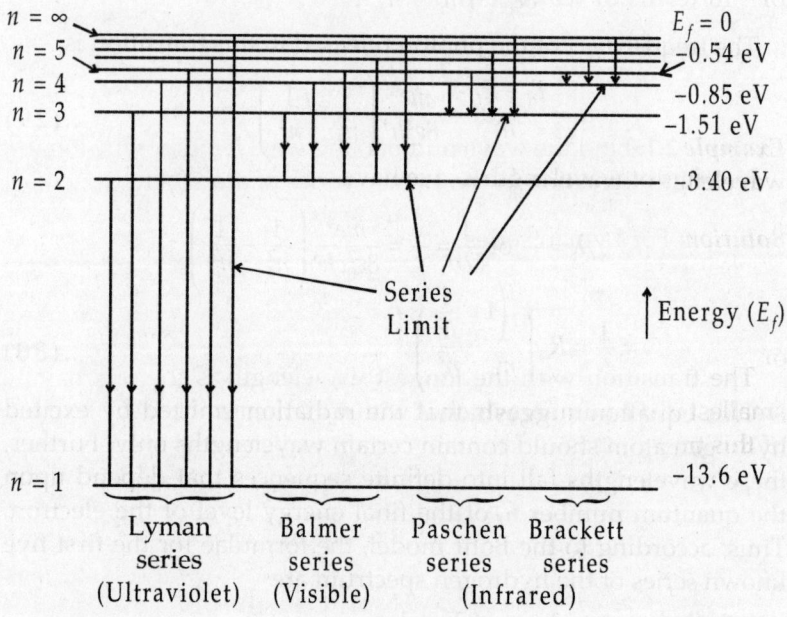

Fig. 2.2: The spectral lines originating in hydrogen spectrum

2.3.3 Bohr Frequency Conditions in Terms of Term Values

Term value is defined as

$$T_n = -\frac{E_n}{hc}\left(=\frac{R_\infty}{n^2}Z^2 \text{ m}^{-1}\right)$$

Now $E_i = h\nu_i$ and $E_f = h\nu_f$

$$T_i = -\frac{E_i}{hc}\left(=\frac{R_\infty Z^2}{n_i^2}\right) \qquad ..(36)$$

$$T_f = -\frac{E_f}{hc}\left(=\frac{R_\infty Z^2}{n_f^2}\right) \qquad ..(37)$$

Thus two terms T_i and T_f combine to produce a spectral line of frequency

$$\nu = \frac{E_i - E_f}{h} = \frac{(T_f - T_i)hc}{h}$$

or $\nu = cT_f - cT_i$ ..(38)

or   in terms of wave number

$$\bar{\nu} = \frac{1}{\lambda} = \frac{\nu}{c} = T_f - T_i$$ ..(39)

Example 2.1: Find the wave number and wavelength of the longest wavelength transition in the Lyman series of atomic hydrogen.

Solution: For Lyman series

$$\bar{\nu} = R_H\left(1 - \frac{1}{n^2}\right); n = 2, 3, \ldots$$

The transition with the longest wavelength is the one having smallest wave number which is with $n = 2$, hence the wave number of this transition is

$$\bar{\nu} = \frac{3}{4} \times R_H = 82258 \text{ cm}^{-1}$$

$$\lambda = \frac{1}{\bar{\nu}} = \frac{1}{82258} \text{ cm} \approx 1.216 \times 10^{-7} \text{ m}$$

$$= 121.6 \text{ nm.}$$

The theoretical explanation of Balmer series in hydrogen spectrum was a great success for Bohr's theory. The success was particularly noteworthy because the Lyman, Brackett and Pfund series were not discovered at the time the Bohr's theory was put forth. However the existence of these series was predicted by the theory and these series were indeed observed experimentally in 1916, 1922 and 1924 respectively at the predicted positions in the spectrum.

2.4 CORRECTION FOR FINITE NUCLEUS MASS (NUCLEAR MOTION AND REDUCED MASS OF THE ELECTRON)

In Bohr's theory, it has been assumed that the nucleus remains stationary and infinitely massive compared with the mass of the electron. But actually the mass of the nucleus is finite, therefore, both nucleus and the electron revolve around their common centre of mass.

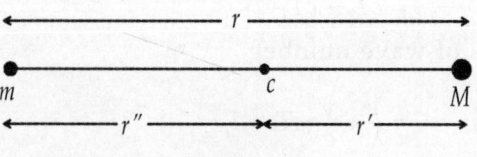

Fig. 2.3

Let c denote the centre of mass of the system; m, the mass of the electron; M, mass of the nucleus; r, distance between electron and nucleus; r', the distance from c to the nucleus and r'', the distance from c to the electron. Then,

$$r = r' + r'' \quad \text{and} \quad mr'' = Mr'$$

∴ $$r'' = \frac{M}{m} r' \quad \text{or} \quad r - r' = \frac{M}{m} r'$$

or $$r' = \frac{r}{1 + \dfrac{M}{m}} = \frac{mr}{m + M} \qquad ..(40)$$

Similarly $$r'' = \frac{Mr}{m + M} \qquad ..(41)$$

The total angular momentum about the centre of mass c is

$$= mr''^2\omega + Mr'^2\omega$$

$$= m\frac{M^2 r^2}{(m+M)^2}\omega + M\frac{m^2 r^2}{(m+M)^2}\omega$$

$$= \frac{mM^2 r^2 + Mm^2 r^2}{(m+M)^2}\omega = \left(\frac{mM}{m+M}\right)r^2\omega \qquad ..(42)$$

Thus the nucleus-electron system is equivalent to a single particle of mass $\mu = \dfrac{mM}{(m+M)}$ which appears to move about the nucleus, with its distance from the nucleus equal to distance of the orbital electron from the nucleus. Also $\mu < m$, therefore μ is called the *reduced mass* of the electron. To take nuclear motion into account, Bohr modified postulate (*iii*) and stated that the electron can move only in those orbits for which the angular momentum of the atom is an integral multiple of $(h/2\pi)$.

$$\mu r^2 \omega = \frac{nh}{2\pi} (n = 1, 2, 3 \ldots) \qquad \qquad ..(43)$$

or $\qquad \mu vr = \frac{nh}{2\pi} \ (\because v = r\omega);$

$$\therefore \qquad E_n = -\frac{2\pi^2 \mu Z^2 e^4}{n^2 h^2} \quad \text{(CGS units)}.$$

2.5 VARIATION OF THE RYDBERG'S CONSTANT

Due to finite nuclear masses, there is a slight variation in the Rydberg constant from atom to atom. For an infinitely heavy nucleus, the Rydberg constant is

$$R_\infty = \frac{2\pi^2 m e^4}{ch^3} \qquad \qquad ..(44)$$

and for a nucleus of mass M,

$$R_M = \frac{2\pi^2 \mu e^4}{ch^3} \qquad \qquad ..(45)$$

where $\qquad \mu = \frac{mM}{m + M} = \frac{m}{1 + \dfrac{m}{M}} \qquad \qquad ..(46)$

(*m* is the mass of the electron)

Thus, $\qquad R_M = \frac{2\pi^2 m e^4}{ch^3} \dfrac{1}{1 + \dfrac{m}{M}} = R_\infty \left(\dfrac{1}{1 + m/M} \right) \qquad ..(47)$

i.e. R_M is less than R_∞ by a factor of $\left(1 + \dfrac{m}{M}\right)$ where M is the nuclear mass. Substituting for the values of m, e, c and h, we obtain

$$R_\infty = 109737 \text{ cm}^{-1} \qquad \qquad ...(48)$$

For hydrogen, $\dfrac{m}{M} = \dfrac{1}{1836}$ and $R_H = 109677 \text{ cm}^{-1}$.

The variation of Rydberg's constant from element to element against the mass number is shown in Fig. 2.4. With increasing mass number, R_M approaches more and more near to R_∞.

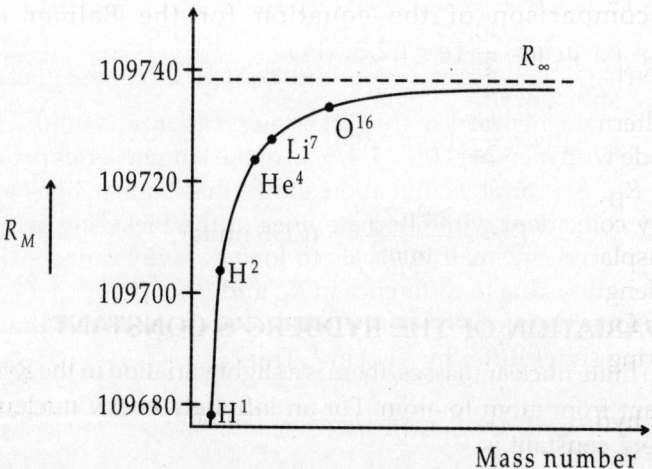

Fig. 2.4: Variation of Rydberg constant R_M with mass number M

2.6 COMPARISON OF H AND He⁺ SPECTRA

Pickering in 1897, discovered a new series in ξ Puppis star which was earlier attributed by Rydberg to some form of hydrogen since it could be represented by a formula similar to the Balmer's formula for hydrogen. Later on, it was shown experimentally by Evans and others that this series arises due to ionized helium.

Putting $Z = 2$ and R_{He} for the Rydberg constant in the Bohr's formula, all the known series of ionized helium can be represented as

$$\nu = 4R_{He}\left(\frac{1}{n_f^2} - \frac{1}{n_i^2}\right)$$

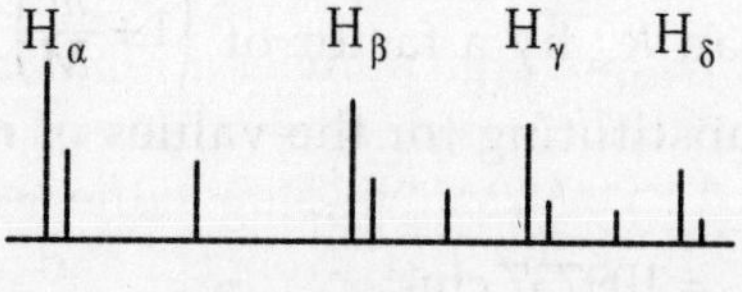

Fig. 2.5

For $n_f = 4$; $n_i = 5, 6, 7$ ...Pickering series

$n_f = 3$; $n_i = 4, 5, 6$ ...Fowler series

$n_f = 2$; $n_i = 3, 4, 5$ ...Lyman in the extreme UV

$n_f = 1$; $n_i = 2, 3, 4$ ...Lyman in the extreme UV

A comparison of the equation for the Balmer series $\left(\nu = R_H \left(\dfrac{1}{2^2} - \dfrac{1}{n^2} \right); n_i = 3, 4, 5 \right)$ with that for the Pickering series shows that alternate members of the Pickering series ($n_i = 6, 8, 10, ..$) must coincide with members ($n_i = 3, 4, 5, ..$) of the Balmer series provided, $R_{He} = R_H$. A careful examination shows that Balmer lines are not exactly coincident with alternate lines of the Pickering series but are displaced by small intervals to lower wave numbers (longer wavelengths), due to difference in R_H and R_{He}.

It is found by a careful study that the series limits of Balmer and Pickering series differ by ~11 cm^{-1}. Thus

For hydrogen: $(\nu_\infty)_H = \dfrac{R_H}{2^2}(n_i \to \infty)$ Balmer

For He^+: $(\nu_\infty)_{He^+} = \dfrac{4R_{He}}{4^2}(n_i \to \infty)$ Pickering

$\Rightarrow \quad R_{He} - R_H = 4[(\nu_\infty)_{He^+} - (\nu_\infty)_H] = 44 \text{ cm}^{-1}$ ..(49)

This result is in excellent agreement with the value obtained from theory.

2.7 BOHR'S CORRESPONDENCE PRINCIPLE

We know that Newtonian mechanics and classical electromagnetic theory are based on thoroughly established experimental facts. Therefore we may expect that the quantum theory should give results that become identical with those of classical physics if the masses and dimensions of the system under consideration are made to approach those of classical system.

It is evident from the energy level diagram of hydrogen atom that energy difference between one level and the next is very small for large values of the quantum number n. Similarly the radii of the orbits are very close for large values of n. Thus, if an electron jumps from an orbit (say $n = 2000$) step by step to the next lower states ($n = 1999$ then $n = 1998$), etc. the steps in energy and radius will appear to proceed almost continuously. Similarly the changes in angular momentum, e.g. from $L = 2000\,\hbar$ to $1999\hbar$, etc. will also appear to be continuous. Thus there is a concordance between classical and quantum physics in the limit of large n. Bohr generalized this concordance in mechanical behaviour of electron to also the electromagnetic behaviour, i.e. emission of radiation. According to him, *In the limiting case of large quantum numbers, the frequencies and the intensities of radiation calculated from classical theory must agree with those of quantum theory.*

This is known as *Bohr's correspondence principle.* It can be easily verified for frequencies of light emitted by an electron in a hydrogen atom.

For a transition from nth orbit to $(n-1)$th orbit, the frequency is

$$\nu = \frac{e^4}{4\pi(4\pi\varepsilon_0)^2}\frac{m}{\hbar^3}\left[\frac{1}{(n-1)^2}-\frac{1}{n^2}\right]$$

$$= \frac{e^4}{4\pi(4\pi\varepsilon_0)^2}\frac{m}{\hbar^3}\frac{(2n-1)}{n^2(n-1)^2} \qquad ..(50)$$

If n is very large

$$\frac{(2n-1)}{n^2(n-1)^2} \simeq \frac{2n}{n^4} = \frac{2}{n^3}$$

then
$$\nu = \frac{e^4}{2\pi(4\pi\varepsilon_0)^2}\frac{m}{\hbar^3}\frac{1}{n^3} \qquad ..(51)$$

This is the frequency according to the quantum theory. To find the frequency according to classical theory, we note that for an accelerated charge, classical electromagnetic theory predicts that the frequency of the emitted light coincides with the frequency of the motion:

$$\nu_{\text{classical}} = \frac{v}{2\pi r} = \frac{n\hbar/mr}{2\pi r}$$

$$= \frac{n\hbar}{2\pi m}\frac{1}{r^2} = \frac{n\hbar}{2\pi m}\left(\frac{e^2 m}{4\pi\varepsilon_0 n^2\hbar^2}\right)^2$$

$$= \frac{e^4}{2\pi(4\pi\varepsilon_0)^2}\frac{m}{\hbar}\left(\frac{1}{n^3}\right) \qquad ..(52)$$

Comparing eqns (51) and (52), we see that for large n, the results of the quantum mechanical calculation agree with that obtained by classical calculation. Hence quantum physics, in the limit of large quantum numbers, gives the same result as provided by classical physics.

2.8 WILSON-SOMMERFELD QUANTIZATION RULES

Planck developed his quantum theory of black body radiation by treating atoms as harmonic oscillators which can radiate or absorb energy in multiples of Planck's constant of action ($h = (6.547 \pm 0.008) \times 10^{-27}$ erg-s)

i.e. $E = h\nu$

where ν is the frequency of the emitted or absorbed radiation. Bohr, on the other hand, quantized the angular momentum of the electron moving in an atomic orbit.

Let us consider a one dimensional simple harmonic oscillator. Its total energy in terms of position coordinate x and linear momentum p_x is

$$E = \text{P.E.} + \text{K.E.} \qquad ..(53)$$

$$= \frac{1}{2}kx^2 + \frac{p_x^2}{2m}$$

or $$\frac{x^2}{2E/k} + \frac{p_x^2}{2mE} = 1 \qquad ..(54)$$

Thus relation between x and p_x is the equation of an ellipse with semimajor axis $a = \sqrt{2E/k}$ and semiminor axis $h = \sqrt{2mE}$. Any instantaneous state of the oscillator is represented by some point on this ellipse plotted on a two dimensional space having coordinates x and p_x (This is known as phase-space plot). During one cycle of oscillation, the point (x, p_x) travels once round the ellipse, so that $\oint p_x dx$ is just the area of the ellipse, i.e.

$$\oint p_x dx = \pi ab$$

$$= \pi\sqrt{2E/k} \times \sqrt{2mE}$$

$$= 2\pi E\sqrt{m/k}$$

$$= E/\nu \left(\text{since frequency } \nu = \frac{1}{2\pi}\sqrt{\frac{k}{m}} \right) \qquad ..(55)$$

Sommerfeld conjectured that orbit of an electron in an atom might be an ellipse (instead of a circle) and to select from the infinity of classically allowed ellipses, he generalized the quantum condition which had been used by Planck and Bohr. According to Planck, the ellipse (arising from a plot of p as a function of x) for a harmonic oscillator had an area equal to an integer times h, i.e.

$$\oint p_x dx = nh \quad \left(\text{since } E = nh\nu \text{ in eqn 55}\right) \qquad ..(56)$$

Now, this rule can be modified to apply to a Bohr orbit. But, it is evident that the above condition is not sufficient to fix unambiguously both axes of the ellipse. Wilson and Sommerfeld, in 1916, introduced a new and more general postulate than the original one of Bohr. This can be stated as follows:

For any physical system in which the coordinates are periodic functions of time, the action integral (for stationary states) $\oint p_i dq_i$ extended over one period of the motion (where q_i is one of the coordinates, p_i is the associated generalized momentum), must be an integral multiple of h, i.e.

$$\oint p_i dq_i = n_i h \qquad ..(57)$$

where n_i is a whole number called the quantum number and the integral is called the *phase-integral*.

The above condition is known as the *Wilson-Sommerfeld quantization* rule. Thus for Sommerfeld's elliptical orbits

$$\oint p_r dr = n_r h \qquad ..(58)$$

$$\oint p_\theta d\theta = kh \qquad ..(59)$$

where the integers n_r and k are known respectively, as the *radial* and *azimuthal quantum numbers*. Again the two quantum conditions (58) and (59) are for one quantum system, therefore

$$n_r + k = n \qquad ..(60)$$

We can deduce the Bohr quantization of angular momentum from the Wilson Sommerfeld rule. We have polar coordinates (r, θ). The momentum associated with the radial coordinate r is $p_r = mdr/dt$ and that associated with the angular coordinate θ is $p_0 = mr^2 d\theta/dt$ (which is constant). For a circular orbit, r is constant, so p_r is zero ($\therefore n_r = 0$). Hence

$$\oint p_\theta d\theta = nh \qquad ..(61)$$

or $$p_\theta \int_0^{2\pi} d\theta = nh$$

or $$p_\theta = \frac{nh}{2\pi} \qquad ..(62)$$

or $$|j| = \frac{nh}{2\pi} \qquad ..(63)$$

i.e. angular momentum of the electron is an integral multiple of $h/2\pi$. This is Bohr's quantization law. The quantum number n in eqn (60) is known as the principal quantum number.

2.9 SOMMERFELD'S EXTENSION OF BOHR'S MODEL

In Bohr's model of the H-atom, the electron's motion was restricted to circular orbits. However, this restriction was quite arbitrary. Further, in spite of striking numerical agreements, Bohr's theory was unable to explain the fine structure of the hydrogen spectrum. The fine structure lines, i.e. observation of several distinct and close components in place of single lines when observed under a high resolution equipment.

Sommerfeld (1916), in an attempt to explain the fine structure extended Bohr's model by suggesting that the electron could move in an elliptical orbit with nucleus at one of the two foci of the elliptical orbit (Fig. 2.6).

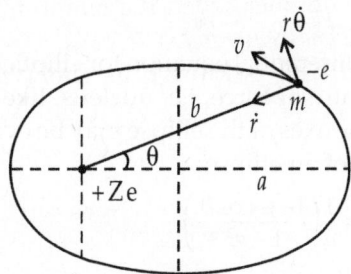

Fig. 2.6: Elliptical electron motion (charge $-e$, mass m) with nucleus (charge $+Ze$) at one focus of the ellipse

The instantaneous position of the electron is described by polar coordinates r and θ which are to be quantized separately (Wilson Sommerfeld quantization).

$$\text{Potential Energy } V = -\frac{Ze^2}{r}$$

and $$\text{Kinetic Energy } k = \frac{1}{2}m\left(\dot{r}^2 + r^2\dot{\theta}^2\right) \qquad ..(64)$$

where $\dot{r}$ is the radial velocity and $r\dot{\theta}$ the angular velocity components. The momentum components are

$$p_r = m\dot{r} \qquad \ldots (65)$$
$$p_\theta = mr^2\dot{\theta} \qquad \ldots (66)$$

The total energy of the electron

$$E = K + V$$

$$= \frac{1}{2}m\dot{r}^2 + \frac{1}{2}mr^2\dot{\theta}^2 - \frac{Ze^2}{r}$$

Now $\qquad \dot{\theta} = \dfrac{p_\theta}{mr^2}$ and $\dot{r} = \dfrac{dr}{dt} = \dfrac{dr}{d\theta}\dfrac{d\theta}{dt} = \dfrac{dr}{d\theta}\dfrac{p_\theta}{mr^2}$

$\therefore \qquad E = \dfrac{1}{2} m \left(\dfrac{dr}{d\theta} \cdot \dfrac{p_\theta}{mr^2} \right)^2 + \dfrac{1}{2}\dfrac{p_\theta^2}{mr^2} - \dfrac{Ze^2}{r}$

Multiplying both sides by $\dfrac{2mr^2}{p_\theta^2}$, we get

$$\dfrac{2mr^2}{p_\theta^2} E = \left(\dfrac{1}{r}\dfrac{dr}{d\theta} \right)^2 + 1 - \dfrac{2mrZe^2}{p_\theta^2}$$

or $\qquad \left(\dfrac{1}{r}\dfrac{dr}{d\theta} \right)^2 = \dfrac{2mE}{p_\theta^2} r^2 + \dfrac{2mZe^2}{p_\theta^2} r - 1 \qquad ..(67)$

This represents differential equation for elliptical orbit of electron-motion about a centre of force, i.e. nucleus, like planetary motion in Kepler orbits. The axes of the ellipse may be evaluated by writing the polar equation of an ellipse:

$$\dfrac{1}{r} = \dfrac{1}{a}\left(\dfrac{1 - \epsilon \cos\theta}{1 - \epsilon^2} \right) \qquad ..(68)$$

with $\qquad \sqrt{1 - \epsilon^2} = \dfrac{b}{a}$

where a is the semimajor axis and b, the semiminor axis.

Taking logarithmic differentiation of r with respect to θ, we have

$$-\dfrac{1}{r^2}\dfrac{dr}{d\theta} = \dfrac{1}{a}\dfrac{\varepsilon \sin\theta}{1 - \epsilon^2}$$

$$\dfrac{1}{r}\dfrac{dr}{d\theta} = -\dfrac{\epsilon \sin\theta}{1 - \epsilon \cos\theta} \qquad \left(\because \text{ from eqn 68 } \dfrac{r}{a} = \dfrac{1 - \epsilon^2}{1 - \epsilon \cos\theta} \right)$$

or $\qquad \left(\dfrac{1}{r}\dfrac{dr}{d\theta} \right)^2 = \dfrac{\epsilon^2 \sin^2\theta}{(1 - \cos\theta)^2} \qquad ..(69)$

Also from eqn (68)

$$(1 - \epsilon \cos\theta) = a(1 - \epsilon^2)/r$$

or $\qquad \epsilon^2 \cos^2\theta = [1 - a(1 - \epsilon^2)/r]^2$

or $\qquad \epsilon^2 \sin^2\theta = \epsilon^2 - \epsilon^2 \cos^2\theta$

$$= \varepsilon^2 - 1 - \dfrac{a^2\left(1 - \varepsilon^2\right)^2}{r^2} + \dfrac{2a\left(1 - \epsilon^2\right)}{r}$$

$$\therefore \quad \frac{\epsilon^2 \sin^2\theta}{\left(1 - \epsilon\cos\theta\right)^2} = \frac{-\left(1 - \epsilon^2\right) - \dfrac{a^2\left(1 - \epsilon^2\right)^2}{r^2} + \dfrac{2a\left(1 - \epsilon^2\right)}{r}}{a^2\left(1 - \epsilon^2\right)^2 / r^2} \qquad ..(70)$$

Consequently from eqns (69 and 70)

$$\left(\frac{1}{r}\frac{dr}{d\theta}\right)^2 = -\frac{r^2}{a^2\left(1 - \epsilon^2\right)} - 1 + \frac{2r}{a\left(1 - \epsilon^2\right)} \qquad ..(71)$$

This equation is comparable with eqn (67). Comparing coefficients of r^2 and r, we get

$$\frac{2mE}{p_\theta^2} = -\frac{1}{a^2\left(1 - \epsilon^2\right)}$$

and
$$\frac{mZe^2}{p_\theta^2} = \frac{1}{a\left(1 - \epsilon^2\right)}$$

Now $1 - \epsilon^2 = b^2/a^2$, hence the above two equations become

$$\frac{2mE}{p_\theta^2} = -\frac{1}{b^2} \qquad ..(72)$$

$$\frac{mZe^2}{p_\theta^2} = \frac{a}{b^2} \qquad ..(73)$$

From eqns (72) and (73)

$$\frac{2mE / p_\theta^2}{m^2 Z^2 e^4 / p_\theta^4} = -\frac{b^2}{a^2}$$

or
$$E = -\frac{m^2 Z^2 e^4 \left(b^2 / a^2\right)}{2 p_\theta^2} \qquad ..(74)$$

From eqn (73)

$$a = \frac{mZe^2 \left(b^2\right)}{p_\theta^2} \qquad ..(75)$$

Equations (74) and (75) describe energy and size-shape relations of the electron's orbit in terms of its angular momentum. Now, applying Wilson Sommerfeld quantization conditions, viz.

$$\oint p_r dr = n_r h \qquad ..(76)$$

$$\oint p_\theta d\theta = kh \qquad ..(77)$$

where n_r and k are *radial* and *azimuthal quantum numbers* respectively and h is Planck's constant. Taking integration over one complete orbit, the second integral yields

$$p_\theta \int_0^{2\pi} d\theta = kh \text{ or } p_\theta = k\frac{h}{2\pi} \quad\quad ..(78)$$

We now designate the orbital angular momentum as I, so

$$|I| = k\frac{h}{2\pi} \quad\quad ..(79)$$

This restriction is the same as that on orbital angular momentum in Bohr's theory. To evaluate the integral in eqn (76), we notice

$$p_r = m\dot{r} = m\frac{dr}{dt} = m\frac{dr}{d\theta}\cdot\frac{d\theta}{dt}$$

$$= \frac{p_\theta}{r^2}\frac{dr}{d\theta} \quad\quad (\because p_\theta = mr^2\dot\theta)$$

$$p_r dr = p_\theta \frac{1}{r^2}\cdot\frac{dr}{d\theta}\left(\frac{dr}{d\theta}\cdot d\theta\right)$$

$$= p_\theta\left(\frac{1}{r}\frac{dr}{d\theta}\right)^2 d\theta$$

$$= p_\theta \frac{\epsilon^2 \sin^2\theta}{(1-\epsilon\cos\theta)^2}d\theta \quad\quad \text{(using eqn (69))}$$

Hence from eqn (76)

$$p_\theta = \int_0^{2\pi} \frac{\epsilon^2 \sin^2\theta}{(1-\epsilon\cos\theta)^2}d\theta = n_r h$$

The value of the definite integral in the above expression is $2\pi\left(\frac{1}{\sqrt{1-\epsilon^2}}-1\right)$. Thus,

$$2\pi p_\theta\left(\frac{1}{\sqrt{1-\epsilon^2}}-1\right) = n_r h$$

or $$2\pi\frac{kh}{2\pi}\left(\frac{1}{\sqrt{1-\epsilon^2}}-1\right) = n_r h$$

or $$\frac{1}{\sqrt{1-\epsilon^2}}-1 = \frac{n_r}{k}$$

or $\qquad \sqrt{1-\epsilon^2} = \dfrac{k}{n_r + k}$

or $\qquad\qquad \dfrac{b}{a} = \dfrac{k}{n_r + k}$ ..(80i)

Since n_r and k are integers, we put

$\qquad n_r + k = n$ (an integer)

Here n is called principal (or total) quantum number. In order to exclude orbits passing through the nucleus, Sommerfeld postulated that the azimuthal quantum number $k \neq 0$, i.e.

$$k = 1, 2, 3, \ldots$$

However radial quantum number

$$n_r = 0, 1, 2, \ldots$$

Hence $\qquad n = 1, 2, 3, 4, \ldots$

For a given value n, the k can assume only the value

$$k = 1, 2, 3 \ldots n$$

Now eqn (80i) becomes

$$\dfrac{b}{a} = \dfrac{k}{n} \qquad\qquad ..(80ii)$$

This is the quantum condition for elliptical orbits.

Now, applying the quantum conditions $p_\theta = kh/(2\pi)$ and $b/a = k/n$, in eqns (74) and (78), we get

$$E = -\dfrac{2\pi^2 m Z^2 e^4}{n^2 h^2} \qquad\qquad ..(81)$$

and $\qquad a = \dfrac{n^2 h^2}{4\pi^2 m Z e^2} = a_0 \dfrac{n^2}{Z} \qquad ..(82)$

and $\qquad b = a\dfrac{k}{n} = a_0 \dfrac{nk}{Z} \qquad\qquad ..(83)$

where a_0 is the radius of the smallest Bohr orbit of hydrogen atom.

From eqn (81), we see that this energy expression is exactly identical with that for the Bohr's circular orbit, i.e. the energy still depends on the principal quantum number n and is independent of the azimuthal quantum number k.

The size and shape of Sommerfeld's orbits are determined by eqns (82) and (83). Let us now consider few cases for the hydrogen atom ($Z = 1$).

For $n = 1$, we take $k = 1$, so we have

$$a = a_0$$
$$b = a_0$$

This is a circular orbit of radius a_0 which is the same as the Bohr orbit. For $n = 2$, we have

(*i*) $k = 2$ so $a = 4a_0$
                       $b = 4a_0$

(*ii*) $k = 1$ so $a = 4a_0$
                       $b = 2a_0$

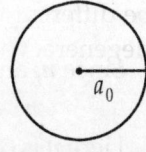

Fig. 2.7 (a)

i.e. for $n = 2$, we have a Bohr's circular orbit of radius $4a_0$ and an elliptical orbit of semimajor axis $4a_0$ and semiminor axis $2a_0$.

Similarly, we can find that for $n = 3$, there are possible three orbits. Thus, for any given value of n, there are n different quantized allowed orbits. One of these is circular (just the one described by the original Bohr theory) and the others are elliptical, all having the same semimajor axis as the radius of the circular orbit but different semiminor axes. The orbit having the lowest k value is most elliptical.

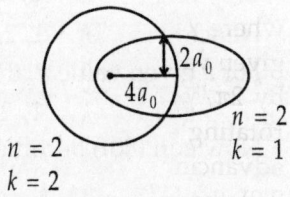

Fig. 2.7 (b)

From eqn 81, we find that Sommerfeld's introduction of elliptical orbits adds no new energy levels (the total energy of the electron is just the same as Bohr's theory expression). Hence, it fails to explain the fine structure.

To designate an orbit of given n and k values, the value of n is followed by one of the letters, $s, p, d, f, g,$ ...according as $k = 1, 2, 3,$ ...respectively. The several orbits characterised by a common value of energy, i.e. same n but different k, are said to be degenerate.

2.10 SOMMERFELD'S RELATIVISTIC CORRECTION

Sommerfeld considered the relativistic variation of mass of an electron in Bohr orbit and thus removed the degeneracy in total energy of the electron for different azimuthal quantum number k values for a given n. For an electron in the innermost orbit of the H atom, v/c equals 10^{-2} or less. Although this gives a relativistic correction to the total energy of the order 10^{-4}, yet it is just of the order of the splitting of energy levels required to explain the observed fine structure of hydrogen spectral lines.

In an elliptic orbit, the velocity of the electron is not constant and is largest near the perihelion. The extent of the relativistic correction depends upon the average velocity of the electron which in turn depends upon the ellipticity of the orbit. Hence, the correction will be different for different azimuthal quantum numbers k's and the degeneracy will be removed. Using the relativistic expression

$$K = m_0 c^2 \left\{ \frac{1}{\sqrt{1 - v^2/c^2}} - 1 \right\} \qquad ..(84)$$

for the kinetic energy of the electron, Sommerfeld obtained the equation for the path of the electron as

$$\frac{1}{r} = \frac{1}{a} \left\{ \frac{1 + \epsilon \cos(\gamma \theta)}{1 - \epsilon^2} \right\} \qquad ..(85)$$

where γ is a constant (< 1). This shows that r does not return to a given value when θ increases by 2π, rather only when it increases by $2\pi/\gamma$ (which is greater than 2π). Thus, the path instead of a rotating ellipse, is like a rosette with the perihelion of the orbit advancing in the same direction as the rotation of the electron (Fig. 2.8), the advance per revolution being $\{(2\pi/\gamma) - 2\pi\}$.

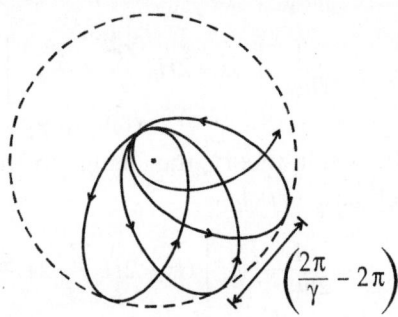

Fig. 2.8: The rosette path of an electron in an atom due to relativistic effect

The level splitting due to different k values (but level having the same n value) may be calculated as follows:

$$\text{mass } m = m_0 \left(1 - \frac{v^2}{c^2} \right)^{1/2} \qquad ..(86)$$

Hamiltonian $H = K + V$

$$= \left\{ \left(p^2 c^2 + m_0^2 c^4 \right)^{1/2} - m_0 c^2 \right\} + V \qquad ..(87)$$

or
$$H = m_0 c^2 \left(1 + \frac{p^2 c^2}{m_0^2 c^4} \right)^{1/2} - m_0 c^2 + V$$

$$= m_0 c^2 \left[1 + \frac{p^2 c^2}{2 m_0^2 c^4} - \frac{p^4 c^4}{8 m_0^4 c^8} \right] - m_0 c^2 + V$$

(using binomial theorem)

$$= \frac{p^2}{2 m_0} - \frac{p^4}{8 m_0^3 c^2} + V$$

$$= H_0 - \frac{p^4}{8 m_0^3 c^2} \qquad ..(88)$$

where $H_0 = \dfrac{p^2}{2 m_0} + V$ is the unperturbed Hamiltonian. Perturbation

$$H' = -\frac{p^4}{8 m_0^3 c^2}$$

$$= -\frac{4 m_0^2}{8 m_0^3 c^2} \left(H_0 - V \right)^2 \qquad (\because p^2 = 2 m_0 \left(H_0 - V \right))$$

$$= -\frac{1}{2 m_0 c^2} \left[H_0^2 + 2 H_0 \frac{Z e^2}{r} + \frac{Z^2 e^4}{r^2} \right] \qquad ..(89)$$

(where Ze = nuclear charge)

Using the perturbation theory, the change in energy

$$\Delta E = \int \psi * H' \psi \, dv$$

$$= -\frac{1}{2 m_0 c^2} \int \psi * \left[H_0^2 + 2 H_0 \frac{Z e^2}{r} + \frac{Z^2 e^4}{r^2} \right] \psi \, dv$$

$$= -\frac{1}{2 m_0 c^2} \left(E_0^2 + 2 Z e^2 E_0 \left\langle \frac{1}{r} \right\rangle + Z^2 e^4 \left\langle \frac{1}{r^2} \right\rangle \right)$$

Now, from Bohr's theory

$$E_0 = -\frac{m_0 Z^2 e^4}{2 \hbar^2 n^2} \qquad ..(90)$$

$$\left\langle \frac{1}{r} \right\rangle - \frac{Z}{a_0 n^2} \quad \text{and} \quad \left\langle \frac{1}{r^2} \right\rangle = \frac{Z^2}{a_0^2 n^3 \left(l + \frac{1}{2} \right)}$$

where a_0 is the Bohr radius = 0.528 Å

$$\therefore \quad \Delta E = -\frac{1}{2m_0c^2}\left[\frac{m_0^2e^8Z^4}{4\hbar^4n^4} - \frac{2m_0Z^2e^4}{2\hbar^2n^2}\cdot\frac{Z^2e^2}{a_0n^2} + \frac{Z^4e^4}{a_0^2n^3\left(l+\frac{1}{2}\right)}\right]$$

$$= -\frac{1}{2m_0c^2}\left[\frac{m_0^2e^8Z^4}{4\hbar^4n^4} - \frac{2m_0Z^4e^4}{2\hbar^2n^4}\frac{4\pi^2m_0^2}{h^2} + \frac{Z^4e^4}{n^3\left(l+\frac{1}{2}\right)}\times\left(\frac{4\pi^2m_0^2}{h^2}\right)^2\right]$$

$$\left(\text{because } a_0 = \frac{h^2}{4\pi^2m_0^2}\right)$$

$$= -\frac{m_0}{c^2}\cdot\frac{2\pi^4}{h^4}\cdot\frac{e^8Z^4}{n^4} + \frac{8\pi^4m_0e^8Z^4}{c^2h^4n^4} - \frac{8\pi^2m_0e^4Z^4}{c^2n^3\left(l+\frac{1}{2}\right)h^4}$$

$$= \frac{6\pi^4m_0e^8Z^4}{n^4ch^3(ch)} - \frac{8\pi^2m_0e^4Z^4}{n^3\left(l+\frac{1}{2}\right)ch^3\cdot ch}$$

or
$$\Delta E = \frac{3R\alpha^2chZ^4}{4n^4} - \frac{Rch\alpha^2Z^4}{n^3\left(l+\frac{1}{2}\right)} \qquad ..(91)$$

$$\left(\because R = \frac{2\pi^2me^4}{ch^3} \text{ and } \alpha = \frac{2\pi e^2}{ch}\right)$$

$$\therefore \quad \Delta T = -\frac{R\alpha^2Z^4}{n^3}\left[\frac{1}{\left(l+\frac{1}{2}\right)} - \frac{3}{4n}\right] \quad \left(\Delta T = \frac{\Delta E}{ch}\right) \qquad ..(92)$$

Here R is the Rydberg constant and α is the fine structure constant.

$l+\frac{1}{2} = k$, the azimuthal quantum number. ΔT is positive since $k \leq n$.

2.11 FINE STRUCTURE OF SPECTRAL LINES

Since a spectral line arises due to the transition of atom from one energy state to another, in the interpretation of fine structure of a spectral line, one must necessarily start with the determination of the different possible energy states. The different energy states of the atom between which transitions can take place are known in spectroscopy by the name the spectral terms. The problem of fine structure is thus essentially a problem of spectral terms when the latter is solved, the former is readily understood (subject to the selection rules). In fact, the appearance of faint lines, or first spectral lines in some series was discovered by Rydberg and also by Kayser and Runge. Even the hydrogen spectrum, the simplest of all the systems, is observed to have a finest structure. At an early date, Michelson studied the Balmer lines with an interferrometer and found that both H_α and H_β are close doublets with separations of only 0.14 Å and 0.8 Å or 0.32 cm^{-1} and 0.329 cm^{-1} respectively. Many subsequent investigations by others have confirmed these results.

2.12 LIMITATIONS OF THE BOHR-SOMMERFELD THEORY

Although the old quantum mechanical put forth by Bohr and Sommerfeld succeeded in predicting the elementary aspects of atomic spectra of hydrogen and hydrogen-like atoms yet there were some limitations:

(i) It could not explain the variation in intensity of the spectral lines.

(ii) The theory is applicable only to one electron atoms. It, however, fails when applied to explain the spectra of atoms having more than one valence electrons, e.g. neutral He-atom (having only two electrons).

(iii) Even in hydrogen-like atoms, the theory is silent on the observed fine-structure of spectral lines.

(iv) No logical reason for assuming different quantum numbers was given. Though, by assuming elliptical orbits in addition to the circular ones and by separately quantizing the angular and the radial motion, Sommerfeld contributed substantially to the problem, however, only with the introduction of electron spin (by Uhlenbeck and Goudsmit), the vector atom model was capable of predicting the exact position of the energy levels of atoms.

(v) The theory did not throw any light on the distribution and arrangement of electrons in atoms.

(vi) The theory could not completely explain the effect of magnetic field on spectral lines, particularly, the anomalous Zeeman effect.

2.13 DE BROGLIE'S MATTER WAVES

The photon concept of electromagnetic radiation led to a dual nature of light. In view of dual nature of light, some physicists were wondering if a theory could be developed in which electron also showed a dual nature of a particle as well as of a wave. Further, according to Bohr's theory, the stable states of electrons in the atom were governed by 'integer' rules. Since the only phenomena involving integers in physics were those of interference and modes of vibration of a stretched string (both of which imply wave motion), so the electrons in their stable states could not be regarded simply as material particles but a certain intrinsic periodicity should be assigned to them.

These consideration led Louis de Broglie to the idea of matter waves. A moving particle has always got a wave associated with it and the particle is controlled by the wave in a manner similar to that in which a photon is controlled by waves.

For a photon of frequency ν, energy $E = h\nu$ and momentum

$$p = \frac{E}{c} = \frac{h\nu}{c} \text{ so that photon wavelength}$$

$$\lambda = \frac{c}{\nu} = \frac{h}{p} \qquad (\because c = \lambda) \qquad \qquad ..(93i)$$

According to de Broglie, this equation applies also to material particles: For a particle of mass m and velocity v, (i.e. momentum $p = mv$), the de Broglie wavelength is given by $\lambda = \dfrac{h}{mv}$,

where $m = \dfrac{m_0}{\sqrt{1 - v^2/c^2}}$ \qquad\qquad ..(93ii)

is its relativistic mass (m_0 being the rest mass).

2.14 THE SCHRÖDINGER EQUATION OF QUANTUM MECHANICS

Matter waves of de Broglie indicated the existence of a more general theory of atomic structure than that of Bohr. Irwin Schrödinger was the first to show that ad-hoc assumptions of Bohr's theory could be replaced by a set of postulates in which there was no a-priori assumption of quantum numbers. He noted that the most characteristic feature of waves is their interference behavior and he modified the classical wave equation such that its solutions included the features of the new mechanics as indicated by the de Broglie equation and the uncertainty principle of Heisenberg. This led Schrödinger to develop the Schrödinger equation. A procedure by which the equation can be constructed is as follows:

Consider the one-dimensional wave equation

$$\frac{\partial^2 \psi}{\partial x^2} = \frac{1}{c^2} \frac{\partial^2 \psi}{\partial t^2} \qquad ..(94)$$

where c is the wave-velocity and $\psi \equiv \psi(x, t)$ is an unknown function representing the amplitude of the wave. A solution of this equation can be written as

$$\psi(x, t) = A_0\, e^{2\pi i \left(\frac{x}{\lambda} - \nu t\right)} \qquad ..(95)$$

where A_0 is a constant and $\nu = c/\lambda$. From de Broglie relationship, we have

$$\lambda = \frac{h}{p_x} \qquad ..(96)$$

and from Planck's hypothesis

$$\nu = \frac{E}{h} \qquad ..(97)$$

where E is the total energy and p_x is the x-component of momentum, consequently

$$\psi(x, t) = A_0\, e^{2\pi i (xp_x - Et)/h} \qquad ..(98)$$

Differentiating ψ once w.r.t. t and rearranging the terms, we get

$$-\frac{h}{2\pi i} \frac{\partial \psi}{\partial t} = E\psi \qquad ..(99)$$

Differentiating eqn. (98) once w.r.t. x and rearranging we get

$$\frac{h}{2\pi i} \frac{\partial \psi}{\partial t} = p_x\, \psi \qquad ..(100)$$

Equations (99) and (100) may be regarded as operator equations where classical linear momentum and the classical energy have been replaced by the following operators:

$$p_x \to -i\hbar \frac{\partial}{\partial x}$$

$$E \to i\hbar \frac{\partial}{\partial t} \qquad ..(101)$$

where
$$\hbar = \frac{h}{2\pi}$$

Now, for a particle of mass m in a conservative potential $V(x)$, the total energy (classically) is given by

$$E = \frac{p_x^2}{2m} + V(x) \qquad ..(102)$$

Using the linear momentum operator from eqn. (101) and replacing $V(x)$ by potential energy operator V_{op}, we can construct the total energy operator as:

$$E_{op} = -\frac{\hbar^2}{2m}\frac{\partial^2}{\partial x^2} + V_{op} \qquad ..(103)$$

Equation (102) can then be written in the operator form as

$$\left(-\frac{\hbar^2}{2m}\frac{\partial^2}{\partial x^2} + V_{op}\right)\psi = i\hbar\frac{\partial}{\partial t}\psi \qquad ..(104)$$

This is the Schrödinger equation in one dimension.

The corresponding equation in three dimensions can be obtained by replacing $\frac{\partial^2}{\partial x^2}$ by

$$\nabla^2 \equiv \frac{\partial^2}{\partial x^2} + \frac{\partial^2}{\partial y^2} + \frac{\partial^2}{\partial z^2} \qquad ..(105)$$

This is known as the Laplacian operator. The $V_{op}(x)$ is replaced by $V_{op}(x, y, z)$ and $\psi(x, t)$ by $\psi(x, y, z, t)$:

$$\left(-\frac{\hbar^2}{2m}\nabla^2 + V_{op}\right)\psi = i\hbar\frac{\partial\psi}{\partial t} \qquad ..(106)$$

This is known as the Schrodinger's time-dependent equation. This is a linear homogeneous second order differential equation and if $\psi_1(x, y, z, t)$, $\psi_2(x, y, z, t)$ ...etc. are the solutions of this equation and a_1, a_2 ...etc. are arbitrary constants depending only on time (real or complex), then

$$\psi = \sum_{n=1}^{\infty} a_n(t)\,\psi_n \qquad ..(107)$$

also represents a solution. (This is known as the principle of super position).

The function $\psi(x, y, z, t)$ is such that it contains all possible meaningful information about the state of the system. Such functions are called state functions. Then, the probability of finding the system in the volume element $dx\,dy\,dz$ at the time t is given by

$$P \equiv \psi^*(x, y, z, t) \cdot \psi(x, y, z, t) \qquad ..(108)$$

Equation (106) may also be written as

$$H\psi(x, y, z, t) = i\hbar \frac{\partial \psi}{\partial t}(x, y, z, t) \qquad ..(109)$$

Here H is known as the *Hamiltonian* (the total energy operator)

$$H = -\frac{\hbar^2}{2m} \nabla^2 + V_{op}(x, y, z) \qquad ..(110)$$

The general solution of the Schrödinger eqn. (109) is given by

$$\psi(x, y, z, t) = \sum_n a_n(t)\, u_n(x, y, z) \qquad ..(111)$$

with $\qquad a_n(t) = \int u_n^*(x, y, z)\, \psi(x, y, z, t)\, dx\,dy\,dz \qquad ..(112)$

Substitution of eqn. (111) into the wave eqn. (109) gives

$$i\hbar \sum_n u_n(\vec{r}) \frac{\partial}{\partial t} a_n(t) = \sum_n a_n(t)\, E\, u_n(r) \qquad ..(113)$$

$$\Rightarrow i\hbar \frac{\partial a_n(t)}{\partial t} = E\, a_n(t) \qquad ..(114)$$

Integrating,

$$a_n(t) = a_n(0)\, e^{-iEt/\hbar} \qquad ..(115)$$

Thus

$$\psi(\vec{r}, t) = \sum_n a_n(0)\, e^{-iEt/\hbar}\, u_n(r) \qquad ..(116)$$

or $\qquad \psi(x, y, z, t) = \psi_n(x, y, z)\, e^{-iEt/\hbar} \qquad ..(117)$

where $\qquad \psi_n(x, y, z) \equiv \sum_n a_n\, u_n(r) \qquad ..(118)$

$\psi_n(x, y, z)$ are known as stationary state functions because the probability of locating the particle is now dependent only on the space coordinates, i.e.

$$P \equiv \psi_n^* \psi_n = \psi_n^*(x, y, z)\, \psi_n(x, y, z) \qquad ..(119)$$

The stationary state functions ψ_n satisfy the time independent Schrödinger equation

$$H\psi_n(x, y, z) = E_n \psi_n(x, y, z) \qquad ..(120)$$

where H is the Hamiltonian operator and E_n are the energy eigenvalues characterised by the set of quantum numbers n. Satisfactory solutions E_n are obtained only if $\psi_n(x, y, z)$ satisfy the following conditions:

1. $\psi_n(x, y, z)$ must vanish at the boundary of the system and $\int \psi_n^* \psi_n \, dx \, dy \, dz$ must be finite for the integration over all configuration space.
2. $\psi_n(x, y, z)$ must be single valued.
3. $\psi_n(x, y, z)$ and its gradient must be continuous.

2.15 SCHRÖDINGER EQUATION FOR SPHERICALLY SYMMETRIC POTENTIAL

In the case of spherically symmetric potential, the potential energy has the form $V(r)$, r being the distance of the particle from the centre of force (the origin). Then Schrödinger equation is given by

$$\nabla^2\psi + \frac{2m}{\hbar^2}[E - V(r)]\psi = 0 \qquad ..(121)$$

The symmetry of the problem permits us to use spherical polar coordinates (r, θ, ϕ).

Let P be a point in space at the position of the particle having Cartesian coordinates (x, y, z). The spherical coordinates are given by

$$x = r \sin \theta \cos \phi$$
$$y = r \sin \theta \sin \phi \qquad ..(122)$$
$$z = r \cos \theta$$

$$\Rightarrow \qquad x^2 + y^2 + z^2 = r^2 \quad \text{or} \quad r = \sqrt{x^2 + y^2 + z^2}$$

$$\tan \phi = \frac{y}{x}; \quad \tan \theta = \frac{\sqrt{x^2 + y^2}}{z} \qquad ..(123)$$

$$\therefore \quad \frac{\partial r}{\partial x} = \sin \theta \cos \phi; \quad \frac{\partial r}{\partial y} = \sin \theta \sin \phi; \quad \frac{\partial r}{\partial z} = \cos \theta$$

$$\frac{\partial \theta}{\partial x} = \frac{\cos \theta \cos \phi}{r}; \quad \frac{\partial \theta}{\partial y} = \frac{\cos \theta \sin \phi}{r}; \quad \frac{\partial \theta}{\partial z} = -\frac{\sin \theta}{r} \qquad ..(124)$$

$$\frac{\partial \phi}{\partial x} = -\frac{\sin \phi}{r \sin \theta}; \quad \frac{\partial \phi}{\partial y} = \frac{\cos \phi}{r \sin \theta}; \quad \frac{\partial \phi}{\partial z} = 0$$

Thus,

$$\frac{\partial \psi}{\partial x} = \frac{\partial \psi}{\partial r} \cdot \frac{\partial r}{\partial x} + \frac{\partial \psi}{\partial \theta} \cdot \frac{\partial \theta}{\partial x} + \frac{\partial \psi}{\partial \phi} \frac{\partial \phi}{\partial x}$$

$$= \sin \theta \cos \phi \, \frac{\partial \psi}{\partial r} + \frac{\cos \theta \cos \phi}{r} \frac{\partial \psi}{\partial \theta} - \frac{\sin \phi}{r \sin \theta} \frac{\partial \psi}{\partial \phi}$$

$$\therefore \quad \frac{\partial}{\partial x} = \sin\theta\cos\phi \frac{\partial}{\partial r} + \frac{\cos\theta\cos\phi}{r}\frac{\partial}{\partial\theta} - \frac{\sin\phi}{r\sin\theta}\frac{\partial}{\partial\phi} \qquad ..(125)$$

Now $\dfrac{\partial^2\psi}{\partial x^2} = \dfrac{\partial}{\partial x}\left(\dfrac{\partial\psi}{\partial x}\right)$

$$= \left(\sin\theta\cos\phi \frac{\partial}{\partial r} + \frac{\cos\theta\cos\phi}{r}\frac{\partial}{\partial\theta} - \frac{\sin\phi}{r\sin\theta}\frac{\partial}{\partial\phi}\right)$$

$$\times\left(\sin\theta\cos\phi \frac{\partial\psi}{\partial r} + \frac{\cos\theta\cos\phi}{r}\frac{\partial\psi}{\partial\theta} - \frac{\sin\phi}{r\sin\theta}\frac{\partial\psi}{\partial\phi}\right) \quad ..(126)$$

similarly

$$\frac{\partial}{\partial y} = \left(\sin\theta\sin\phi \frac{\partial}{\partial r} + \frac{\cos\theta\cos\phi}{r}\frac{\partial}{\partial\theta} - \frac{\cos\phi}{r\sin\theta}\frac{\partial}{\partial\phi}\right) \quad ..(127)$$

so that

$$\frac{\partial^2\psi}{\partial y^2} = \frac{\partial}{\partial y}\left(\frac{\partial\psi}{\partial y}\right) = \left(\sin\theta\sin\phi \frac{\partial}{\partial r} + \frac{\cos\theta\cos\phi}{r}\frac{\partial}{\partial\theta} - \frac{\cos\phi}{r\sin\theta}\frac{\partial}{\partial\phi}\right)$$

$$\times\left(\sin\theta\sin\phi \frac{\partial\psi}{\partial r} + \frac{\cos\theta\cos\phi}{r}\frac{\partial\psi}{\partial\theta} - \frac{\cos\phi}{r\sin\theta}\frac{\partial\psi}{\partial\phi}\right) \quad ..(128)$$

Also, $\quad \dfrac{\partial\psi}{\partial z} = \dfrac{\partial\psi}{\partial r}\cdot\dfrac{\partial r}{\partial z} + \dfrac{\partial\psi}{\partial\theta}\cdot\dfrac{\partial\theta}{\partial z}$

$$= \left(\cos\theta \frac{\partial\psi}{\partial r} + \frac{(-\sin\theta)}{r}\cdot\frac{\partial\psi}{\partial\theta}\right) \qquad ..(129)$$

$$\therefore \quad \frac{\partial}{\partial z} = \cos\theta \frac{\partial}{\partial r} - \frac{\sin\theta}{r}\frac{\partial}{\partial\theta}$$

and $\quad \dfrac{\partial^2\psi}{\partial z^2} = \dfrac{\partial}{\partial z}\left(\dfrac{\partial\psi}{\partial z}\right) = \left(\cos\theta \dfrac{\partial}{\partial r} - \dfrac{\sin\theta}{r}\cdot\dfrac{\partial}{\partial\theta}\right)$

$$\times\left(\cos\theta \frac{\partial\psi}{\partial r} - \frac{\sin\theta}{r}\frac{\partial\psi}{\partial\theta}\right) \qquad ..(130)$$

Adding eqns (126), (128) and (130) and simplifying we get

$$\nabla^2\psi = \frac{\partial^2\psi}{\partial x^2} + \frac{\partial^2\psi}{\partial y^2} + \frac{\partial^2\psi}{\partial z^2}$$

$$= \left[\frac{1}{r^2}\frac{\partial}{\partial r}\left(r^2\frac{\partial}{\partial r}\right) + \frac{1}{r^2\sin\theta}\frac{\partial}{\partial\theta}\left(\sin\theta\frac{\partial}{\partial\theta}\right)\right.$$

$$\left. + \frac{1}{r^2\sin^2\theta}\frac{\partial^2}{\partial\phi^2}\right]\psi$$

Consequently, from eqn. (121) we get Schrödinger's equation in spherical polar coordinates as

$$\frac{1}{r^2}\frac{\partial}{\partial r}\left(r^2\frac{\partial\psi}{\partial r}\right) + \frac{1}{r^2\sin\theta}\frac{\partial}{\partial\theta}\left(\sin\theta\frac{\partial\psi}{\partial\theta}\right) + \frac{1}{r^2\sin^2\theta}\frac{\partial^2\psi}{\partial\phi^2}$$

$$+ \frac{2m}{\hbar^2}[E - V(r)]\psi = 0. \qquad ..(131)$$

2.16 SCHRÖDINGER EQUATION FOR THE HYDROGEN ATOM

A hydrogen atom consists of an electron circulating around a proton and the Schrödinger wave equation for the electron is

$$\nabla^2\psi + \frac{8\pi^2 m_e}{h^2}(E - V)\psi = 0 \qquad ..(132i)$$

or $$\nabla^2\psi + \frac{2m_e}{\hbar^2}(E - V)\psi = 0$$

We know that, in the ground state, the system has spherical symmetry and in such a situation, the Laplacian operator ∇^2 becomes

$$\nabla^2 = \frac{1}{r^2}\left[\frac{\partial}{\partial r}\left(r^2\frac{\partial}{\partial r}\right) + \frac{1}{\sin\theta}\frac{\partial}{\partial\theta}\left(\sin\theta\frac{\partial}{\partial\theta}\right) + \frac{1}{\sin^2\theta}\frac{\partial^2}{\partial\phi^2}\right]$$

$$..(132ii)$$

using this, eqn. (132i) yields

$$\frac{1}{r^2}\left[\frac{\partial}{\partial r}\left(r^2\frac{\partial}{\partial r}\right) + \frac{1}{\sin\theta}\frac{\partial}{\partial\theta}\left(\sin\theta\frac{\partial}{\partial\theta}\right) + \frac{1}{\sin^2\theta}\frac{\partial^2}{\partial\phi^2}\right]\psi$$

$$+ \frac{8\pi^2\mu}{h^2}(E - V)\psi = 0 \qquad ..(133)$$

Here μ represents the reduced mass of the combined electron mass and proton mass:

$$\mu = \left(\frac{m_e m_p}{m_e + m_p}\right) \qquad ..(134)$$

The potential V is

$$V = \frac{-e^2}{4\pi\epsilon_0 r} \qquad ..(135)$$

2.16.1 Energy and Wave Function for the Ground State

It is appropriate to solve for the energy E_0 and the wave function ψ_0 of the ground state ($n = 1$) first, since this is the simplest electronic energy state of any atom. The solution for ψ_0 turns out to be completely spherically symmetric and relatively simple. Instead of investigating a general solution, we use a trial solution for ψ_0 that is radially symmetric and decays exponentially:

$$\psi_0 = e^{-\alpha r} \qquad\qquad ..(136)$$

where α is a constant to be determined, using this, Schrödinger's equation becomes

$$\left(\alpha^2 + \frac{8\pi^2 \mu E_0}{h^2}\right) e^{-\alpha r} + \left(\frac{2\pi \mu e^2}{\epsilon_0 h^2} - 2\alpha\right) \frac{e^{-\alpha r}}{r} = 0 \quad ..(137)$$

For this equation to be valid for all values of r, both factors in parenthesis must separately be equal to zero. Equating the right hand terms to zero leads to a solution for α of

$$\alpha = \frac{\pi \mu e^2}{\epsilon_0 h^2} = \frac{1}{a_0} \qquad\qquad ..(138)$$

i.e. α represents the reciprocal of the value of the first Bohr radius (with m_e replaced by the reduced mass μ). Using this value of α within the left hand parentheses of eqn. (137) we obtain

$$E_1 = -\frac{\mu e^4}{8 \epsilon_0^2 h^2} \qquad\qquad ..(139)$$

This value is exactly the same as that obtained by Bohr's theory except that the reduced electron-proton mass μ is used instead of electron-mass m_e.

Equation (136) gives only the radial dependence of the wave function ψ_0. To be complete, it has to include a normalization constant C:

$$\psi_0 = C\, e^{-\alpha r} = C\, e^{-r/a_0} \qquad\qquad ..(140)$$

To satisfy the normalization condition for a spherically symmetric volume element $4\pi r^2 dr$ (which would be a shell of radius r and thickness dr), we have

$$= \int_0^\infty C^2 e^{-2r/a_0}\, 4\pi r^2\, dr = 1 \qquad\qquad ..(141)$$

This leads to $C = 1/(\pi^{1/2} a_0^{3/2})$.

Thus, the ground state wave function for the hydrogen can be expressed as

$$\psi_0 = \left(\frac{1}{\pi^{1/2} a_0^{3/2}} \right) e^{-r/a_0} \qquad \ldots (142)$$

This wave function, being spherically symmetric, has only r-dependence. This is depicted in Fig. 2.9.

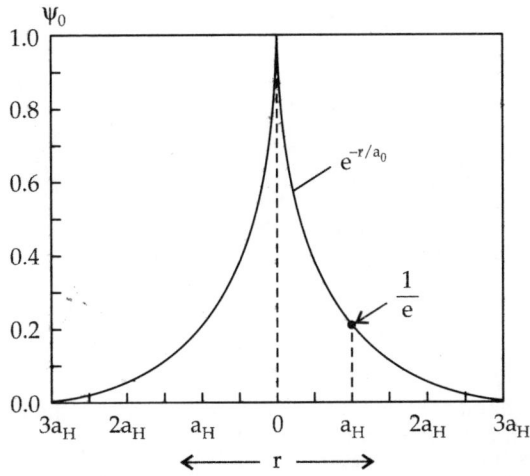

Fig. 2.9: Wave function ψ_0 for the ground state of the hydrogen atom

It has a maximum at $r = 0$ and is reduced by a factor of $1/e$ at $r = a_0$ (the first Bohr radius).

This wave function does not tell us where we would actually find the electron. The probability of finding the electron at any radial distance r within a spherically symmetric shell of thickness dr and volume $4\pi r^2\, dr$ is obtained by considering $|\psi_0|^2\, dV$, i.e.

$$|\psi_0|^2\, dV = \psi_0^* \psi_0\, dV = \psi_0^* \psi_0\, 4\pi r^2\, dr$$
$$= 4\pi r^2\, C^2\, e^{-2r/a_0}\, dr \qquad ..(143)$$

Thus, the probability of finding the electron within a shell of thickness dr is given by

$$4\pi r^2\, C^2\, e^{-2\alpha r} = 4\alpha^3\, r^2\, e^{-2\alpha r} \qquad ..(144)$$

It can be found that the maximum value of this probability occurs at $r = a_0$, the *first Bohr radius*.

Thus, the Schrödinger equation yields a solution for the lowest energy state $n = 1$ (ground state) of the hydrogen atom that is consistent with the Bohr theory. It suggests a "smeared out" distribution in which the electron spends a very high percentage of time near the Bohr-radius but is almost never in that exact solution-contrary to what was suggested by Bohr's theory (Fig. 2.10).

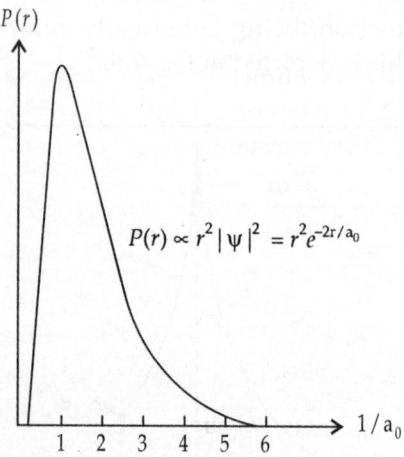

Fig. 2.10: Probability of finding an electron at any specific distance from the centre of the nucleus in units of r/a_0

2.17 EXCITED STATES OF HYDROGEN

The ground state of hydrogen atom was assumed to be spherically symmetric. It was therefore possible to use a trial solution $\psi_0 = e^{-\alpha r}$ that had no angular dependence, i.e. it as independent of θ and ϕ. However, the excited state wave functions are much more complicated than the ground state wave function $\psi_0 = (1/(\pi^{1/2} a_0^{3/2})) e^{-r/a_0}$, consequently, the complete Schrödinger equation (133) must be solved. In this case, a trial solution for ψ can be written as

$$\psi(r,\,\theta,\,\phi) = CR\,(r)\,\Theta(\theta)\,\Phi(\phi) \qquad ..(145)$$

where C is a constant, substituting this solution in Schrödinger's eqn. in spherical polar coordinates (i.e. eqn. (133)) we get

$$\frac{1}{\Phi}\frac{\partial^2\Phi}{\partial\phi^2} + \frac{\sin\theta}{\Theta}\frac{\partial}{\partial\theta}\left(\sin\theta\frac{\partial\Theta}{\partial\theta}\right)$$

$$+ \frac{\sin^2\theta}{R}\frac{\partial}{\partial r}\left(r^2\frac{\partial R}{\partial r}\right) + \sin^2\theta\,\frac{8\,\pi^2\,\mu\,r^2}{h^2}\,(E-V) = 0 \quad ..(146)$$

Rearranging

$$\frac{\sin^2\theta}{R}\frac{\partial}{\partial r}\left(r^2\frac{\partial R}{\partial r}\right) + \frac{\sin\theta}{\Theta}\frac{\partial}{\partial\theta}\left(\sin\theta\frac{\partial\Theta}{\partial\theta}\right)$$

$$+ \frac{2\mu r^2\sin^2\theta}{\hbar^2}(E-V) = -\frac{1}{\Phi}\frac{\partial^2\Phi}{\partial\phi^2} \quad ..(147)$$

The l.h.s. of this eqn is a function of r and θ whereas r.h.s. is a function of ϕ only, this can be correct only if both sides are equal to the same constant. Let the constant be $+ m^2$, i.e.

$$-\frac{1}{\Phi}\frac{\partial^2\Phi}{\partial\phi^2} = m^2 \quad ..(148)$$

and

$$\frac{1}{R}\frac{\partial}{\partial r}\left(r^2\frac{\partial R}{\partial r}\right) + \frac{2\mu r^2}{\hbar^2}(E-V)$$

$$= \frac{m^2}{\sin^2\theta} - \frac{1}{\Theta\sin\theta}\frac{\partial}{\partial\theta}\left(\sin\theta\frac{\partial\Theta}{\partial\theta}\right) \quad ..(149)$$

Again, we have an equation in which different variables appear on each side, requiring that both sides be equal to some same constant. This constant is $l(l + 1)$. Thus, we have following set of three differential equations

$$\frac{1}{\Phi}\frac{\partial^2\Phi}{\partial\phi^2} = -m^2 \quad ..(150)$$

$$\frac{m^2}{\sin^2\theta} - \frac{1}{\Theta\sin\theta}\frac{\partial}{\partial\theta}\left(\sin\theta\frac{\partial\Theta}{\partial\theta}\right) = l(l+1) \quad ..(151)$$

$$\frac{1}{R}\frac{\partial}{\partial r}\left(r^2\frac{\partial R}{\partial r}\right) + \frac{2\mu r^2}{\hbar^2}(E-V) = l(l+1) \quad ..(152)$$

Here V is the coulomb potential

$$V = \frac{e^2}{4\pi\epsilon_0 r} \quad ..(153)$$

Out of the above three separate equations, only eqn. (152) contains the potential function, it follows that the radial function $R(r)$ is the only one which depends on the form of the central field. The angular functions $\Theta(\theta)$ and $\Phi(\phi)$ remain the same for all central fields. Further, eqn. (150) has a solution of the form

$$\Phi = e^{im\phi} \quad ..(154)$$

This function must have the same value at $\phi = 0, 2\pi, 4\pi, 6\pi \ldots$ since that would be one complete revolution around the atom. This would occur only if the constant m has the value 0 or an integer. Thus, we have the following restriction on m:

$$m = 0, \pm 1, \pm 2, \pm 3 \ldots \qquad ..(155)$$

Solving eqns (151) and (152) is somewhat more complex and here, we simply provide the solutions.

Equation (151) has the form of Legendre's differential equation which has solutions only if l is a positive integer with a value greater than or equal to $|m|$. This integer l is called the *azimuthal quantum number*. The solution to Legendre's equation involve terms known as *Legendre polynomials*. These polynomials are of the form $P_l^m (\cos \theta)$ and can be generated by the function

$$\Theta(\theta) = P_l^{|m|} (x) = \frac{(1 - x^2)^{1/2|m|}}{2^l \, l!} \frac{d^{|m|+l}}{d_x^{|m|+l}} (x^2 - 1) \quad ..(156)$$

where $x \equiv \cos \theta$.

Some of these functions are

$$P_0^0(x) = 1 \qquad\qquad P_1^1(x) = (1 - x^2)^{1/2}$$
$$P_1^0(x) = x \qquad\qquad P_2^1(x) = 3x(1 - x^2)^{1/2} \qquad ..(157)$$
$$P_2^0(x) = \frac{1}{2} (3x^2 - 1) \quad P_2^2(x) = 3(1 - x^2)$$

The radial part of the equation viz. eqn. (152) is of the form known as *Laguerre differential equation* and its solutions are given by

$$R(\rho) = \rho^l \, e^{-\rho/2} \, L_{n+l}^{2l+1} (\rho) \qquad\qquad ..(158)$$

where $$\rho \equiv \left(\frac{2r}{n \, a_0} \right)$$

For solutions of the Schrödinger equation, n is referred to as the *principal quantum number* and can have the values $n \geq l + 1$.

The functions $L_{n+l}^{2l+1} (\rho)$ are known as *associated Laguerre polynomials* and can be generated by the following equation.

$$L_{n+l}^{2l+1} (\rho) = \left(\frac{d}{d\rho} \right)^{2l+1} \left[e^\rho \left(\frac{d}{d\rho} \right)^{n+l} \left(\rho^{n+l} \, e^{-\rho} \right) \right] \quad ..(159)$$

As with the Legendre functions, a few examples of the low order Laguerre polynomials are:

$$L_1^1(\rho) = -1 \qquad\qquad L_2^2(\rho) = 2$$

$$L_2^1(\rho) = 2\rho - 4 \qquad L_3^3(\rho) = -6 \qquad ..(160)$$

The radial equation (152) contains the energy E, from which the energy eigenvalues can be obtained as

$$E_n = -\frac{\mu e^4}{8 \, \epsilon_0^2 \, h^2} \left(\frac{1}{n^2}\right) \qquad ..(161)$$

This is identical to the Bohr equation for the energy of various electronic levels or states of the hydrogen atom with μ in place of m_e.

2.18 ALLOWED QUANTUM NUMBERS AND WAVE FUNCTIONS

We have following restrictions on various quantum numbers:

$$n \geq 1 \qquad\qquad ..(162)$$
$$n \geq l + 1 \qquad\qquad ..(163)$$
$$l \geq 0 \qquad\qquad ..(164)$$
$$l \geq |m| \qquad\qquad ..(165)$$
$$m = 0, \pm 1, \pm 2, ... \pm l \qquad\qquad ..(166)$$

We are familiar with the quantum number n from the Bohr theory. n is *principal quantum number*. The *angular momentum quantum number* l can have n values ranging from 0 to $n - 1$. (These values were given the names s, p, d, f, g ...respectively for increasing values of l). The quantum number m is referred to as the *magnetic quantum number* (Table 2.1). From eqn. (166) there are $2l + 1$ values of m for every value of l. Thus, using the above rules, we can write down all the allowed eigenfunctions describing the various possible quantum states (eigenstates) for the hydrogen atom using the notation $\psi_{n,l,m}$ for each state. Thus, there is one possible state for $n = 1$ viz. $\psi_{1,0,0}$, four states for $n = 2$ viz. $\psi_{2,0,0}$, $\psi_{2,1,1}$, $\psi_{2,1,0}$ and $\psi_{2,1,-1}$, nine states for $n = 3$ etc. suggesting that there are n^2 states for every possible value of n, (taking into account all allowed values of l and m) i.e. there are several states all with the same energy for any value of $n > 1$. These energy states are therefore said to be *degenerate*.

We can now write the entire wave function for various states of the hydrogen atom as

$$\psi_{n,l,m} = C\rho^l e^{-\rho/2} L_{n+l}^{2l+1}(\rho) \, P_l^{|m|}(\cos\theta) \, e^{im\phi} \qquad ..(167i)$$

where $$\rho = \frac{2r}{na_0} \qquad\qquad ..(167ii)$$

TABLE 2.1: The four quantum numbers

Symbol	Name	Meaning	Equation	Values
n	Total quantum number	Quantizes the total energy of the electron	$E_n = \dfrac{1}{2}\dfrac{\mu Z^2 e^4}{2\left(4\pi\varepsilon_0\right)^2 n^2 \hbar^2}$	$n = 1, 2, \ldots$
l	Orbital quantum number	Quantizes the magnitude of the total orbital angular momentum	$L = \sqrt{l(l+1)}\,\hbar$	$l = 0, 1, 2, \ldots (n-1)$ (n values of l)
m	Magnetic orbital quantum number	Quantizes the z-component of of the total orbital angular momentum. This is called spatial quantization	$L_z = m\hbar$	$m = -l\ldots, 0, \ldots, +l$ ($2l+1$ values of m)
m_s	Magnetic spin quantum number	Quantizes the z-component of the spin	$s_z = m_s \hbar$	$m_s = -\tfrac{1}{2}, +\tfrac{1}{2}$

The normalization constant C is given by

$$C = -\left\{\left(\frac{2Z}{na_0}\right)^3 \frac{(n-l-1)!}{2n\left[(n+l)!\right]^3}\right\}^{1/2} \left[\left(\frac{2l+1}{4\pi}\right)\frac{(l-|m|)!}{(l+|m|)!}\right]^{1/2} \quad ..(168)$$

Here Z represents charge on the nucleus.

Thus, we can write $\psi_{nlm}(r, \theta, \phi) = R_{nl}(r)\,\Theta_{lm}(\theta)\,\Phi_m(\phi)$

where $\quad \Phi_m(\phi) = \dfrac{1}{\sqrt{2\pi}}\, e^{im\phi}$ $\qquad\qquad ..(169i)$

$$\Theta_{lm}(\theta) = \left[\left(\frac{2l+1}{2}\right)\frac{(l-|m|)!}{(l+|m|)!}\right]^{1/2} P_l^{|m|}(\cos\theta) \quad ..(169ii)$$

and

$$R_{nl}(r) = -\left\{\left(\frac{2Z}{na_0}\right)^3 \frac{(n-l-1)!}{2n\left[(n+l)!\right]^3}\right\}^{(1/2)} \cdot e^{-\frac{Zr}{na_0}}\left(\frac{2Zr}{na_0}\right)^l L_{n+l}^{2l+1}\left(\frac{2Zr}{na_0}\right)$$

$$...(169iii)$$

TABLE 2.2: The hydrogen like wavefunctions

(Here $\rho = Zr/a_0$, $a_0 = \dfrac{4\pi\epsilon_0 \hbar^2}{\mu e^2}$)

n	l	m	$\psi_{nlm}(r, \theta, \phi)$
1	0	0	$\dfrac{1}{\sqrt{\pi}}\left(\dfrac{Z}{a_0}\right)^{3/2} e^{-\rho}$
2	0	0	$\dfrac{1}{4\sqrt{2\pi}}\left(\dfrac{Z}{a_0}\right)^{3/2}(2-\rho)e^{-\rho/2}$
2	1	0	$\dfrac{1}{4\sqrt{2\pi}}\left(\dfrac{Z}{a_0}\right)^{3/2}\rho e^{-\rho/2}\cdot\cos\theta$
2	1	± 1	$\dfrac{1}{8\sqrt{\pi}}\left(\dfrac{Z}{a_0}\right)^{3/2}\rho e^{-\rho/2}\cdot\sin\theta\cdot e^{\pm i\phi}$

2.19 PERIODIC TABLE OF ELEMENTS

The periodic table is built up of elements (atoms) that have various numbers of protons, neutrons and electrons. In developing the periodic table of elements, as more protons and electrons are added to produce new elements, the electrons are added sequentially into shells according to their n, l, m and s quantum numbers beginning with the lowest allowed values and working towards higher values, with the maximum number of electrons determined by the number of protons in the nucleus (maintaining charge neutrality). From equations (162) to (166) we know the restrictions on quantum numbers n, l and m. We also have the restrictions on s of $\pm 1/2$, and also the Pauli exclusion principle indicating that no two electrons can occupy the same state (i.e. all four quantum numbers cannot be the same). With these restrictions, Table 2.3 shows how the various elements evolve according to the way electrons fill the possible quantum states.

TABLE 2.3: Quantum numbers and numbers of electrons for various shells and subshells

Shell	K	L		M			N				O					P		
n	1	2		3			4				5					6		
Subshell	1s	2s	2p	3s	3p	3d	4s	4p	4d	4f	5s	5p	5d	5f	5g	6s	6p	6d
l	0	0	1	0	1	2	0	1	2	3	0	1	2	3	4	0	1	2
Number of electrons	2	2	6	2	6	10	2	6	10	14	2	6	10	14	18	2	6	10

2.19.1 Aufbau Principle

The distribution of electrons in an atom takes places according to the following two principles

(a) The Pauli exclusion principle which implies that the maximum number of electrons in s, p, d, f, g subshells are respectively, 2, 6, 10, 14, 18, etc.

(b) The electrons in a given atom occupy the lowest possible quantum state i.e. one with the lowest energy. The energy of the sublevels increases in the following order:

$1s < 2s < 2p < 3s < 3p < 4s < 3d <$
$4p < 5s < 4d < 5p < 6s < 4f$
$< 5d < 6p < 7s < 5f < 6d$

A simple mnemonic for this sequential order of energy levels is shown in Fig. 2.11.

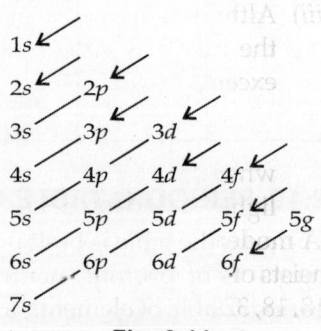

Fig. 2.11

i.e. write the subshells for given shells in rows and read downward along successive diagonals. We can build up the ground state configuration of the atoms in the periodic table by the above 'aufbau' (building up) process beginning with the hydrogen configuration and add (successively) one proton (to the nucleus) and one electron, always placing the additional electron in the unoccupied state of lowest energy. The electron configuration of the atoms of a few elements are

Hydrogen	$(Z = 1)$	:	$1s^1$
Helium	$(Z = 2)$	:	$\boxed{1s^2}$
Lithium	$(Z = 3)$	:	$\boxed{1s^2}\, 2s_1$
Neon	$(Z = 10)$	:	$\boxed{1s^2}\,\boxed{2s^2\, 2p^6}$
Sodium	$(Z = 11)$	:	$\boxed{1s^2}\,\boxed{2s^2\, 2p^6}\, 3s^1$

Here a rectangle denotes a closed shell. For atoms with $Z > 18$, the energy ordering of the subshells as deduced from detailed study of the atomic spectra is given according to in Fig. 2.11.

2.19.2 The Periodic Table

In the periodic table, the elements are arranged in the order of increasing atomic number. The periodic table was first constructed by Mendeleev in 1871 and perfected by Mosley and others from a study of X-ray spectra. It was observed that:

(i) Chemical properties of elements repeat after a certain number of elements

(ii) The atomic number Z characterizes the nuclear charge as well as the number of electrons in a neutral atom.

(iii) Although an increase in Z usually goes hand in hand with the increase in the atomic weight, there are a number of exceptions, e.g.

$$\left(^{40}_{18}\, Ar - ^{39}_{19}\, K\right); \left(^{128}_{52}\, Te - ^{127}_{53}\, I\right)$$

where the element of higher atomic weight precedes the lighter element.

A modern version of the periodic table is shown in Table 2.4. It consists of 7 horizontal rows which are called "periods" having 2, 8, 8, 18, 18, 32 and 17 elements respectively. Across each period, there is a steady transition from an active metal to an inert gas. As a

TABLE 2.4: Periodic table of elements

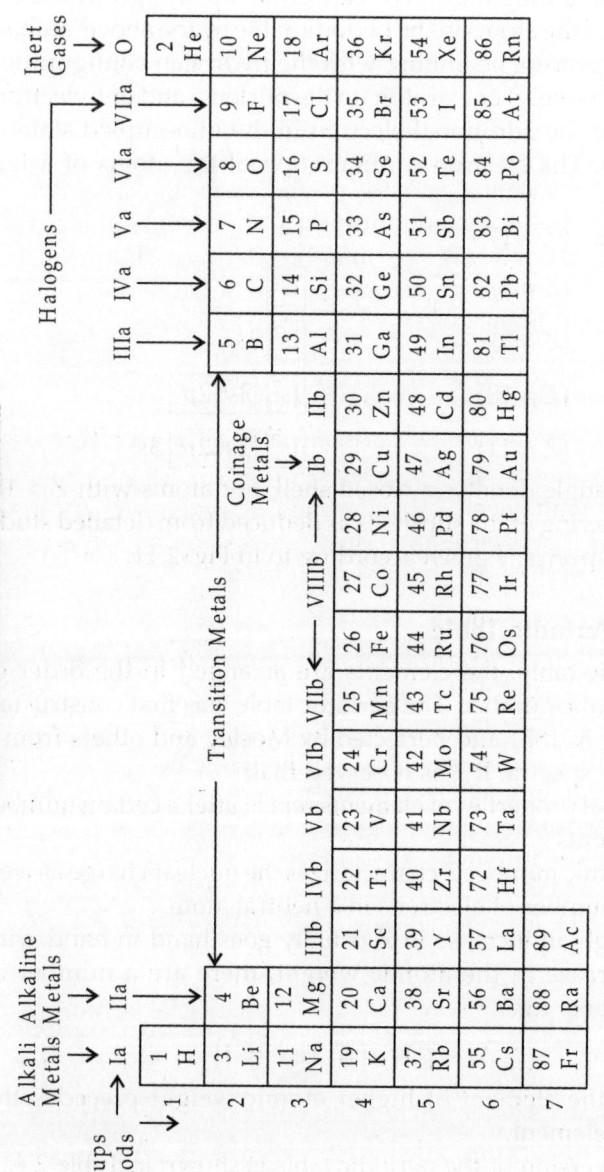

consequence, elements with similar properties occur in vertical columns which are called "groups". For instance, group I contains hydrogen and the alkali metals which are highly active and the last group VIII consists of noble gases all of which are inert.

In each period after the third, a series of "transition elements" appear between the group II and group III elements. The elements of each series show chemical resemblance to one another but are different from the elements of the preceding period. The transition elements of period 6 include fourteen "rare earth elements called "lanthanides" which are virtually indistinguishable in their properties. A similar group of closely related metals called "actinides" occur in period 7.

2.19.3 Periodicity in the Properties of Elements

The periodicity in the properties of elements is connected with the periodicity in the filling of the outer shell which can contain a maximum of eight electrons and which determines the optical as well as chemical properties of atoms. The order in which the shells and subshells are filled is shown in Fig. 2.12.

The first group elements (hydrogen and alkali metals) contain only one electron in the outer shell. This means that the optical terms (with the exception of the s-term) have a doublet structure and the elements are monovalent.

The elements of the second group (alkaline earths) beryllium, magnesium, calcium, etc., have two valence electrons, and hence their spectral terms should be singlets and triplets, while the valency of these elements should be equal to two.

The third group elements have three electrons in their outer shells and hence maximum splitting of their optical terms should be equal to four (quarters) and their maximum valency is equal to three.

In the seventh group comprising halogens, on the contrary, the outer shell lacks one electron. Hence, in addition to the maximum positive valency of seven, they may turn out to be monovalent in ionic compounds, i.e. they may have a negative valency of one.

Finally, in the group of inert gases, the last outer shell is completely filled, while the new one has not yet been occupied by electrons. Hence these elements are said to belong to VIII a (or zeroth) group.

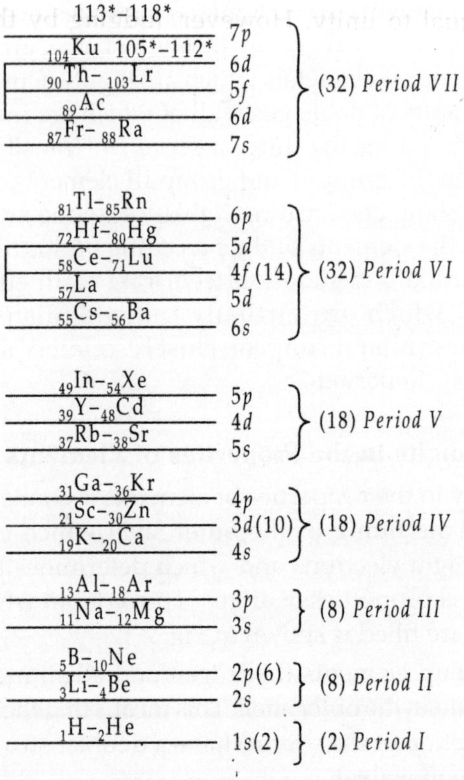

$$
\left.
\begin{array}{l}
\underline{}\;\;113^*-118^* \\[-2pt]
\underline{}\;_{104}\text{Ku}\;\;105^*-112^* \qquad 7p \\[-2pt]
\underline{}\;_{90}\text{Th}-_{103}\text{Lr} \qquad\qquad 6d \\[-2pt]
\underline{}\;_{89}\text{Ac} \qquad\qquad\qquad 5f \\[-2pt]
\underline{}\;_{87}\text{Fr}-_{88}\text{Ra} \qquad\qquad 6d \\[-2pt]
\qquad\qquad\qquad\qquad\qquad 7s
\end{array}
\right\}\;(32)\;\textit{Period VII}
$$

Above items correspond to:

```
        113*-118*
    104Ku 105*-112*        7p  ⎫
    90Th-103Lr             6d  ⎪
    89Ac                   5f  ⎬ (32) Period VII
    87Fr-88Ra              6d  ⎪
                           7s  ⎭

    81Tl-85Rn              6p  ⎫
    72Hf-80Hg              5d  ⎪
    58Ce-71Lu              4f(14) ⎬ (32) Period VI
    57La                   5d  ⎪
    55Cs-56Ba              6s  ⎭

    49In-54Xe              5p  ⎫
    39Y-48Cd               4d  ⎬ (18) Period V
    37Rb-38Sr              5s  ⎭

    31Ga-36Kr              4p  ⎫
    21Sc-30Zn              3d(10) ⎬ (18) Period IV
    19K-20Ca               4s  ⎭

    13Al-18Ar              3p  ⎫
    11Na-12Mg              3s  ⎬ (8) Period III

    5B-10Ne                2p(6) ⎫
    3Li-4Be                2s  ⎬ (8) Period II

    1H-2He                 1s(2) ⎬ (2) Period I
```

Fig. 2.12: Schematic diagram of the filling of the energy levels in the atoms of the Mandeleev Periodic Table. The subshells s and p can lie in the outer shell. The d-subshell can lie in the first inner shell or the following shells. The f subshell can lie from the second inner shell onwards. The atomic numbers with asterisks show the elements that have not been discovered so far.

However, there are a number of exceptions to the general rule of the existence of eight elements in each period. The first exceptions are hydrogen ($Z = 1$) and helium ($Z = 2$) which from the first period. This period contains only two elements instead of eight. This is due to the fact that the K-shell does not include the p-subshell. Consequently, these elements have dual properties in some respects.

As a matter of fact, hydrogen must repeat the chemical and physical properties of alkali metals, in accordance with the number of electrons in the outer shell (e.g. formation of H^+Cl^- and Na^+Cl^-). It is well known that the maximum splitting of spectral terms in both of them (hydrogen and alkali metals) is equal to two, and their valency is equal to unity. However, judging by the number of missing electrons, hydrogen is similar to halogens, (there is one electron less than the number required for filling the outer shell). Hence, it may combine with a free second electron, forming a negatively charged ion like halogen.

According to the number of electrons in the outer shell (two) helium should resemble alkaline earths of the second group. Both in helium and alkaline earths, the spectral terms should be either singlets (spin is equal to zero), or triplets (spin equal to unity). However in its chemical properties, helium is a typical inert gas, since the outer K-shell is completely filled.

The noble gases (^{2}He, ^{10}Ne, ^{18}Ar, ^{36}Kr, ^{54}Xe and ^{86}Rn) are monoatomic and chemically inert. The reason is as follows. In each of the noble gas atoms (except He) a p subshell is just completed, and the atom can have its first excitation only by going to the succeeding s-subshell. But every s-subshell has a particularly large energy gap from the preceding p-subshell. As a result, these atoms are particularly difficult to excite. Further, the total orbital and spin angular momenta of the electrons in completed subshells are zero, thus zero total magnetic dipole moment and the effective charge distributions are perfectly symmetrical. Thus, there is no tendency of interaction with other atoms to produce chemical compounds and very little tendency to condense into liquids or solids.

It is to be noted here that ^{2}He (having configuration $1s^2$) is a noble gas atom because, for it, the first unfilled $2s$ subshell is an s-subshell and has a large energy gap from the preceding $1s$ subshell and since, in ground state the shell is closed, it produces no external fields. On the other hand, an atom like ^{20}Ca ($1s^2$, $2s^2$, $2p^6$, $3s^2$, $3p^6$, $4s^2$) is not a noble gas atom even though it has completed subshells. The reason is that in its first excited state, an electron goes to a $3d$ subshell which is quite near in energy to the ground state.

In the periodic table, a filling of the inner $3d$-subshell takes place from scandium ($Z = 21$) to nickel ($Z = 28$). Iron ($Z = 26$), cobalt ($Z = 27$) and nickel ($Z = 28$) have identical properties and if we take the chemical or physical properties as a criterion for forming groups, these elements are combined into the same group viz. VIIIB. In particular, these elements have typical *ferromagnetic properties* which are due to uncompensated spins of $3d$-electrons in the inner shell.

For the first three rows of periodic table, the properties such as valency, ionization energy, etc. of the elements change uniformly from element to element. However, for succeeding rows, this is not true.

Transitions Metals: In the fourth row, the elements from ^{21}Sc to ^{30}Zn have quite similar chemical properties and almost same ionization energies. They form the first transition series. This observation can be explained as follows. These elements occur during the filling of the 3d subshell whose radius is considerably less than that of the 4s-subshell which is completely filled for the transition elements (except ^{24}Cr and ^{29}Cu). The filled 4s subshell shields the 3d electrons from external influence. Hence the chemical properties of these elements, which depend on the electrons in the outer subshells of their atoms, are quite similar, independent of their 3d electrons. The properties of ^{24}Cr and ^{29}Cu are somewhat different because they have only a single 4s electron. Similar transition series occur during the filling of 4d and 5d subshells in 5th and 6th periods.

Lanthanides and Actinides: Lanthanides (^{58}Ce to ^{71}Lu) are the elements in which the 4f subshell is filling. This subshell lies deep within the outermost 6s subshell, which is *completed* in all the rare-earths. Thus, 4f electrons have virtually no effect on the chemical properties of rare-earths. Outer configuration being same i.e. 5s^2 5p^6 6s^2, the chemical properties of the rare-earths are almost identical. The same is true for actinides (^{90}Th to ^{103}Lr) which occur in the periodic table during the filling of the 5f subshell inside the filled 7s subshell. They too have the same outer configuration (6s^2 6p^6 7s^2) and hence identical chemical properties.

2.20 CENTRAL FIELD APPROXIMATION

In a single electron atom, the form of the interaction energy is $-Ze^2/(4\pi\varepsilon_0 r)$ and this with the kinetic energy of the electron leads to ground state energy of the form

$$-\frac{m_0 e^4 Z^4}{8\varepsilon_0^2 h^2} = -13.6 \text{ eV} = 110000 \text{ cm}^{-1} \qquad ..(170)$$

In many electron systems, the electrostatic potential energy has got the form $\Sigma - Ze^2/4\pi\varepsilon_0 r_j$ (for jth electron it is $- Ze^2/(4\pi\varepsilon_0 r_j)$). Apart from this attractive force, there are electrostatic repulsion terms between each pair of electrons given by

$$\sum_{j=2}^{N} \sum_{k=1}^{j-1} \frac{e^2}{4\pi\varepsilon_0 r_{jk}} \qquad ..(171)$$

This repulsive contribution cancels part of attraction between nucleus and electrons. In the first instance, it can be assumed that each electron (in an atom) moves in a spherically symmetric potential $V(r)$, that is produced by the nucleus and the other electrons. This is termed as *central field approximation*.

The difference between a single electron atom, i.e. H-atom, and a many electron atom is that, in case of hydrogen atom, the levels with same n but different l have same energy, but in a many electron atom, the degeneracy is removed. This is due to non-central part of interaction (arising because of eqn 171). Thus in other atoms, electrons having small angular momentum penetrate very close to the nucleus, as a result of which, nuclear charge is partly screened.

This results in an increase in the value of $V(r)$. Mathematically, this can be understood as follows:

The attractive potential terms are

$$\sum_{j} -\frac{Ze^2}{4\pi\varepsilon_0 r_j}$$

Part of it is neutralized by the central part of repulsive interaction

$$\sum_{j=2}^{N} \sum_{k=1}^{j-1} \frac{e^2}{4\pi\varepsilon_0 r_{jk}}$$

and let it be represented by $\sum_{j} C(r_j)$, i.e. non-central part left is

$$\sum_{j=2}^{N} \sum_{k=1}^{j-1} \frac{e^2}{4\pi\varepsilon_0 r_{jk}} - \sum_{j} C(r_j) \qquad \qquad ..(172)$$

and the central part is

$$\sum_{j} V(r_j) = \sum_{j} \left(\frac{-Ze^2}{4\pi\varepsilon_0 r_j} + C(r_j) \right) \qquad \qquad ..(173)$$

As a result, the Hamiltonian of the system is

$$\hat{H} = \hat{H}_0 + \hat{H}' \qquad \qquad ..(174)$$

where $\hat{H}_0 = \sum_{j} \left[-\frac{\hbar^2}{2m} \nabla_j^2 + V(r_j) \right] \qquad \qquad ..(175)$

and
$$\hat{H}' = \sum_{j=1}^{N} \sum_{k=1}^{j-1} \frac{e^2}{4\pi\varepsilon_0 r_{jk}} - \sum_j \frac{Ze^2}{4\pi\varepsilon_0 r_j} - \sum_j V(r_j) \qquad ..(176)$$

(using eqns (172) and (173))

Under central field approximation $\hat{H}' << \hat{H}_0$ and we are left with the eigen value equation

$$\hat{H}_0 \,\psi \equiv \sum_j \left\{ -\frac{\hbar^2}{2m} \nabla_j^2 + V(r_j) \right\} \psi = E\psi \qquad ..(177)$$

where
$$\hat{H}_0 = \sum_j \hat{H}_j, \; \psi = \prod_j \psi_j \qquad ..(178)$$

and
$$E = \sum_j E_{nj} l_j \qquad ..(179)$$

Equations (177), (178) and (179) are for a many electron atoms, under central field approximation. For a single electron atom,

$$\left(-\frac{\hbar^2}{2m} \nabla^2 + V(r) \right) \psi_{nlm_l m_s} = E_{nl} \psi_{nlm_l m_s} \qquad ..(180)$$

i.e. energy of a state depends on n and l and is degenerated with respect to m_l and m_s. (as also in eqn (179)).

The above central field approximation is used in Hartree's self consistent field method for solution of problems of atomic structure. Hartree method neglects the exchange property of wave functions. This is taken care of in the Hartree-Fock method.

Questions

2.1 What are Bohr's postulates? Derive an expression for allowed energies of the hydrogen atom. Draw an energy level diagram showing the observed transitions. What are limitations of Bohr's theory?

2.2 Deduce an expression for the series spectra of a hydrogen-like atom taking into account the finite mass of the nucleus.

2.3 Write postulates of Bohr's theory and derive an expression for the Rydberg's constant. Explain why the value of this constant for helium is different from hydrogen. How has this difference been used to determine m/M_H and e/m?

2.4   Discuss the Bohr-Sommerfeld theory of elliptical orbits of hydrogen atom and compare its results with those of Bohr's theory of circular orbits.

2.5   Give a brief account of Sommerfeld's relativity correction so as to explain the fine structure of hydrogen spectral lines.

2.6   Prove that for very large quantum numbers, the quantum theory frequency and the classical orbital frequency become equal.

2.7   The series limit wavelength of Balmer series in hydrogen spectrum is experimentally found to be 3646 Å. Calculate the Rydberg's constant.

2.8   The wavelength of the second member of Balmer series of hydrogen is 4861 Å. Calculate the wavelength of the first member.   (**Ans:** 6562 Å)

Spectral Lines, Line Width and Line Width Broadening Mechanisms

3.1 ORIGIN OF SPECTRAL LINES ACCORDING TO QUANTUM MECHANICS

An atom dropping from an energy level E_m to a lower level E_n causes emission of a radiation of frequency ν, given by

$$\nu = \frac{E_m - E_n}{h} \qquad ...(1)$$

For quantum mechanical treatment, we consider only motion of an electron in the x-direction.

The time dependent wave function Ψ_n (of an electron) in a state of quantum number n and energy E_n is the product of a time independent wave function ψ_n and a time varying function whose frequency is

$$\nu = \frac{E_n}{h} \qquad ...(2)$$

i.e. $\qquad \Psi_n = \psi_n e^{-i(E_n/\hbar)t}$

so $\qquad \Psi_n^* = \psi_n^* e^{i(E_n/\hbar)t} \qquad ...(3)$

The expectation value of the position of electron

$$< x > = \int\limits_{-\infty}^{+\infty} \Psi_n^* \, x \, \Psi_n dx \qquad ...(4)$$

$$= \int\limits_{-\infty}^{+\infty} \psi_n^* \, x \psi_n \, e^{i[(E_n/\hbar) - (E_n/\hbar)]t} dx$$

$$= \int\limits_{-\infty}^{+\infty} \psi_n^* \, x \psi_n \, dx \qquad ..(5)$$

which is constant in time, since ψ_n^* and ψ_n are time independent. In this case, the electron does not radiate.

We now consider an electron changing from one energy state to another, say, in particular an atom being in its ground state at $t = 0$, and an excitation process (say a beam of radiation or collisions with other particles) acts upon it. In this condition, the atom undergoes a transition from the state of energy E_m. The wave function Ψ of an electron capable of existing in states n or m may be written

$$\Psi = a\Psi_n + b\Psi_m \qquad ..(6)$$

where $a \times a$ is the probability that the electron is in state ψ_n and $b \times b$ in state ψ_m. Further, since electrons being fermions, they satisfy anticommutation relation

$$a^* \times a + b^* \times b = 1 \qquad ..(7)$$

At $t = 0$, $a = 1$ and $b = 0$ (by assumption). When the electron is in state m (the excited state), $a = 0$ and $b = 1$, and finally $a = 1$ and $b = 0$.

When the electron is in either of the states m or n, there is no radiation, while in the midst of the transition, electromagnetic waves are produced.

Hence

$$\bar{x} = \int\limits_{-\infty}^{+\infty} \left(a^* \, \psi_n^* + b^* \, \psi_m^* \right) x \left(a\psi_n + b\psi_m \right) dx$$

$$= \int\limits_{-\infty}^{+\infty} x \left(a^2 \psi_n^* \psi_n + b^* \, a\psi_m \psi_n + a^* \, b\psi_n^* \psi_m + b^2 \psi_m^* \psi_m \right) dx \quad ..(8)$$

Here $a^* a = a^2$ and $b^* b = b^2$.

The first and the last integrals are constants, and only the second and third integrals give a non-vanishing contribution to a time variation of $\bar{x}$. Because x is position vector only, its presence in each integral has no significance.

$$\bar{x} = a^2 \int\limits_{-\infty}^{+\infty} x\psi_n^* \psi_n dx + b^* \, a \int\limits_{-\infty}^{+\infty} x\psi_m^* e^{+i(E_m/\hbar)t} \, \psi_n e^{-i(E_n/\hbar)t} \, dx$$

$$+ a^* \, b \int\limits_{-\infty}^{+\infty} x\psi_n^* e^{+i(E_n/\hbar)\,t} \psi_m e^{-i(E_m/\hbar)\,t} dx + b^2 \int\limits_{-\infty}^{+\infty} x\psi_m^* \psi_m dx \quad ..(9)$$

In case of a bound system of two states $\psi_n^* \psi_m = \psi_m^* \psi_n$ and $a \times b = b \times a$. So, we can combine the time varying terms of eqn (1) into the single term.

$$a^* b \int_{-\infty}^{+\infty} x \psi_n^* \psi_m \left[e^{(i/\hbar)(E_m - E_n)t} + e^{-(i/\hbar)(E_m - E_n)t} \right] dx$$

Now $e^{i\theta} + e^{-i\theta} = 2\cos\theta$

hence the above term becomes

$$2a^* b \cos\left(\frac{E_m - E_n}{\hbar} \right) t \int_{-\infty}^{+\infty} x \psi_n^* \psi_m dx$$

The coefficient of this integral is a time varying coefficient which oscillates sinusoidally at the frequency

$$\nu = \frac{E_m - E_n}{h} \qquad \qquad ..(10)$$

Since $\cos\left(\dfrac{E_m - E_n}{\hbar} \right) t = \cos 2\pi \left(\dfrac{E_m - E_n}{h} \right) t$

$$= \cos 2\pi \nu t \qquad \qquad \left(\because \ \hbar = \frac{h}{2\pi} \right)$$

Hence $\bar{x} = a^2 \int_{-\infty}^{+\infty} x \psi_n^* \psi_n dx + b^2 \int_{-\infty}^{+\infty} x \psi_m^* \psi_m dx$

$$+ 2a^* b \cos 2\pi \nu t \int_{-\infty}^{+\infty} x \psi_n^* \psi_m dx \quad ..(11)$$

When the electron is in either state ψ_n or ψ_m, then b^2 or a^2 are zero (stationary states). These terms do not contribute. When the electron is undergoing a transition between these states, its average position oscillates with the frequency ν. This frequency is identical with that permitted by Bohr's theory. So, eqn (10) could be derived using quantum mechanics without any specific assumptions. The allowed stationary states n and m correspond to solutions of the time independent Schrodinger equations, and for a spectroscopic process to occur, a cause must exist which may allow a state to change from one stationary state to another.

3.2 LINE INTENSITIES, OSCILLATOR STRENGTHS AND THE THOMAS–REICHE–KUHN SUM RULE

According to dipole approximation, the intensity of a transition between a pair of states a and b is proportional to the quantity $| r_{ba} |^2$ where

$$r_{ba} = \langle \psi_b | r | \psi_a \rangle \qquad \qquad ..(12)$$

Consequently, the relative intensities of a series of transitions from a given initial state 'a' to various final states 'k' are determined by the quantities $|r_{ka}|^2$. While discussing intensities of lines, it is customary to introduce a related dimensionless quantity f_{ka} called the *oscillator strength*, defined by

$$f_{ka} = \frac{2m\omega_{ka}}{3\hbar}|r_{ka}|^2 \qquad ..(13)$$

where $\qquad \omega_{ka} = (E_k - E_a)/\hbar \qquad\qquad ..(14)$

This definition implies that $f_{ka} > 0$ for absorption ($\because E_k > E_a$), and $f_{ka} < 0$ for emission. The oscillator strengths, defined by eqn (13) obey the sum rule, due to Thomas, Reiche and Kuhn:

$$\sum_k f_{ka} = 1 \qquad\qquad ..(15)$$

where the sum is overall states, including the continuum. This sum rule can be proved as follows. Let f_{ka}^x be defined as

$$f_{ka}^x = \frac{2m\omega_{ka}}{3\hbar}|x_{ka}|^2 \qquad ..(16)$$

$$= \frac{2m\omega_{ka}}{3\hbar}\langle\psi_a|x|\psi_k\rangle\langle\psi_k|x|\psi_a\rangle$$

Now,

$$p = \langle\psi_k|\,p\,|\psi_a\rangle = m\langle\psi_k|\,\dot{r}\,|\psi_a\rangle \qquad ..(17)$$

and using Heisenberg's equation of motion

$$\dot{r} = \frac{1}{i\hbar}\left[r, H_0\right] \qquad ..(18)$$

Further

$$\langle\psi_k|\dot{r}|\psi_a\rangle = \frac{1}{i\hbar}\langle\psi_k|rH_0 - H_0r|\psi_a\rangle$$

$$= \frac{1}{i\hbar}(E_a - E_k)\langle\psi_k|\,r\,|\psi_a\rangle$$

$$= \frac{i(E_k - E_a)}{\hbar}\langle\psi_k|r|\psi_a\rangle \qquad ..(19)$$

or $\qquad\qquad p_{ka} = im\,\omega_{ka}r_{ka} \qquad\qquad ..(20)$

or $\qquad\qquad x_{ka} \equiv \langle\psi_k|x|\psi_a\rangle = \frac{1}{im\omega_{ka}}\langle\psi_k|p_x|\psi_a\rangle$

or $\qquad\qquad x_{ka} = \frac{-i}{m\omega_{ka}}\langle\psi_k|p_x|\psi_a\rangle \quad (i^2 = -1) \qquad ..(21)$

$\therefore$ $$x_{ak} = \left\langle \psi_a | x | \psi_k \right\rangle = \frac{i}{m\,\omega_{ka}} \left\langle \psi_a | p_x | \psi_k \right\rangle \qquad ..(22)$$

Hence $$f_{ka}^x = \frac{2i}{3\hbar} \left\langle \psi_a | p_x | \psi_k \right\rangle \left\langle \psi_k | x | \psi_a \right\rangle$$

$$= -\frac{2i}{3\hbar} \left\langle \psi_a | x | \psi_k \right\rangle \left\langle \psi_k | p_x | \psi_a \right\rangle$$

or $$f_{ka}^x = \frac{i}{3\hbar} \Big\{ \left\langle \psi_a | p_x | \psi_k \right\rangle \left\langle \psi_k | x | \psi_a \right\rangle - \left\langle \psi_a | x | \psi_k \right\rangle \left\langle \psi_k | p_x | \psi_a \right\rangle \Big\}$$
$$..(23)$$

Now, using the closure property $\left(viz \sum_k | k \rangle\langle k | = 1 \right)$, we get

$$\sum_k f_{ka}^x = \frac{i}{3\hbar} \left\langle \psi_a | p_x x - x p_x | \psi_a \right\rangle \qquad ..(24)$$

But since $[x_1\, p_x] = i\hbar$, hence

$$\sum_k f_{ka}^x = \frac{1}{3} \qquad ..(25)$$

Using same arguments (as above), we get

$$\sum_k f_{ka}^y = \frac{1}{3} \text{ and } \sum_k f_{ka}^z = \frac{1}{3}$$

Hence $$\sum_k f_{ka} = \frac{1}{3} + \frac{1}{3} + \frac{1}{3} = 1 \qquad ..(26)$$

This proves the sum rule.

We know that under dipole approximation, the transition rate for spontaneous emission of a photon is given by

$$W_{ab} = \frac{4\alpha}{3c^2} \omega_{ba}^3 | r_{ba} |^2 \qquad ...(27)$$

where α is the fine structure constant. In terms of oscillator strengths, we have

$$W_{ka} = \frac{2\hbar\alpha}{mc^2} \omega_{ka}^2 | f_{ka} | \qquad ...(28)$$

For hydrogenic atoms, the oscillator strengths and transition probabilities decrease as the principal quantum number n of the upper level increases.

Table 3.1 gives average oscillator strengths for some transitions in hydrogenic atoms.

TABLE 3.1: Average oscillator strengths for some transitions in hydrogenic atoms

Initial Level	Final Level	$n = 1$	$n = 2$	$n = 3$	$n = 4$
1s	np	–	0.416	0.079	0.029
2s	np	–	–	0.435	0.103
2p	ns	– 0.139	–	0.014	0.003
2p	nd	–	–	0.696	0.122

After: *Physics of Atoms and Molecules*, B. H. Bransden and C. J. Joachain, Pearson Education Ltd. (2003).

3.3 LINEWIDTH

Imagine that we have a classical monochromatic light source. Then, the light we see at any moment comes from a number of atoms, each making a transition between the same pair of energy levels, but the emission from any one atom is in no way related to that from any other atom. A careful spectroscopic analysis shows us that the light is not really monochromatic in the strict sense of the word, it contains components of various wavelengths within a certain range called the *linewidth*. If the linewidth is much less than the average wavelength, one uses the term *quasimonochromatic* for such radiation.

3.4 NATURAL LINEWIDTH

Imagine that an atom/molecule emits a spectral line by changing its state from level A to level B, with energies E_A and E_B respectively. A wave of frequency $\omega_0 = (E_A - E_B)/\hbar$ is emitted. However, no energy level except the ground state is an exact stationary state because of fluctuations of the environmental electromagnetic field. As a result an atom in level A will decay to a lower level after an average time T_A. According to the uncertainty principle, the value of E_A is uncertain to the extent $\delta E \approx h/T_A$. The corresponding frequency width of the emitted wave is

$$\delta\omega = 2\pi(T_A^{-1} + T_B^{-1}) \qquad ..(29)$$

This is called the natural linewidth, it is generally smaller than the Doppler and collision linewidths.

3.5 COLLISION BROADENING

When we consider atoms, e.g. that of a real gas, there will always be collisions between the various atoms. According to the kinetic theory of gases, a particular atom will be free for an average line

$$T_1 = (4NvA)^{-1} \qquad ..(30)$$

between collisions, where N is the number of molecules per unit volume and A, their collision cross-section. Both N and v depend on temperature T and pressure P, through the relation

$$Nv = P(3/mk_BT)^{1/2} \qquad ..(31)$$

So $\qquad\qquad T_1 = bT^{1/2}/P \qquad ..(32)$

where b is a constant.

Now consider what happens if an emitting atom suffers a collision. It may be supposed that the shock of the collision will, at the very least, destroy phase correlation between the emitted waves before and after the collision. The emission from all the atoms in the gas will, therefore, appear like a series of uncorrelated bursts of radiation each of average duration τ_1. The actual durations will have a Poisson distribution of mean value τ_1. Therefore, the probability of there being a burst of length between τ and $\tau + \delta\tau$ is

$$p(\tau) = \tau_1^{-1}\exp(-\tau/\tau_1) \qquad ..(33)$$

So, the emitted wave consists of wave trains with frequency ω_0 and random phase, starting at random moments and having durations statistically distributed according to eqn (33).

The spectral intensity of such an emitted wave is the mean spectral intensity of the individual wave trains, the spectral phases being random, i.e.

$$J(\omega) = \int\limits_0^\infty \left[\int\limits_{-\tau/2}^{\tau/2} \exp\left(i\omega_0 t\right) \exp(-i\omega t)dt \right]^2 p(\tau)d\tau \qquad ...(34)$$

In this equation, the integral in square brackets is the Fourier transform of a harmonic wave lasting for a duration τ; the outer integral is the statistical average. The value of the inner integral is

$$\frac{\tau\sin\left[(\omega - \omega_0)\tau/2\right]}{(\omega - \omega_0)\tau/2}$$

Thus, eqn (34) becomes

$$J(\omega) = \frac{4}{\tau_1(\omega - \omega_0)^2} \int\limits_0^\infty \exp(-\tau/\tau_1)\sin^2\left[(\omega - \omega_0)\tau/2\right]d\tau$$

$$= \frac{2\tau_1^2}{[1 + (\omega - \omega_0)^2 \tau_1^2]} \qquad ...(35)$$

The function (eqn 35) is known as a *Lorentzian function* with a half width $2\tau_1^{-1}$.

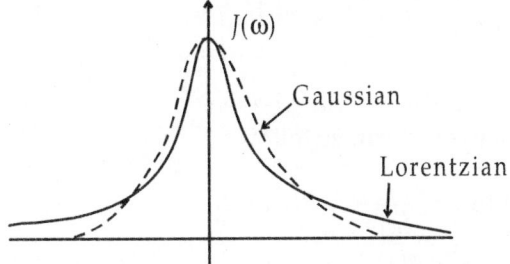

Fig. 3.1: Comparison between Lorentzian and Gaussian functions

3.6 EMISSION LINEWIDTH OF A RADIATING ELECTRON AND HOMOGENEOUS BROADENING

The electron has an electromagnetic field associated with it. While undergoing a radiative transition from a higher to a lower level in an atom, the magnetic field vector remains the same but the electric field does change: it decays, consequently it is described not by an infinitely long wavetrain but by a single pure frequency ω_0, because of Bohr's frequency condition. It has a finite starting point at $t = 0$ and then decays exponentially with a time constant

$$\tau_0 = \frac{2}{\gamma_0}$$

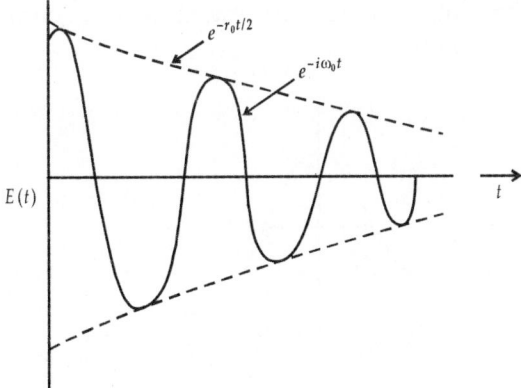

Fig. 3.2: Time variation of electric field of an oscillating electron undergoing radiative decay

From electromagnetic theory, the time dependence of electric field vector is given by

$$E(t) = \begin{cases} E_0 e^{-\gamma_0 t/2} e^{-i\omega_0 t} & \text{for } t \geq 0 \\ 0 & \text{for } t < 0 \end{cases} \quad ..(36)$$

However, we can obtain frequency dependent expression by taking Fourier transform as follows:

$$E(\omega) = \frac{1}{(2\pi)^{1/2}} \int\limits_{-\infty}^{\infty} E(t) e^{i\omega t} dt$$

$$= \frac{E_0}{\sqrt{2\pi}} \int\limits_{0}^{\infty} e^{i[(\omega-\omega_0)+i\gamma_0/2]t} dt$$

$$= -\frac{E_0}{\sqrt{2\pi}} \frac{1}{i[(\omega-\omega_0)+i\gamma_0/2]} \quad ...(37)$$

The intensity distribution per unit frequency $I(\omega)$ of this wave is proportional to $|E(\omega)|^2$ and is given by

$$I(\omega) = I_0 \frac{\gamma_0/2\pi}{(\omega-\omega_0)^2 + \gamma_0^2/4} \quad ...(38)$$

The normalisation condition is ensured by

$$I_0 = \int\limits_{0}^{\infty} I(\omega) d\omega \quad ..(39)$$

where I_0 is the total intensity integrated over the entire frequency width of emission line.

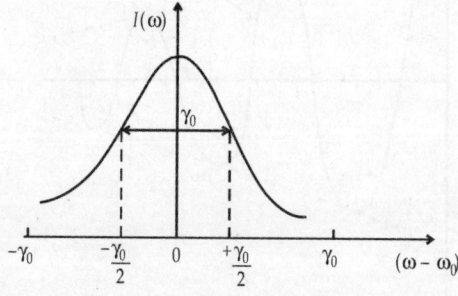

Fig. 3.3: Variation of $I(\omega)$ as a function of ω for an electron undergoing transition from one level to a lower level

The form of the line shape is known as a *Lorentzian distribution* and is symmetrical with respect to $\omega = \omega_0$. The full width at half maximum [FWHM] is

$$\Delta\omega\,|_{FWHM} = 2(\omega - \omega_0) = \gamma_0 = \frac{1}{\tau_0} = 2\pi\Delta\nu_c \qquad ..(40)$$

where $\Delta\nu_c$ is the classical linewidth. This type of emission broadening occurs when every atom of the same species making the same transition produces an identical line shape and width. Such a situation leads to Lorentzian line shape variation and is known as *homogeneous broadening* (Collision broadening is a type of homogeneous broadening).

3.7 DOPPLER BROADENING

Spectral lines are not infinitely narrow and in condensed media, may spread over several thousand cm^{-1}.

An important broadening process in gaseous samples is the Doppler effect, in which radiation is shifted in frequency when the source is moving towards or away from the observer. When a source emitting radiation of frequency ν recedes with speed v, the observer detects radiation of frequency

$$\nu' = \frac{\nu}{1 + v/c} \qquad ..(41)$$

where c is the speed of the radiation. A source approaching the observer appears to be emitting radiation of frequency

$$\nu' = \frac{\nu}{1 - v/c} \qquad ..(42)$$

Molecules reach high speeds in all directions in a gas, and a static observer detects the corresponding Doppler shifted range of frequencies. Some molecules approach the observer and some move away; some quickly and some slowly. The detected spectral line is the absorption or emission profile, arising from all the resulting Doppler shifts. The profile reflects the Maxwellian distribution of molecular speeds parallel to the line of sight, which is a Gaussian curve $(\sim e^{-x^2})$. The Doppler-line-shape is, therefore, also Gaussian.

We shall now derive an expression for width of a spectral line due to Doppler broadening.

Let v be the velocity of an atom and θ, the angle between v and direction of observation. Then, the change in frequency due to Doppler effect is given by

$$\frac{\Delta v}{v_0} = \frac{v - v_0}{v_0} = \frac{v \cos \theta}{c} \equiv \frac{v}{c} \qquad ..(43)$$

where v_0 is the centre frequency when $v = 0$ and v is the observed frequency.

Assuming a Maxwellian distribution of velocities, the probability that an atom of mass m has a velocity component between u and $u + du$ is

$$P(u)du = \left(\frac{m}{2\pi kT}\right)^{1/2} \exp\left(-\frac{mu^2}{2kT}\right)du \qquad ..(44)$$

at a temperature T. Consequently, the probability that the atom emits in the frequency range v and $v + dv$ is given by

$$P(v)dv = \frac{c}{v_0}\left(\frac{m}{2\pi kT}\right)^{1/2} \exp\left\{-\frac{mc^2}{2kT}\left(\frac{v - v_0}{v_0}\right)^2\right\}dv \qquad ..(45)$$

Since the intensity at the frequency v is proportional to this probability, we can write for the line shape

$$g_D(v) = \frac{c}{v_0}\left(\frac{m}{2\pi kT}\right)^{1/2} \exp\left\{-\frac{mc^2}{2kT}\left(\frac{v - v_0}{v_0}\right)^2\right\} \qquad ..(46)$$

It can be seen that this relation satisfies the condition

$$\int_{-\infty}^{+\infty} g_D(v)dv = 1 \qquad ..(47)$$

The plot of the relation (46) shows a Gaussian line shape.

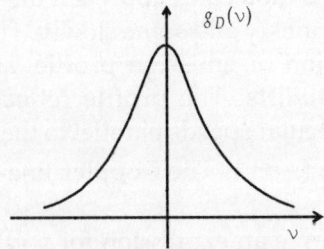

Fig. 3.4: Gaussian line shape

To find the frequency at which the intensity drops to half of its maximum value, we set

$$\exp\left\{-\frac{mc^2}{2kT}\left(\frac{\nu-\nu_0}{\nu_0}\right)^2\right\} = \frac{1}{2} \qquad ..(48i)$$

or
$$-\frac{mc^2}{2kT}\left(\frac{\nu-\nu_0}{\nu_0}\right)^2 = \ln 2 \qquad ..(48ii)$$

or
$$(\nu-\nu_0) = \left(\frac{2kT}{mc^2}\right)^{1/2}\nu_0(\ln 2)^{1/2} \qquad ..(48iii)$$

Therefore,

$$\Delta\nu_D = 2\nu_0\left(\frac{2kT\ln 2}{mc^2}\right)^{1/2}$$

or
$$\Delta\nu_D = \frac{2\nu_0}{c}\times\left(\frac{2kT}{m}\ln 2\right)^{1/2} \qquad ..(49)$$

In terms of the wavelength

$$\Delta\lambda = 2\frac{\lambda_0}{c}\times\left(\frac{2kT}{m}\ln 2\right)^{1/2} \qquad ..(50)$$

In practice, temperature and pressure cause both Doppler and collision broadening in various degrees, and observed spectral lines are rarely strictly Gaussian or Lorentzian, rather, from the point of view of simple optical coherence theory, they can be considered as having a single empirical width.

3.8 LIFETIME BROADENING

It is found that spectral lines are still not infinitely sharp even if Doppler broadening has been largely eliminated.* When the Schrodinger equation is solved for a system that is changing with time, it is found that it is impossible to specify the energy levels exactly. If, on average, a system survives in a state for a time τ, the lifetime of the state, its energy levels are blurred to an extent $\sim\delta E$ where

$$\delta E = \hbar/\tau \qquad ..(51)$$

(From Heisenberg's uncertainty principle).

---

*A novel approach to eliminate Doppler broadening has been possible by use of lasers of extremely high monochromaticity and radio frequency techniques.

Lifetime broadening is often called uncertainty broadening. Expressing the energy spread in wave numbers (through $\delta E = hc\,\delta\bar{\nu}$) and using the values of the fundamental constants, we get a practical form of relation

$$\delta\bar{\nu} \simeq \frac{5.3 \text{ cm}^{-1}}{\tau/ps} \qquad\qquad ..(52)$$

All states are subject to some lifetime broadening and the shorter the lifetimes of the states involved in a transition, the broader the spectral lines.

A study of line broadening mechanisms and, hence, the line shape function is useful in predicting the operation characteristics of lasers.

Questions

3.1 Explain quantum mechanically the origin of spectral lines.

3.2 What do you understand by linewidth of a monochromatic light source? What is natural linewidth?

3.3 Explain Doppler broadening and derive an expression for it.

Angular Momentum of Electrons, Selection Rules and Spectroscopic Notations

4.1 ORBITAL MAGNETIC MOMENT AND BOHR MAGNETON

The orbital quantum number l in a one electron atom determines the magnitude of electron's angular momentum $|L|$. Now, an electron moving in a Bohr orbit constitutes a current I which will have a magnetic moment μ equal to the product of I and the area of the Bohr orbit, i.e.

$$\mu_e = IA = I \cdot \pi r_n^2 = -\frac{e}{cT} \times \pi r_n^2$$

$$= \frac{-ev}{c \times 2\pi r_n} \times \pi r_n^2 = -\frac{e}{2c} v r_n (emu) \qquad \ldots(1)$$

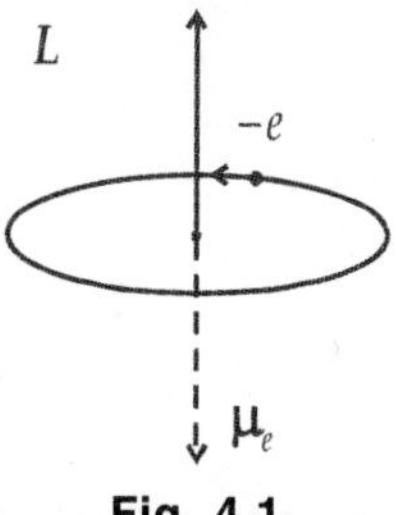

where c is the speed of light, v is the velocity of the electron and $2\pi r_n/v$ is the period ($= T$) of the orbit. Since angular momentum $L = m v r_n$, so electron's magnetic moment

$$\mu_l = -\frac{e}{2mc}(L) \qquad \ldots(2)$$

The quantity $(-e/(2mc))$ is a constant and is known as the *gyromagnetic ratio*.

The minus sign means that μ_l is in the direction opposite to that of L. The subscript l is used here with μ to designate "orbital" to distinguish it from the one corresponding to electron spin. The eqn (2) is usually written as

$$\mu_l = -g_l \left(\frac{e}{2mc} \right) L \qquad \ldots(3)$$

where $g_l = 1$.

Fig. 4.1

The quantity g_l is called the orbital (spectroscopic) *splitting factor.* It is introduced here to preserve the symmetry with further equations involving g factors which are different from one. From quantum mechanics:

$$|L| = \sqrt{l(l+1)}\,\hbar = \sqrt{l(l+1)}\,\frac{h}{2\pi} \qquad ..(4)$$

$$\therefore \qquad |\mu_l| = \frac{eh}{4\pi mc}\sqrt{l(l+1)} \quad (\text{CGS units}) \qquad ..(5)$$

The quantity $\left(\dfrac{eh}{4\pi mc}\right)$ forms a natural unit for the measurement of atomic magnetic moments and is called *Bohr magneton.* It is designated by μ_B, i.e.

$$1\mu_B = \frac{eh}{4\pi mc} = \frac{4.8\times10^{-10}\ \text{esu} \times 6.63\times10^{-27}\ \text{erg-s}}{4\times3.14\times9.1\times10^{-28}\text{g}\times3\times10^{10}\ \text{cm/s}}$$

$$= 0.928 \times 10^{-20}\ \text{erg/Gauss} \qquad ..(6)$$

Thus $\qquad |\mu_l| = \mu_B\sqrt{l(l+1)} \qquad ..(7)$

In the MKS system

$$\mu_B = \frac{eh}{4\pi m} = \frac{1.6\times10^{-19}\ \text{C}\times6.63\times10^{-34}\ \text{J-s}}{4\times3.14\times9.1\times10^{-31}\ \text{kg}}$$

$$= 0.928\times10^{-23}\ \text{A-m}^2 \ \text{ or } \ \left(\frac{\text{J}}{\text{wb}-\text{m}^2}\right)$$

and $\qquad \dfrac{|\mu_l|}{|L|} = \left(\dfrac{e}{2m}\right)$ and $|\mu_l| = \dfrac{e\hbar}{2m}\sqrt{l(l+1)}$ (MKS units) $\quad ..(8)$

4.2 ANGULAR MOMENTUM OF MANY-ELECTRON ATOMS

The optical spectrum of an element is the characteristic of the valence electrons, i.e. electrons lying outside the closed shells of the atom, e.g. alkali metals (possessing one valence electron) show one type of spectrum whereas alkaline earths (having two valence electrons) show another type of spectrum.

The small letters l, s and j depict the state of an electron, while the capital letters L, S and J depict the state of an atom as a whole and depends on the resulting angular momentum of the atom obtained by adding contributions from the free electrons, since core electrons, i.e. electrons in the closed shells, do not contribute to the total angular momentum of the atom.

There are two different ways in which we might sum the orbital and spin momentum of the valence electrons:

1. First sum the orbital contributions and then the spin contributions separately. Finally add the total orbital and total spin contributions to reach the grand total angular momentum.

   Symbolically

   $$\sum_i I_i = L \qquad\qquad ..(9)$$

   $$\sum_i S_i = S \qquad\qquad ..(10)$$

   and $\quad L + S = J \qquad\qquad ..(11)$

   Capital letters without suffixes denote total momentum. Here

   $$|I_i| = \sqrt{l_i(l_i + 1)}\,\hbar \qquad\qquad ..(12)$$

   and $\quad |S_i| = \sqrt{s_i(s_i + 1)}\,\hbar \qquad\qquad [\hbar = h/(2\pi)]$
   $$..(13)$$

   i is the number of valence electrons. The summation in equations (9), (10) and (11) is the vector sum.

2. Sum the orbital and spin momenta of each electron separately, finally summing the individual totals to form the grand total:

   $$I_i + S_i = J_i \text{ and } \sum_i J_i = J \qquad\qquad ..(14)$$

   $$|J| = |L - S|,.., |L + S|$$

   The first method, known as *Russell-Saunder's* or LS coupling, gives results in accordance with the spectra of small and medium sized atoms, while the second (called *j-j coupling*) applies better to large atoms. Symbolically, these are expressible as

   Russell-Saunders coupling:

   $$(s_1, s_2, ..)(l_1, l_2, ..) = (S, L) = J \qquad\qquad ..(15)$$

   j-j coupling:

   $$(l_1, s_1)(l_2, s_2)...= (J_1, J_2..) = J \qquad\qquad ..(16)$$

   $$|J| = (j_1 + j_2), (j_1 + j_2 - 1), ...|j_1 - j_2|$$

   Here $\quad J = \sqrt{J(J+1)}\,\hbar \qquad \left(\hbar = \dfrac{h}{2\pi}\right)$

4.3 THE VECTOR MODEL OF THE ATOM

4.3.1 The Four Quantum Numbers

The principal quantum number n may be taken as a measure of how the total energy of an electron in an atom is quantised. Therefore, the energy levels are completely specified by the principal quantum number. It may have values

$$n = 1, 2, 3, \ldots$$

For a given energy specified by n, there is a range of possible values for the total angular momentum. Quantum mechanically, this range is essentially the same and only certain discrete values are allowed determined by l, the orbital quantum number where

$$l = 0, 1, 2, \ldots, (n - 1) \qquad \ldots(17)$$

and the eigen value of the square of the angular momentum L^2 is $\hbar^2 l(l + 1)$. The number l describes the way in which the angular momentum of the electron is quantised and has a small effect on the energy.

Given l, there is then a range of discrete values of the z-component of the angular momentum (L_z) which determines the orientation of the angular momentum vector L. This is specified by the magnetic quantum number m_l. It can take the values

$$m_l = 0, \pm 1, \pm 2, \ldots \pm l \qquad \ldots(18)$$

These are $(2l + 1)$ values and z-component of angular momentum

$$L_z = \hbar m_l \qquad \ldots(19)$$

In presence of an external magnetic field, the l-level may split up into $(2l + 1)$ values of m_l. The quantum number m_l is a measure of the angle between the electron's angular momentum vector and an applied magnetic field. In addition, there is a spin quantum number s due to spin of the electron. It can take only two values $+\frac{1}{2}$ on $-\frac{1}{2}$.

4.3.2 Space Quantisation

In Sommerfeld's theory, the motion of the electron round the nucleus has two degrees of freedom resulting in the quantum numbers n and k, the former giving the major axis or the size of the elliptic orbits and the latter giving the eccentricity or the shape of the orbits. Description of the electron motion in an atom in terms of these two quantum numbers implies essentially the motion to be confined to the orbital plane. The orbital plane of the electron may have different orientations in space. The orbital angular momentum vector $\vec{p_l}$ is perpendicular to the plane of the orbit. The orientation of vector $\vec{p_l}$

depends on the direction of an external applied electric or magnetic field. In order to study the orientation of vector $\vec{p_l}$ in space we have to consider a special direction fixed in space.

According to classical mechanics, the vector $\vec{p_l}$ may have orientation in any arbitrary direction with respect to a fixed direction in space (e.g. the Z-direction usually taken as the direction of the externally applied field). Thus in actual case the situation is different in case of electronic orbits and some quantum conditions are to be satisfied in regard to orientations of these orbits in space relative to a preferred direction. This requires the third quantum number m_l. This quantum number does not change the size or shape of the Bohr Sommerfeld orbit but simply determines their orientations relative to the reference direction. Because of the restrictions on the orientations of the electron orbit, the orbits are said to be space-quantized.

Let p_ϕ denote the component of orbital angular momentum p_l along the field direction $\vec{B}$ then

$$p_\phi = p_l \cos \alpha$$

where α is the angle which $\vec{p_l}$ makes with the direction of $\vec{B}$ (i.e. with the z-axis) the quantum condition to be satisfied is

$$\int p_\phi \, d_\phi = m_l \, h$$

where h is Planck's constant.

$$2\pi \, p_l \cos \alpha = m_l \, h$$

$$p_l \cos \alpha = m_l \, \hbar$$

since
$$p_l = l\hbar$$

$$\therefore l \cos \alpha = m_l$$

Since $\cos \alpha$ lies between -1 and $+1$, the permitted values of m_l are

$$m_l = 0 \pm 1, \pm 2, \ldots \pm l \qquad (2l + 1 \text{ values})$$

i.e. there are $(2l + 1)$ permitted orientations of the electron orbit in space or $2l + 1$ values for the component of lh along any chosen direction.

4.3.3 Space Quantisation Diagrams

The space quantization of an electron orbit is represented by means of the orbit normal and this is treated as a vector. We know that the number of possible m_l values is limited by the value of l. In Fig. 4.2, the space quantization diagram for s-electron ($l = 0$) shows the possible orientations of p_ϕ. Similar diagrams for $p(l = 2)$ and $d(l = 3)$ electrons show four and six possible orientations respectively.

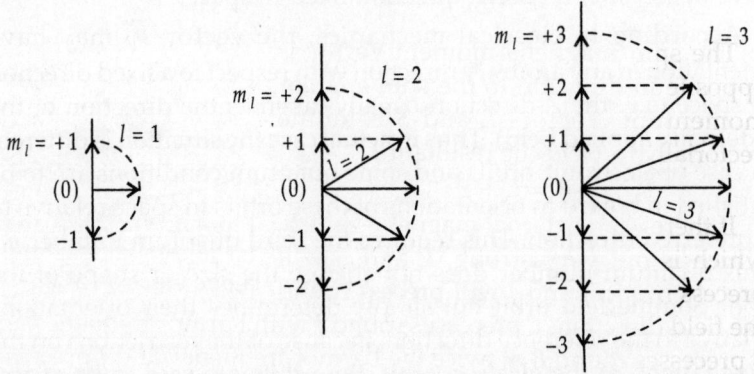

Fig. 4.2: Space quantization diagrams for Bohr-Sommerfeld orbits with l = 1, 2 and 3

$$m_l h = 2\pi\, p_\phi$$

$$m_l = \frac{2\pi\, p_\phi}{h}$$

$$l \cos \alpha = 2\pi\, p_\phi / h$$

By drawing the orientated vector of length $l = 2\pi p_\phi/h$, the projection will always be just the magnetic quantum number m_l.

4.3.4 The Vector Model

The magnitude L of the orbital angular momentum L of an atomic electron is given by

$$L = \sqrt{l(l+1)}\,\hbar$$

The magnetic moment due to the orbital motion of the electron is given by

$$\mu_l = \left(\frac{e}{2m_e}\right)L \qquad ..(20)$$

As the charge on the electron is negative, the direction of the orbital magnetic moment vector is opposite to that of the angular momentum vector.

The spin angular momentum S has a magnitude S given by

$$S = \sqrt{s(s+1)}\,\hbar \qquad ..(21)$$

The magnetic moment due to the spin of the electron is given by

$$m_s = 2\left(\frac{e}{2m}\right)S \qquad ..(22)$$

The spin magnetic moment vector is opposite in direction to the spin angular momentum. Vectors L and S combine vectorially to yield the resultant J.

$$J = L + S$$

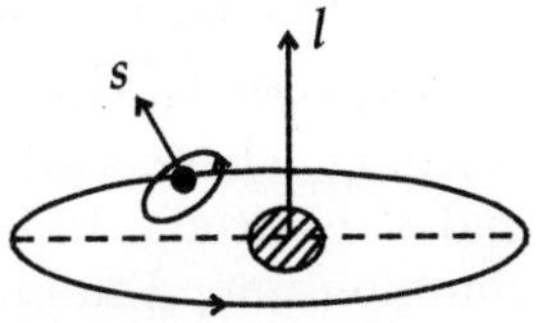

If there is an external magnetic field B, which is not very strong, L and S will precess round J and J will precess round the field direction. L precesses round J with Larmor frequency and S precesses round B at twice the Larmor frequency and J precesses with a compromise frequency. This we will further consider in Chapter 7 while discussing anomalous Zeeman effect.

Fig 4.3: Vectors l and s for an orbital electron in a one electron atom

4.4 TWO-ELECTRON ATOM

For elements with more than one valence electron, e.g. Ca or Mg, the development of energy levels would require an extension of vector model of the one electron atom. Prior to this, we need to distinguish between equivalent and non-equivalent electrons. If two electrons have the same n and l quantum numbers, they are said to be *equivalent electrons*, otherwise they are *non-equivalent*.

In case of a one-electron atom, orbital moment l^* and spin moment s^* couple to form the total moment j^*.

$$j^* = l^* + s^*$$

In case of a two-valence electron atom, there arise two possibilities. The first case is that of Russell-Saunders coupling (also called L-S coupling) and the other is j-j coupling.

In LS coupling, l_1^* and l_2^* couple to form L^* and s_1^* and s_2^* couple to form S^*, both L^* and S^* in turn couple together to form J^*, the total angular momentum of the atom. The quantum conditions are such that

$$L^* = \sqrt{L(L+1)}\,\hbar$$

$$S^* = \sqrt{S(S+1)}\,\hbar \qquad ..(23)$$

and

$$J^* = \sqrt{J(J+1)}\,\hbar$$

Here, the magnitude of the quantum numbers L and S are determined by equations

$$L = (l_1 + l_2), (l_1 + l_2 - 1), (l_1 + l_2 - 2), .., (l_1 - l_2) \quad ..(24)$$

$$S = (s_1 + s_2), (s_1 + s_2 - 1), (s_1 + s_2 - 2), .., (s_1 - s_2) \quad ..(25)$$

In Russell-Saunders coupling, the interaction between L^* and S^* is strong, whereas in case of j-j coupling, the interaction between l^* and s^* is strong. In fact, the Russell-Saunders coupling is more common among atoms than j-j coupling. Figure 4.4 illustrates a few examples of this coupling.

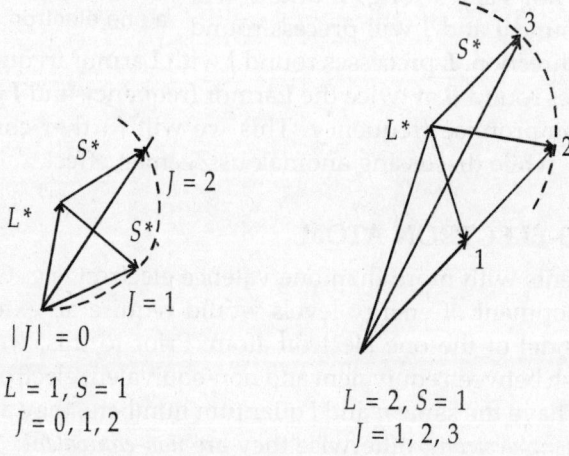

Fig 4.4: Vector diagrams for *LS* coupling

In the j-j coupling l^* and s^* values of each electron combine to form a j^* and the values j_1^* and j_2^* combine to give a resultant J^*. For j-j coupling, the vectors are defined by

$$\left. \begin{array}{c} l^* = \sqrt{l(l+1)}\,\hbar \\[2mm] s^* = \sqrt{s(s+1)}\,\hbar \\[2mm] j^* = \sqrt{j(j+1)}\,\hbar \end{array} \right] \quad ..(26)$$

where $\quad l_1^* + s_1^* = j_1^*$

$$l_2^* + s_2^* = j_2^*$$

and $\qquad J^* = j_1^* + j_2^* \qquad\qquad\qquad ..(27)$

$$|J^*| = \sqrt{J(J+1)}\,\hbar\,(\equiv J^*)$$

Hence $J = (j_1 + j_2), (j_1 + j_2 - 1), ..., (j_1 - j_2)$

and J takes integral values only. Figure 4.5 shows examples of j-j coupling for a two electrons atom.

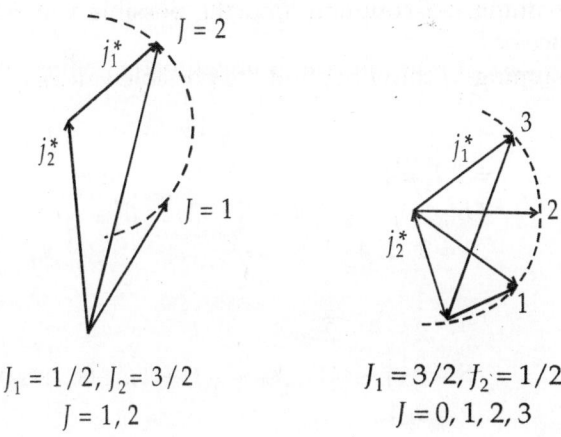

$J_1 = 1/2, J_2 = 3/2$ $J_1 = 3/2, J_2 = 1/2$

$J = 1, 2$ $J = 0, 1, 2, 3$

Fig 4.5: Vector diagrams for *j-j* coupling

How the component vectors couple together to form the resultant vector J^* for an atom having two non-equivalent electrons is illustrated in Fig. 4.6.

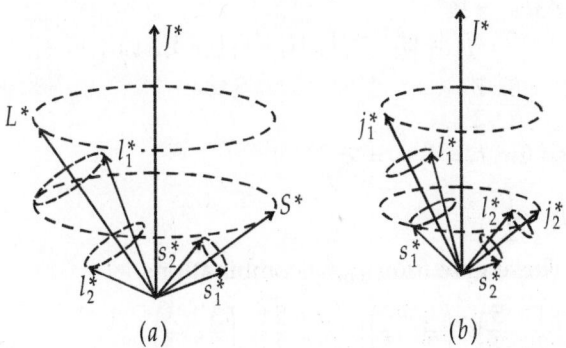

(a) (b)

Fig 4.6: The vector diagram of (a) *L-S* and (b) *j-j* coupling for an atom having two non-equivalent electrons

The multiplicity of the levels is defined by

$$M = (2S + 1)$$

                                              ..(28)

Example **4.1:** The quantum numbers of two electrons in a two valence electron atom are

$$n_1 = 6 \qquad l_1 = 3 \qquad s_1 = \tfrac{1}{2}$$
$$n_2 = 5 \qquad l_2 = 1 \qquad s_2 = \tfrac{1}{2}$$

(a) Assuming *L-S* coupling, find the possible values of *L* and hence of *J*

(b) Assuming *j-j* coupling, find the possible values of *J*.

Solution:

(a) Given $l_1 = 3, l_2 = 1$

$$\therefore \qquad L = |l_1 - l_2|; \ |l_1 - l_2| + 1; \ ... \ (l_1 + l_2)$$
$$= 2, 3, 4$$

$$s_1 = \frac{1}{2}, \ s_2 = \frac{1}{2}$$

$$\therefore \qquad S = |s_1 - s_2|, \ |s_1 - s_2| + 1, ..., (s_1 + s_2)$$
$$= 0, 1.$$

Hence, the *J* values are

$$J = |L - S|, .., (L + S)$$

For $\quad S = 0$ and $L = 2, 3, 4$, we have

$$J = 2, 3, 4$$

For $\quad S = 1$ and $L = 2, 3, 4$, we have

$$J = 1, 2, 3; \ 2, 3, 4 \text{ and } 3, 4, 5.$$

(b) $l_1 = 3, s_1 = \tfrac{1}{2}$

$$\therefore \qquad j_1 = |l_1 - s_1|, \ |l_1 - s_1| + 1, \ ..., (l_1 + \ s_1)$$

$$= \frac{5}{2}, \frac{7}{2}$$

Also for, $l_2 = 1, s_2 = \tfrac{1}{2}$

$$J_2 = \frac{1}{2}, \frac{3}{2}$$

These give four (j_1, j_2) combinations

$$\left(\frac{1}{2}, \frac{5}{2}\right); \ \left(\frac{1}{2}, \frac{7}{2}\right); \ \left(\frac{3}{2}, \frac{5}{2}\right); \ \left(\frac{3}{2}, \frac{7}{2}\right)$$

$$\left(\frac{1}{2}, \frac{5}{2}\right) \Rightarrow J = 2, 3$$

$$\left(\frac{1}{2}, \frac{7}{2}\right) \Rightarrow J = 3, 4$$

$$\left(\frac{3}{2}, \frac{5}{2}\right) \Rightarrow J = 1, 2, 3, 4$$

$$\left(\frac{3}{2}, \frac{7}{2}\right) \Rightarrow J = 2, 3, 4, 5$$

4.5 TERM SYMBOLS

The term symbol for a particular atomic state is written as follows:

Term symbol $= {}^{2S+1}L_J$

where the numerical superscript gives the multiplicity of the state (due to degeneracy), the numerical subscript gives the total angular momentum quantum number J. The value of the orbital quantum number L is expressed by a letter:

For $L = 0, 1, 2, 3, 4, \ldots$

Symbol $= S, P, D, F, G, \ldots$ respectively in analogy with the single electron states $s, p, d, f, \ldots$, etc., for $l = 0, 1, 2, \ldots$

Given a term symbol for a particular atomic state, we can immediately deduce the various total angular momenta of that state. For example,

1. ${}^3S_1$: $\Rightarrow$ $2S + 1 = 3$ hence $S = 1$
   and $L = 0$ and $J = L + S = 1$.
2. ${}^2P_{3/2}$: $\Rightarrow$ $L = 1, J = 3/2,$
   $2S + 1 = 2$ $\Rightarrow$ $S = \frac{1}{2}$.

Sometimes, the value of the principal quantum number is also mentioned, e.g. $2^2P_{3/2}$. It is to be noted that although the term symbol tells only the total spin, total orbital and total angular momenta of the whole atom, it does not tell anything about the state of individual electrons. However, this is not important in spectroscopy and we are concerned only with the energy state of the atom as a whole.

In a one electron systems, the ground state is always singlet, (Why)?:

The value of J is given by $L + S$ to $L - S$ with a difference of 1 in successive values . Since in one electron system $S = \pm\frac{1}{2}$ hence $J = L + \frac{1}{2}$ and $L - \frac{1}{2}$, i.e. J has only two values and the state is said to be a doublet. For the ground state $L = 0$, $J = +\frac{1}{2}$ or $-\frac{1}{2}$. But since J is always positive (net angular momentum of the atom) so, the possibility of $-\frac{1}{2}$ is ruled out. Therefore, the ground state of single electron system is always singlet.

4.6 THE VECTOR MODEL AND SPIN ORBIT INTERACTION

Due to spin orbit interaction, there arises a first order energy shift

$$\Delta E'' = \frac{t^2}{2m_0^2 c^2} \left\langle \frac{1}{r}\frac{dV}{dr} s \cdot l \right\rangle \qquad ..(29)$$

where s is the spin and l, the orbital angular momentum and

$$V = -\frac{Ze^2}{4\pi\varepsilon_0 r} \qquad ..(30)$$

$\xi s \cdot l$ acts as a small perturbation

where $\quad \xi \equiv \frac{\hbar^2}{2m_0^2 c^2} \left\langle \frac{1}{r}\frac{dV}{dr} \right\rangle \qquad ..(31)$

We can write

$$H = H_0 + \xi s \cdot l \qquad ..(32)$$

where $H_0 \equiv -\frac{\hbar^2}{2m}\nabla^2 - \frac{Ze^2}{4\pi\varepsilon_0 r}$ is the interaction free hamiltonian.

The eigen functions of H are the product of space and spin functions, i.e.

$$U_{n,l,m_l,m_s} = R_{nl} Y_{l,m_l} \chi(m_s) \qquad ..(33)$$

For a given n these are all degenerate with respect to l, m_l and m_s. U_{n,l,m_l,m_s} is an eigen function of l^2, s^2, l_z and s_z.

∴ These operators commute with H_0.

Now, in order to find the first order energy shift arising from the perturbation $\xi s \cdot l$, we have to use a representation for the wave functions in which $s \cdot l$ is diagonal. The functions U_{n,l,m_l,m_s} will not suffice because $s \cdot l$ does not commute with l_z or s_z. Then, we have to write

$$U_{n,l,j,m_j} = \sum_{m_l m_s} U_{n,l,m_l,m_s} C_{m_l,m_s,j} \phi_{mj} \qquad ..(34)$$

or in Dirac notation

$$|n,l,j,m_j\rangle = \sum_{m_l,m_s} |n,l,m_l,m_s\rangle \langle l,s,m_l,m_s \,|\, l,s,j,m_j\rangle \qquad ..(35)$$

where the coefficients

$$C_{m_l,m_s,j} \equiv \langle l,s,m_l,m_s \,|\, l,s,j,m_j\rangle$$

are called (Clebsch-Gordon (or $C.G.$) coefficients. We here need not know these coefficients, rather, here, we use the fact that $<n, l, j, m_j>$ is an eigen function of l^2, s^2, j^2 and j_z.

Thus, $\qquad \Delta E_1 = \langle n, l, j, m_j | \xi s \cdot l | n, l, j, m_j \rangle$ ..(36)

Now,

$$j^2 = (l + s) \cdot (l + s)$$
$$= l^2 + s^2 + 2l \cdot s$$

hence $\quad l \cdot s = \dfrac{1}{2}\left(j^2 - l^2 - s^2\right)$ ..(37)

and consequently,

$$\Delta E_1 = \frac{\xi}{2}\langle n, l, j, m_j | j^2 - l^2 - s^2 | n, l, j, m_j \rangle$$

$$= \frac{\xi}{2}\left[j(j+1) - l(l+1) - s(s+1) \right]$$ ..(38)

The labels of $U_{n,\,l,\,j,\,m_j}$ indicate that these functions are simultaneous eigen functions of l^2 (and s^2), j^2 and j_z (because $s \cdot l$ commutes with l^2, s^2, j^2, j_z). Here each level labelled by j is $(2j + 1)$ fold degenerate.

Main results for the hydrogen spectrum are summarised in Table 4.1.

TABLE 4.1

Characteristics	Remarks
1. The wave function is reparable into radial and angular parts	Because $V(r)$ is a central field
2. The constants of the motion are l^2, s^2, l_z and s_z	Because s and l are assumed not to interact with each other
3. There is a degeneracy in m_l and m_s	Because no axis in space has been physically established
4. The electric dipole between rules are $\Delta l = \pm 1$ $\Delta m_l = 0$ (π polarisation) $\Delta m_l = \pm 1$ (σ polarisation)	Because there is change of parity

Fine Structure

The fine structure arises due to perturbation. The hamiltonian

$$H = H_0 + H_1 + H_2$$

where $H_1 = \xi s \cdot l$

$$H_2 = \text{(due to other relativistic effects)}$$

TABLE 4.2

Characteristics	Remarks
1. $H_1 = \xi s \cdot l$ is treated as a small perturbation	Because s and l are now interacting
2. l^2, s^2, j^2 and j_z are constants of the motion (and not l_z and s_z)	There is a torque on l and s but no torque on j
3. The corresponding zeroth order wave functions are $\lvert n, l, j, m_j \rangle$ and not $\lvert n, l, m_l, m_s \rangle$.	The perturbation does not mix up states of different j and m_j and $s \cdot l$ is diagonal in the $\lvert n, l, j, m_j \rangle$ representation

4.7 HYDROGEN FINE STRUCTURE

Due to spin-orbit interaction, there is a magnetic interaction energy

$$\Delta E_{ls} = -\mu_s \cdot \mathbf{B} \qquad\qquad ..(39)$$

where

$$\mu_s = -g_s \frac{e}{2mc} s \qquad\qquad ..(40)$$

and

$$\mathbf{B} = \frac{1}{mc}\frac{1}{r}\frac{dV}{dr} l \qquad\qquad ..(41)$$

$$\therefore \qquad \Delta E_{ls} = \frac{e}{m^2 c^2}\frac{dV}{r dr} s \cdot l \qquad (\because g_s = 2) \qquad ..(42i)$$

This is the interaction energy in a frame in which the electron is at rest. Taking into account the *Thomas precession*, i.e. transforming to the frame in which nucleus is at rest, we have

$$\Delta E_{ls} = \frac{1}{2}\frac{e}{m^2 c^2}\frac{1}{r}\frac{dV}{dr} s \cdot l \qquad\qquad ..(42ii)$$

Since

$$j = l + s$$

$$\therefore \qquad j \cdot j = l^2 + s^2 + 2l \cdot s$$

or

$$l \cdot s = \frac{1}{2}\big[j(j+1) - l(l+1) - s(s+1) \big] \hbar^2$$

$$\left(\because |j| = \big[j(j+1) \big]^{1/2} \hbar, \text{ etc. and } \hbar = \frac{h}{2\pi} \right)$$

Thus,

$$\Delta E_{ls} = \frac{eh^2}{16\pi^2 m^2 c^2}\big\{ j(j+1) - l(l+1) - s(s+1) \big\}\frac{1}{r}\frac{dV}{dr} \quad ..(43)$$

Since

$$V(r) = -\frac{Ze}{r} \qquad\qquad \therefore \quad \frac{dV}{dr} = \frac{Ze}{r^2}$$

$$\therefore \qquad \Delta E_{ls} = \frac{Ze^2h^2}{16\pi^2m^2c^2}\left\{j(j+1)-l(l+1)-s(s+1)\right\}\left\langle\frac{1}{r^3}\right\rangle \quad ..(44)$$

The average value $\left\langle\dfrac{1}{r^3}\right\rangle$ taking into account the hydrogen radial eigen function $R_{nl}(r)$ is given by

$$\left\langle\frac{1}{r^3}\right\rangle = \frac{Z^3}{a_0^3 n^3 l\left(l+\frac{1}{2}\right)(l+1)} \quad \text{(if } l>0) \qquad ..(45)$$

Here $a_0 = h^2/(4\pi^2\,me^2)$ is the Bohr radius, i.e. atomic radius in ground state of H-atom.

Hence $\Delta E_{ls} = \dfrac{Ze^2h^2}{16\pi^2m^2c^2}\dfrac{Z^3}{a_0^3 n^3 c\left(l+\frac{1}{2}\right)(l+1)}$

$$\times \{j(j+1)-l(l+1)-s(s+1)\}$$

$$\equiv \frac{R\alpha^2 hcZ^4}{2n^3 l\left(l+\frac{1}{2}\right)(l+1)}\left\{j(j+1)-l(l+1)-s(s+1)\right\} \quad ..(46)$$

where $\alpha = \dfrac{2\pi^2 e^2}{hc}$ is fine structure constant and $R = \dfrac{2\pi^2 me^4}{ch^3}$ is Rydberg constant (taking nucleus to be at rest).

The term shift due to above derived spin orbit interaction is

$$\Delta T_{ls} = -\frac{\Delta E_{ls}}{hc}$$

$$= -\frac{R\alpha^2 Z^4}{2n^3 l\left(l+\frac{1}{2}\right)(l+1)}\left\{j(j+1)-l(l+1)-s(s+1)\right\} \text{ cm}^{-1} \qquad ..(47)$$

For H atom, we have a single electron, $s = \dfrac{1}{2}$ and $j = l \pm \dfrac{1}{2}$.

Further, we have seen that due to relativistic effect,

$$\Delta T_r = \frac{R\alpha^2 Z^4}{n^3}\left[\frac{1}{l+1/2}-\frac{3}{4n}\right] \qquad ..(48)$$

Hence, the net term-shift due to spin-orbit interaction and the relativistic effect for a H atom, is

$$\Delta T = \Delta T_{ls} + \Delta T_r$$

$$= -\frac{R\alpha^2 Z^4}{2n^3 l(l+1/2)(l+1)}\left[j(j+1)-l(l+1)-s(s+1)\right]$$

$$+\frac{R\alpha^2 Z^4}{n^3}\left[\frac{1}{(l+1/2)}-\frac{3}{4n}\right]$$

$$= \frac{R\alpha^2 Z^4}{n^3}\left\{\frac{1}{\left(l+\dfrac{1}{2}\right)}-\frac{j(j+1)-l(l+1)-s(s+1)}{2l\left(l+\dfrac{1}{2}\right)(l+1)}-\frac{3}{4n}\right\} \quad ..(49)$$

Now $j(j+1) - l(l+1) - s(s+1) = l$ for $j = l + 1/2$

$$= -(l+1) \text{ for } j = l - 1/2$$

Therefore,

$$\left.\begin{aligned}
\Delta T &= \frac{R\alpha^2 Z^4}{n^3}\left(\frac{1}{(l+1)}-\frac{3}{4n}\right) \\[2mm]
\text{and} \qquad \Delta T &= \frac{R\alpha^2 Z^4}{n^3}\left(\frac{1}{l}-\frac{3}{4n}\right)
\end{aligned}\right\} \qquad ..(50)$$

These two can be written together as

$$\Delta T = \frac{R\alpha^2 Z^4}{n^3}\left(\frac{1}{j+\dfrac{1}{2}}-\frac{3}{4n}\right) \qquad ..(51)$$

This expression is similar to expression obtained for Sommerfeld's relativistic expression with k replaced by $(j + 1/2)$. The above expression was independently obtained by Dirac and is known as *Dirac equation*.

Taking $R = 1.097 \times 10^5 \text{ cm}^{-1}$

$$\alpha = 1/137$$

and $Z = 1$, we obtain

$$\Delta T = \frac{5.84}{n^3}\left(\frac{1}{k}-\frac{3}{4n}\right): \text{ Sommerfeld's formula} \qquad ..(52)$$

$$\Delta T = \frac{5.84}{n^3}\left(\frac{1}{j+\dfrac{1}{2}}-\frac{3}{4n}\right): \text{ Dirac's formula} \qquad ..(53)$$

These are term-shifts from the Bohr level.

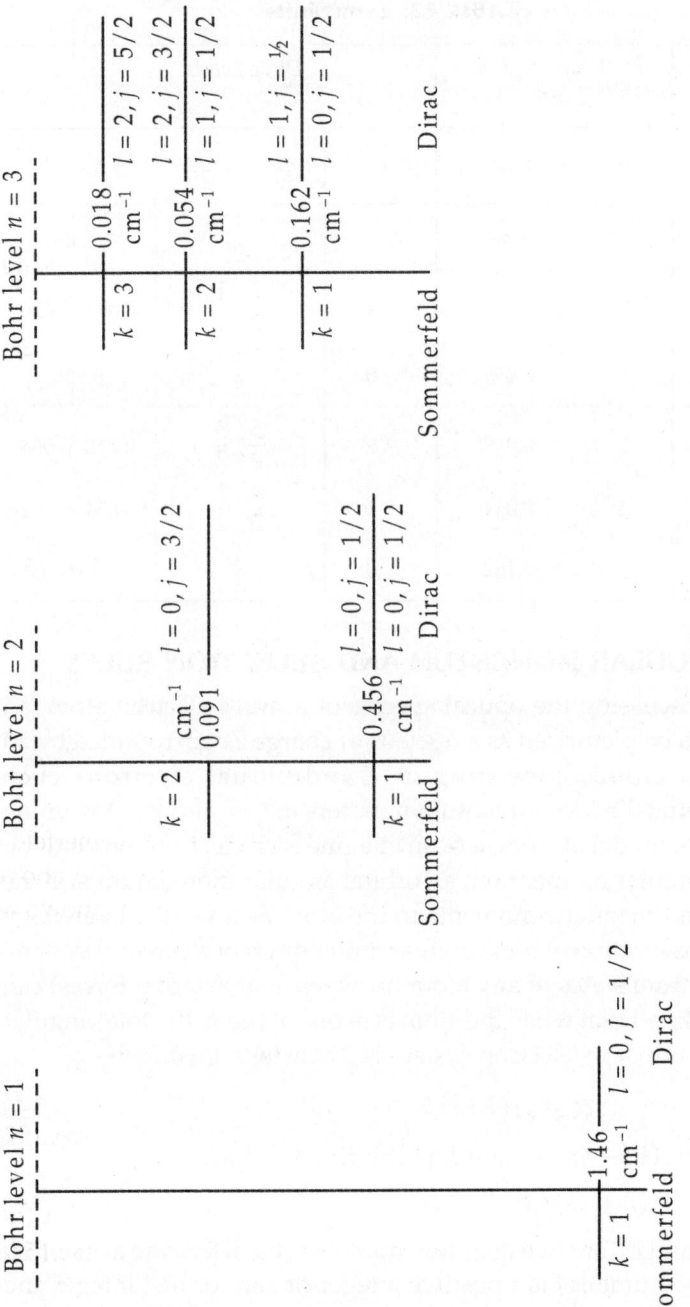

Fig. 4.7: Energy levels according to Sommerfeld's theory and Dirac's theory

TABLE 4.3: Term Shifts

Bohr Levels	Sommerfeld Levels		Dirac Levels		
n	l	$\Delta T(\text{cm}^{-1})$	l	$j\left(=l\pm\dfrac{1}{2}\right)$	$\Delta T(\text{cm}^{-1})$
1	1	1.46	0	$\dfrac{1}{2}$	1.46
2	2	0.091	1	$\dfrac{3}{2},\dfrac{1}{2}$	0.091, 0.456
	1	0.456	0	$\dfrac{1}{2}$	0.456
3	3	0.018	2	$\dfrac{5}{2},\dfrac{3}{2}$	0.018, 0.054
	2	0.054	1	$\dfrac{3}{2},\dfrac{1}{2}$	0.054, 0.162
	1	0.162	0	$\dfrac{1}{2}$	0.162

4.8 ANGULAR MOMENTUM AND SELECTION RULES

While discussing the optical spectra of a multi-electron atom, an atom can be picturised as a nucleus of charge Ze surrounded by an electron cloud, consisting of Z individual electrons each characterized by four quantum numbers n, l, m_l and m_s. Assuming the vector model of atom, we can imagine each electron contributing a spin angular momentum, an orbital angular momentum and the associated magnetic moments to the atom as a whole. In analogy with conservation of total angular momentum of a classical system, the quantum states of any atom (in absence of external forces) can be so defined that when the atom is in one of them, the total angular momentum of its electron system is a constant, given by

$$A_J = \sqrt{J(J+1)}\,\hbar \qquad\qquad ..(54)$$

and also its z-component has the fixed value

$$(A_J)_z = M_J\hbar \qquad\qquad ..(55)$$

where J and M_J are two quantum numbers characterizing a quantum state. The number J is a positive integer or zero or half integer and M_J has one of the $2J + 1$ values $-J, -J + 1, .., 0, 1, .., J$.

For example, for $J = 0$, $M_J = 0$; for $J = \frac{1}{2}$, $M_J = +\frac{1}{2}$ or $-\frac{1}{2}$, for $J = 1$, $M_J = 1$ or 0 or –1.

For each of the electronic quantum states, there is an integral value of the quantum number m_l and m_s is either $+\frac{1}{2}$ or $-\frac{1}{2}$; further, for a chosen z-axis,

$$M_J = \Sigma(m_l + m_s) \qquad \qquad ..(56)$$

where the summation is to be taken over all electrons in the atom. It is to be noted here that M_J, and hence also J has integral value when the atom contains an even number of electrons and half integral value when it contains an odd number of electrons.

4.8.1 The Selection Rules

It is experimentally found that all the possible combinations of permitted energy states of atom do not actually appear as spectral lines. Selection rules give the reason for such a state of affairs. For the vector atom model, three selection rules have been devised which are described here under.

(*a*) **The Selection Rule for L**

Most of the observed spectral lines are due to transition between states, in which a single electron jumps from one orbit to another, and in such cases, the selection rule is

$\Delta L = \pm 1$

i.e. only those lines are observed for which the value of L changes by ± 1.

(*b*) **The Selection Rule for J**

Spectral lines arise only when transitions take place between states for which

$\Delta J = \pm 1$ or 0

$0 \rightarrow 0$ being excluded.

(*c*) **The Selection Rule for S**

It is given by

$\Delta S = 0$

i.e. states with different S (hence different multiplicities) do not combine with one another. Theory and experiment however show that this selection rule is adhered to less and less strictly as the atomic number increases. Hence it is only an approximate rule holding good in case of light atoms. In the presence of a magnetic field,

$\Delta m_L = 0$ or ± 1

The magnetic spin quantum number m_S remains unchanged, i.e.

$\Delta m_S = 0$

In consequence

$\Delta m_J = 0$ or ± 1

These rules furnish invaluable help in the allocation of observed spectral series to the proper quantum number. With their aid, energy level diagrams can be constructed both for the natural complex multiplet lines and for the Zeeman effect of such lines.

Transitions which contravene these rules are sometimes observed, but the intensities of such "forbidden lines" (as they are called), are usually weak in comparison with the normal lines.

4.8.2 The Intensity Rules

Intensity rules have been devised to supplement the selection rules, in order to predict also the intensity of the lines that are observed. These are as follows:

(a) Those transitions are strong, giving rise to intense lines, in which L and J change in the same sense; the transitions are the weaker, the more the change in direction of L and J is different.

(b) A transition in the decreasing sense $(L \to L - 1)$ is stronger than a transition in the increasing sense $(L \to L + 1)$

(c) The case of oppositely directed transitions does not occur, in general, either in X-ray spectra or in doublet spectra; because it would lead to a final state in which $(J - L)$ would be two units greater than in the initial state, which is forbidden.

Hence, we have the following cases:

$\Delta L = -1, \Delta J = -1$; most intense line (a)

$\Delta L = -1, \Delta J = 0$; less intense (a)

$\Delta L = +1, \Delta J = +1$; weaker (b)

$\Delta L = +1, \Delta J = 0$; weakest (a and b)

$$\left.\begin{array}{l} \Delta L = -1, \Delta J = +1 \\ \Delta L = +1, \Delta J = -1 \end{array}\right\} \text{ on line } (c)$$

It can be seen both theoretically and experimentally that the total intensity of all the lines coming to, or starting from, any given J term of a multiplet is proportional to $2J + 1$.

4.9 SPECTROSCOPIC TERMS AND SPECTRAL NOTATIONS

We now define some terms associated with spectroscopy:

State: The state of an atom is specified by listing four quantum numbers n, l, m_l and m_s. It signifies the state of motion of all the electrons. The state of lowest energy is termed as the ground state. If several states have the same energy, they are said to be degenerate.

Energy Level: It is a collection of states having the same energy in the absence of (external) electric or magnetic field. It is characterized by quantum number J.

Sublevel: An external field splits an energy level into two or more sublevels. Each is characterized by one or more magnetic quantum numbers.

Term: A collection of levels characterized by an orbital angular momentum and multiplicity comprises a 'spectroscopic term', e.g. a ^3D term is the weighted average energy of the $^3D_3$, $^3D_2$ and $^3D_1$ levels.

Equivalent Electrons: Subshells with the same n and l are termed equivalent orbitals. The electrons in equivalent orbitals are called 'equivalent electrons'.

Statistical Weight: It is the number of distinct states in a specified collection, e.g. statistical weight of a level is $(2J + 1)$, for a term it is $(2S + 1)(2L + 1)$ and for a single electron is $2n^2$.

Spectral lines are categorized according to the following conventions:

Component: A transition between two sublevels is termed a component.

Line: A transition between two levels is a 'line'. It is thus a blend of components.

Multiplet: A collection of transitions between two terms is called a multiplet. It consists of a number of lines.

Resonance Lines: Among the lines arising from transitions between the ground level and higher levels, the line of lowest frequency is called the 'resonance line'.

4.9.1 Term Symbol from Electronic Configuration

Example 4.2: Consider carbon atom. Its electronic configuration is $1s^2\, 2s^2\, 2p^2$. It has two equivalent electrons in the outer p-subshell. For these electrons,

$$l_1 = l_2 = 1 \quad \text{and} \quad s_1 = s_2 = \tfrac{1}{2}$$

so, $\qquad L = 2, 1, 0 \quad$ and $\quad S = 1, 0$

The terms corresponding to these values are $(^{2S+1}L_J)$ or ^3D, ^3P, ^3S and ^1D, ^1P, ^1S according as $S = 1$ and $S = 0$ respectively. Corresponding J values are

$$^3D: J = 3, 2, 1 \qquad ^1D: J = 2$$
$$^3P: J = 2, 1, 0 \qquad ^1P: J = 1$$
$$^3S: J = 1 \qquad\qquad ^1S: J = 0$$

Hence, the complete list of states is

$$^3D_3,\, ^3D_2,\, ^3D_1 \quad ; \quad ^1D_2$$
$$^3P_2,\, ^3P_1,\, ^3P_0 \quad ; \quad ^1P_1$$
$$^3S_1 \quad ; \quad ^1S_0$$

Following rules provide a suitable guideline to determine various terms of an atom:

1. Electrons enter unfilled orbitals to give maximum multiplicity (Hund's rule).

2. For terms resulting from equivalent electrons, those with highest multiplicity will be most stable.

3. Among the states having same configuration and same multiplicity, the most stable state is the one with largest value of J.

4. For states with given L and S values, there arise two possibilities:

   (*i*) if the orbital is less than half filled, the state with smallest J is most stable and

   (*ii*) if the orbital is half or more than half filled, the state with largest J will be most stable.

5. For equivalent electrons the ground state will be the one with $L = \Sigma m_l$ maximum provided no two electrons have same m_l value.

Example 4.3: Consider the ground state of flourine atom:

(i) F : $1s^2\, 2s^2\, 2p^5$

(ii) Ignore the closed shells and couple the $l's$ of the valence incomplete orbitals to find L:

$$\text{(incomplete orbital is } p \therefore l = 1)$$

(iii) Then couple spins and find S

$$S = s = \tfrac{1}{2} \qquad\qquad (\therefore 2p^5 \leftrightarrow 2p_x^2\, p_y^2\, p_z^1)$$

Multiplicity is $\therefore$ 2.

(iv) Couple L and S to find J.

$$L = 1 \text{ and } S = \tfrac{1}{2} \qquad \therefore J = 3/2 \text{ and } \tfrac{1}{2}$$

(v) Express the term symbol as

$$2^2P_{3/2},\ 2^2P_{½}$$

(vi) Using rule (4), the ground state, i.e. most stable state, is $2\,^2P_{3/2}$.

Example 4.4: Consider the Ti atom. Its outermost configuration is $3d^2$.

$$L = \Sigma m_l = 2 + 1 = 3 \qquad\qquad (F \text{ state})$$

$$S = \Sigma s_i = \frac{1}{2} + \frac{1}{2} = 1$$

Hence multiplicity is = 3.

J values are (for $L = 3$, $S = 1$):

$$J = 4,\, 3,\, 2.$$

Since the shell is less than half filled, the ground state is $^3F_2$.

Example 4.5: Write the term symbol for ground state of Mn atom $(3d^5)$.

$$L = \Sigma m_l = 2 + 1 + 0 - 1 - 2 = 0 \Rightarrow S \text{ state}$$

Spin
$$S = \sum_i s_i = \frac{1}{2} + \frac{1}{2} + \frac{1}{2} + \frac{1}{2} + \frac{1}{2} = \frac{5}{2}$$

$$\Rightarrow \qquad 2S + 1 = 6 \text{ (Multiplicity)}$$

$$L = 0,\, S = \frac{5}{2} \quad \therefore J = \frac{5}{2}$$

Hence the ground state is $^6S_{5/2}$.

Example 4.6: Determine the possible terms of a one-electron atom corresponding to $n = 3$ and compute the angle between l and s vectors for the term $^2D_{5/2}$.

Solution: For $n = 3$, l can take the values

$$l = 0,\, 1,\, 2$$

and for an electron $s = ½$

$\therefore$ Multiplicity $(2s + 1) = 2$

The possible values of j are
$$j = l \pm s$$
$$l = 0; j = 1/2$$
$$l = 1; j = 3/2, 1/2$$
$$l = 2; j = 5/2, 3/2.$$

Hence, the possible terms are
$$^2S_{1/2}, {}^2P_{3/2}, {}^2P_{1/2}, {}^2D_{5/2}, {}^2D_{3/2}$$

For the state $^2D_{5/2}$, we have
$$. l = 2; s = 1/2, j = 5/2$$

From vector model,
$$l \cdot s = \frac{1}{2}\left(j^2 - l^2 - s^2\right)$$

or
$$\cos(l, s) = \frac{j(j+1) - l(l+1) - s(s+1)}{2\sqrt{l(l+1)}\sqrt{s(s+1)}}$$

$$= \frac{\dfrac{35}{4} - 6 - \dfrac{3}{4}}{2\sqrt{6 \times 3/4}} = 0.4714$$

$\therefore$ angle $(l, s) = \cos^{-1}(0.4714) = 61.9°.$

4.10 RADIATION SHIFT FOR THE H_α LINE (LAMB SHIFT)

According to Dirac's quantum mechanical theory the term shift from Bohr level is given by

$$\Delta T = \frac{R_\infty \alpha^2 Z^4}{n^3}\left[\frac{1}{j + \dfrac{1}{2}} - \frac{3}{4n}\right]$$

$$\equiv \frac{5.84}{n^3}\left(\frac{1}{j + \dfrac{1}{2}} - \frac{3}{4n}\right) \qquad \qquad ..(59)$$

The only difference between the Sommerfeld and Dirac's theory is that the latter predicted a double degeneracy of most levels because the energy depends on n and j but not on l (generally there are two values of l corresponding to same j). Figure 4.8 shows theoretical structure of H_α as predicted by Dirac's theory. The selection rules are

$$\Delta l = \pm 1$$
$$\Delta j = 0, \pm 1 \text{ but } j = 0 \leftrightarrow j = 0 \qquad \qquad ..(60)$$

These rules allow five transitions as shown in the Fig. 4.8.

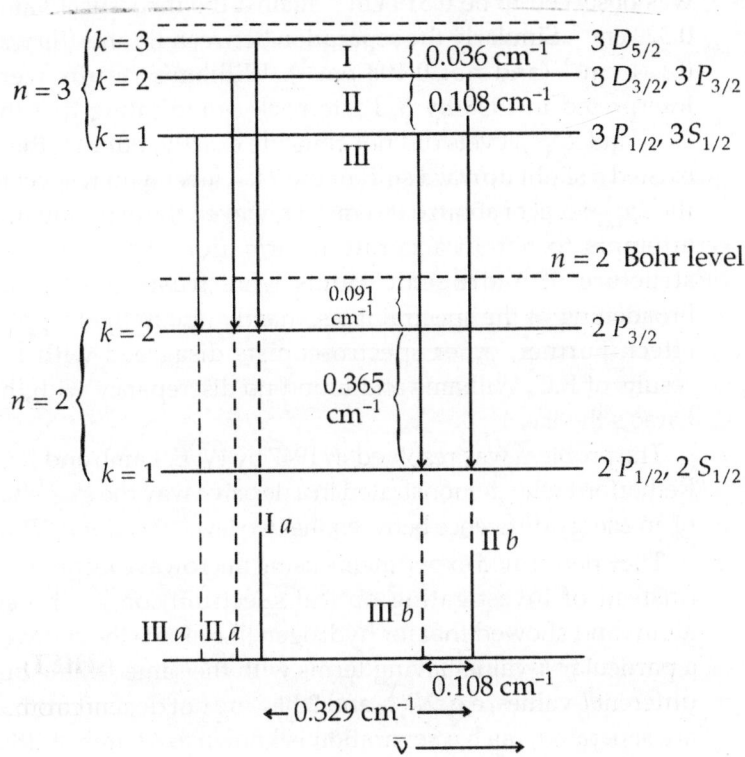

Fig. 4.8: Theoretical structure of H_α as predicted by Dirac's theory

Comparison with Experiment

Many spectroscopic studies of the fine structure of atomic hydrogen were made to test the Dirac's theory but no definite conclusion was reached till 1940. Although there were some evidence strongly supporting the theory, the measurements performed by R.C. Williams in 1938 indicated two main differences between theory and experiment as mentioned below:

(i) The component Ia was observed to be weaker than the component IIb, and the component IIIb was found to be much more stronger than as predicted theoretically. These deviations could be explained as due to unequal excitation of the $n = 3$ levels.

(*ii*) The separation between the two main components I*a* and II*b* was observed to be 0.319 cm^{-1} against the theoretical value 0.329 cm^{-1}. Similarly the separation between II*b* and III*b* was 0.134 cm^{-1} (and not 0.108 cm^{-1}). William's results were interpreted in 1938 by S. Pasternack as indicating that the $2s_{1/2}$ and $2p_{1/2}$ levels did not coincide exactly, but that there existed a slight upward shift of the $2s_{1/2}$ level with respect to the $2p_{1/2}$ level of about 0.03 cm^{-1}. However, the experimental attempts to obtain accurate information about the fine structure of hydrogenic atoms were frustrated by the broadening of the spectral lines, mainly due to the Doppler effect. Further, other spectroscopists disagreed with the results of R.C. Williams and found no discrepancy with the Dirac's theory.

The problem was resolved in 1947 by W. E. Lamb and R.C. Retherford who demonstrated in a decisive way the existence of an energy difference between the two levels $2S_{1/2}$ and $2P_{1/2}$.

They performed experiments using microwave techniques (instead of investigating optical spectrum) on hydrogen atoms and showed that for hydrogen like atoms the states of a particular *n*-value having terms with the same *j* value but different *l*-values, e.g. $2P_{1/2}$ and $2S_{1/2}$, are not degenerate but are separated. Such a separation is known as "Lamb shift".

Lamb-Retherford Experiment

Lamb and Retherford used microwave techniques to stimulate a direct radio frequency transition between the $2S_{1/2}$ and $2P_{1/2}$ levels. There is no selection rule on the principal quantum number *n* for electric dipole transitions. In particular, these transitions can occur between the levels having same *n* value. This fact was pointed out as early as 1928 by W. Grotrian, who suggested that it should be possible with radio waves to induce such transitions among the excited states of the hydrogen atom. Because the frequencies of radio waves are much smaller than those corresponding to optical lines (such as H_α line), the Doppler broadening (which is proportional to frequency) is considerably reduced in radio-frequency experiments and can, in fact, be neglected in the experiment of Lamb and Retherford.

The plan of the experiment is shown in Fig. 4.9.

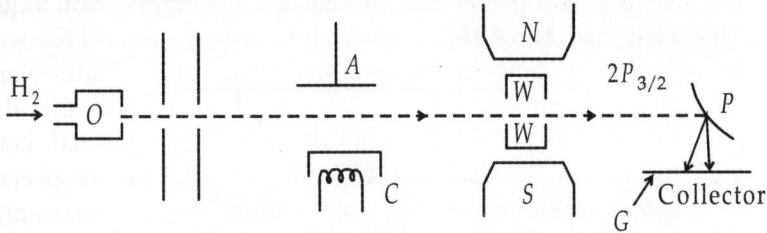

Fig. 4.9: Lamb-Retherford experiment

(After *Atomic and Molecular Spectra* by Raj Kumar, K. Nath and R. Nath, Meerut, India (2003)).

Molecular hydrogen H_2 entering an oven O is dissociated into atomic hydrogen and then passed through slits. This beam of hydrogen atoms is passed through a vacuum diode in which electrons are being emitted from heated cathode C (and accelerated towards anode A). Some of the normal atoms ($1S_{1/2}$) passing through this region collide with the electrons and are excited into $2S_{1/2}$, $2P_{1/2}$ and $2P_{3/2}$ states.

These excited atoms proceed toward a tungsten plate P and collide with it. During this transit, the atoms in the $2P_{1/2}$ and $2P_{3/2}$ states return to their ground state $1S_{1/2}$ but those in the metastable state $2S^*_{1/2}$ cannot do so (because of the selection rule $\Delta l \neq 0$).

These metastable atoms return to their ground state by colliding with plate P from which, therefore electrons are emitted. The beam of electrons so emitted are collected and passed on to a galvanometer whose reading is a measure of the metastable atomic beam density.

Any mechanism, which causes the metastable atoms to undergo a transition to the $2P_{3/2}$ state, will result in a fall in the galvanometer reading which is sensitive only to metastable atoms. Such transitions were induced by passing the atoms through a wave guide (WW) in which microwaves of controllable frequency were generated.

It was observed that, at a certain frequency, the metastable atomic beam intensity suddenly reduced. It was conjectured that the microwaves of this frequency were absorbed by the $2S^*_{1/2}$ atoms which were excited to the $2P_{3/2}$ state from which they decayed at once to the ground state. Consequently, the atoms reaching the plate P were in their ground state and could not eject electrons from it. Hence, this frequency was a measure of the term difference between $2S_{1/2}$ and $2P_{3/2}$ states. Such measurements showed that this difference was not 0.365 cm^{-1} (as predicted by Dirac's theory) but 0.0353 cm^{-1} less. This meant that the state $2S_{1/2}$ is higher than $2P_{1/2}$ by 0.0353 cm^{-1} (*see* Fig. 4.10).

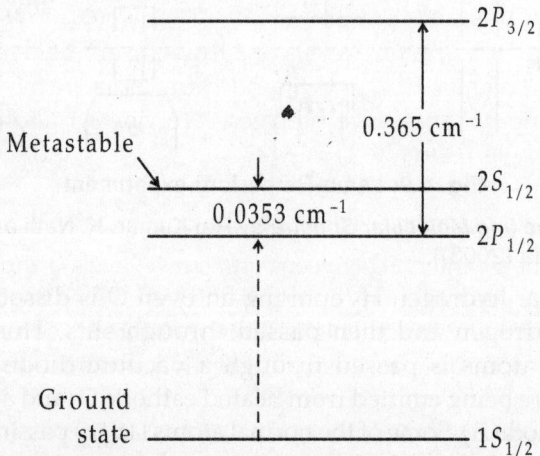

Fig. 4.10: Lamb shift (0.0353 cm^{-1})

Soon after the discovery of this shift, Bethe showed that a revised theory of interaction between matter and radiation causes all S terms to be raised by an amount which agreed well with the experimental value for hydrogen. Hence, Lamb shift is also called as radiation shift.

4.11 LANDE'S INTERVAL RULE

This rule determines the separation of fine structure lines in L-S coupling. Under L-S coupling, the spin orbit interaction energy is

$$\Delta E_{sl} = a(L \cdot S) \qquad ..(61)$$

where a is an interaction constant.

We have

$$J = L + S$$

$$\Rightarrow \qquad J \cdot J = L \cdot L + S \cdot S + 2L \cdot S$$

or $\qquad |J|^2 = |L|^2 + |S|^2 + 2L \cdot S$

or $\qquad L \cdot S = \frac{1}{2}\left\{ |J|^2 - |L|^2 - |S|^2 \right\}$

$$\therefore \qquad \Delta E_{sl} = \frac{a}{2}\left[J(J+1) - L(L+1) - S(S+1) \right]\hbar^2 \qquad ..(62)$$

where $\qquad \hbar^2 = h/(2\pi)$

because $\quad |{}^*J| = \sqrt{J(J+1)}\hbar$ etc., we can write

$$\Delta E_{sl} = A\big[J(J+1) - L(L+1) - S(S+1)\big] \qquad ..(63)$$

where A is a constant.

The various fine structure levels of a Russell Saunders multiplet have the same values of L and S, and differ only in the value J. Hence, the energy difference between two fine structure levels (corresponding to J and $(J + 1)$ is

$$E_{J+1} - E_J = A[(J + 1)\,(J + 2) - J(J + 1)]$$
$$= 2A(J + 1) \qquad\qquad ..(64)$$

Thus, the energy interval between consecutive levels (J and $J + 1$) of a fine structure multiplet is proportional to $J + 1$, i.e. to the larger of the two J-values involved. This is known as *Lande's interval rule*. For example, according to this rule, the fine structure levels $^3P_0$, $^3P_1$, $^3P_2$ have separations in the ratio 1 : 2, and the levels $^3D_1$, $^3D_2$, $^3D_3$ have separations in the ratio 2 : 3.

4.12 NORMAL AND INVERTED TERMS

A normal term is one in which fine structure levels are arranged in ascending order with increasing J whereas an inverted term is one in which fine structure levels are arranged in ascending order with decreasing J.

The experimental studies have revealed that normal term appears when the atom of the element concerned has got the electron configuration which involves less than half filled subshell of electrons. The inverted term appears when the atom has got the electron configuration with subshell having more than half filled electrons. The order of the levels is governed by the spin orbit interaction. The interaction energy is given by

$$\Delta E = \cos(L^*S^*)\ \Sigma a_i l_i^* s_i^* \cos(l_i^* L^*) \cos(s_i^* S^*)$$

When this interaction energy is negative, the terms are inverted and when this energy is positive, the terms are normal.

4.13 SPECTRAL TERMS OF EQUIVALENT ELECTRONS

In an atom, equivalent electrons are those which have the same n and l values. In this case, some of the spectroscopic terms derived for non-equivalent electrons will not be allowed, which is a consequence of Pauli's exclusion principle.

Equivalent electrons, though having same n and l values, must differ in their values of m_l and m_s. Terms of equivalent electrons can easily be obtained using a scheme proposed by G. Breit. The method is illustrated by considering the example of two equivalent p-electrons (p^2).

For the two electrons, $l_1 = 1, l_2 = 1, s_1 = 1/2, s_2 = 1/2$. Therefore $m_{l_1} = 1, 0, -1; m_{l_2} = 1, 0, -1; m_{s_1} = 1/2, -1/2$. The m_{l_1} and m_{l_2} values are tabulated one in a row and the other in a column as shown in Fig. 4.11 (a). We calculate M_L values ($\equiv m_{l_1} + m_{l_2}$) as given in the figure. Similarly compute the M_s values.

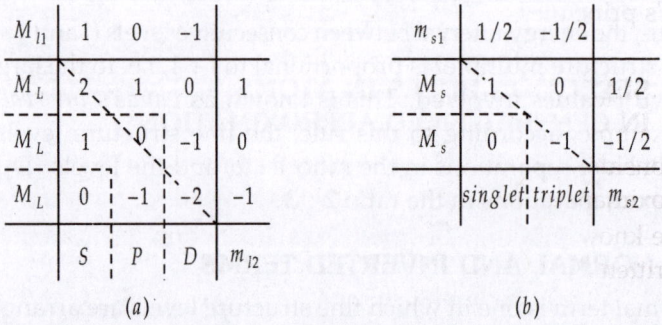

Fig. 4.11: Magnetic quantum numbers for two equivalent p electrons

Figure 4.11 (b). The M_L values are grouped by dotted line as 2, 1, 0, -1, -2 (corresponding to $L = 2$ and hence a D term). $M_L = 1, 0, -1$ corresponds to $L = 1$ giving a P term. $M_L = 0$ corresponds to $L = 0$ and therefore an S term.

Similarly $M_S = 1, 0 -1$ corresponds to a multiplicity of three, giving a triplet term and $M_S = 0$ giving a singlet term. If the electrons are non-equivalent all the combinations are allowed giving $^1S, ^1P, ^1D$ $^3S, ^3P$ and ^3D.

In the case of equivalent electrons, the situation is different. For M_L values 2, 0, -2 (along the diagonal), m_{l_1} and m_{l_2} are equal. Similar is the case with M_S values. Hence if m_{s_1} and m_{s_2} are equal ($M_S = 1$ and -1), the values of $M_L = 2, 0$ - 2 are forbidden. Since the values in the lower left of the array are the mirror images of those in the upper right half, one of these groups must also be left out from the calculations. Leaving the lower left half, the remaining values are

(i) $M_S = 1$, $M_L = 1, 0, -1$

(ii) $M_S = -1$, $M_L = 1, 0, -1$

On the other hand, if we allow the m_l values to be alike, the values of $M_S = 1$ and -1 are forbidden. The remaining values are

(iii) $M_S = 0$, $M_L = 2, 1, 0, -1, -2$

(iv) $M_S = 0$, $M_L = 1, 0, -1$

(v) $M_S = 0$, $M_L = 0$

The rows (i), (ii) and (iv) together give the complete set of quantum numbers for a ^3P term. The row (iii) gives a ^1D term and row (v) gives a ^1S term. Thus the allowed terms in case of two equivalent p electrons are ^1S ^1D and ^3P. The terms ^1P, ^3S and ^3D are excluded by Pauli's principle.

4.14 SELECTION RULES IN ATOMS IN CENTRAL FIELD APPROXIMATION

We consider here the case of hydrogen atom, but in the central field approximation, the results will also hold good for other atoms.

We know that the normalized wave function for hydrogen can be written as

$$\psi_{nlm} = N_r \, R_{nl}(r) \, N_\theta \, P_l^{|m|}(\cos\theta) \cdot \frac{1}{\sqrt{2\pi}} \, e^{im\phi} \qquad ..(66)$$

where N_r and N_θ are normalization factors for 'r' and 'θ' respectively. For a transition between states (n_1, l_1, m_1) and $(n_2\, l_2\, m_2)$ in the electric dipole approximation, we need the integral

$$<n_2 l_2 m_2 \,|\vec{r}|\, n_1 l_1 m_1> \equiv \iiint \psi^*_{n_2 l_2 m_2} \, \vec{r} \, \psi_{n_1 l_1 m_1} \, dx \, dy \, dz \qquad ..(67)$$

Using the spherical polar coordinates, we have

$$\left.\begin{array}{l} z = r \cos\theta \\ x = r \sin\theta \cos\phi \\ y = r \sin\theta \sin\phi \end{array}\right\} \Rightarrow dx \, dy \, dz = r^2 \sin\theta \, d\theta \, d\phi \qquad ..(68)$$

We define

$$u \equiv x + iy = r \sin\theta \, e^{i\phi}$$
$$v \equiv x - iy = r \sin\theta \, e^{-i\phi} \qquad ..(69)$$

In terms of u and v, r is given by

$$\vec{r} = \frac{u+v}{2}\,\hat{i} + \frac{u-v}{2i}\,\hat{j} + z\hat{k} \qquad ..(70)$$

where $\hat{i}, \hat{j}$ and $\hat{k}$ are the three unit vectors in Cartesian coordinates. Thus eqn. (67) can be written as product of three simple integral over ϕ, θ and r coordinates respectively.

(1) The ϕ factor of the transition integral

For the z-component, we know that to

$$= \frac{1}{2\pi} \int_0^{2\pi} e^{-im_2\phi}\, e^{im_1\phi}\, d\phi = 0 \text{ if } m_1 \neq m_2$$

$$= 1 \text{ if } m_1 = m_2 \qquad\qquad ..(71)$$

(using eqn. (66) for $\psi_{n_1 l_1 m_1}$ and $\psi_{n_2 l_2 m_2}$)

But, from eqn. (70) we see that both the x and y components of $\vec{r}$ involve u and v, so we consider the evaluation of integral depending on u. From eqn. (70), we get the following integral that depends on ϕ in eqn. (67)

$$\frac{1}{2\pi} \int_0^{2\pi} e^{-im_2\phi}\, e^{i\phi} \cdot e^{im_1\phi}\, d\phi = \frac{1}{2\pi} \int_0^{2\pi} e^{i(m_1+1)\phi}\, e^{-im_2\phi}\, d\phi$$

$$= 0 \text{ if } m_1 - m_2 + 1 \neq 0$$
$$= 1 \text{ if } |m_2| = |m_1| + 1$$
or $\qquad\qquad |m_2| = |m_1| - 1 \qquad\qquad ..(72)$

The integral containing v and depending on ϕ can be written as

$$\frac{1}{2\pi} \int_0^{2\pi} e^{-i\phi}\, e^{-im_2\phi}\, e^{im_1\phi}\, d\phi = \frac{1}{2\pi} \int_0^{2\pi} e^{-im_2\phi}\, e^{i(m_1-1)\phi}\, d\phi$$

$$= 0 \text{ if } m_1 - m_2 - 1 \neq 0$$
$$= 1 \text{ if } |m_2| = |m_1| - 1$$
or $\qquad\qquad |m_2| = |m_1| + 1 \qquad\qquad ..(73)$

From equations (71), (72) and (73), we find that the factor depending on ϕ vanishes and integral of eqn. (67) reduces to zero, unless

$$\Delta m \equiv |m_2| - |m_1| = 0 \text{ or } \pm 1 \qquad\qquad ..(74)$$

i.e. transition will take place only if the condition (74) is satisfied for the two states.

2. The θ factor of the transition integral

From eqn. (66), (67) and (68) we find that since the z component of $\vec{r}$ is $r \cos\theta$, the integral containing z and depending on θ can be written as

$$N_\theta(l_1, m_1)\, N_\theta(l_2, m_2) \int_0^{\pi} P_{l_2}^{|m_2|}(\theta) \cdot \cos\theta\, P_{l_1}^{|m_1|}(\theta) \sin\theta\, d\theta \qquad ..(75)$$

(Note that factor $\sin \theta$ comes r.h.s. of eqn. (3)). But from eqn. (71) we know that the z-component of transition moment vanishes unless $m_1 = m_2$ and the integral in eqn. (75) reduces to

$$N_\theta(l_1, m_1) \; N_\theta(l_2, m_2) \int_0^\pi P_{l_2}^{|m|} \, P_{l_1}^{|m|} \cdot \cos \theta \, \sin \theta \, d\theta \qquad ..(76)$$

For the associated Legendre polynomials, we have the following recursion relation

$$(2l + 1) \cos \theta \; P_l^{|m|} = (l + |m|) \; P_{l-1}^{|m|} + (l - |m| + 1) \; P_{l+1}^{|m|}$$

$$..(77)$$

so that

$$(2l_1 + 1) \cos \theta \; P_{l_1}^{|m|} = (l_1 + |m|) \; P_{l_1 - 1}^{|m|} + (l_1 - |m| + 1) \; P_{l_1 + 1}^{|m|}$$

or $\qquad P_{l_1}^{|m|} \cdot \cos \theta = \left(\dfrac{l_1 + |m|}{2l_1 + 1} \right) P_{l_1 - 1}^{|m|} + \left(\dfrac{l_1 - |m| + 1}{2l_1 + 1} \right) P_{l_1 + 1}^{|m|} \qquad ..(78)$

substituting $\cos \theta = t$ (i.e. $\sin \theta \, d\theta = dt$) we get the following from the orthonormality of Legendre polynomials (after multiplying both sides of eqn. (78) by $P_{l_2}^m$ and integrating):

$$\int_0^\pi P_{l_2}^{|m|}(\theta) \, P_{l_1}^{|m|}(\theta) \, \cos \theta \, \sin \theta \, d\theta$$

$$= \frac{1}{N_\theta(l_1) N_\theta(l_2)} \left[\frac{l_1 + |m|}{2l_1 + 1} \int_{-1}^{+1} P_{l_2}^{|m|}(t) \, P_{l_1 - 1}^{|m|}(t) \, dt \right.$$

$$\left. + \frac{l_1 - |m| + 1}{2l_1 + 1} \int_{-1}^{+} P_{l_2}^{|m|}(t) \, P_{l_1 + 1}^{|m|}(t) \, dt \right] \qquad ..(79)$$

$$= \frac{1}{N_\theta^2(l_1 - 1, m)} \frac{l_1 + |m|}{(2l_1 + 1)} \quad \text{if } l_2 = l_1 - 1$$

$$= \frac{1}{N_\theta^2(l_1 + 1, m)} \frac{l_1 - |m| + 1}{2l_1 + 1} \quad \text{if } l_2 = l_1 + 1 \qquad ..(80)$$

$$= 0 \text{ otherwise}$$

The integral involving u and depending on θ must have $\Delta m = \pm 1$ and comes out to be

$$N_\theta(l_1, m_1) \; N_\theta(l_2, m_2) \int_0^\pi P_{l_2}^{|m_2|}(\theta) \sin \theta \, P_{l_1}^{|m_1|}(\theta) \, \sin \theta \, d\theta \qquad ..(81)$$

This can be evaluated by using the recursion relation

$$(2l_1 + 1) \sin \theta \ P_{l_1}^{|m_1|} = P_{l_1+1}^{|m_1|-1} - P_{l_1-1}^{|m_1|+1} \qquad ..(82)$$

using this and substituting $t = \cos \theta$, we get the following relations from orthogonality of $P_l^{|m|}(\theta)$:

$$\int_0^\pi P_{l_2}^{|m_2|} P_{l_1}^{|m_1|} \sin^2 \theta \ d\theta = \frac{1}{N_\theta(l_1, m_1) \, N_\theta(l_2, m_2)} \times \frac{1}{(2l_1 + 1)}$$

$$\left[\int_{-1}^{+1} P_{l_2}^{|m_2|}(t) \, P_{l_1+1}^{|m_1|-1}(t) \, dt - \int_{-1}^{+1} P_{l_2}^{|m_2|}(t) \, P_{l_1-1}^{|m_1|+1}(t) \, dt - \right]$$

$$= \frac{1}{N_\theta^2(l_1 + 1, m_1 - 1)} \frac{1}{(2l_1 + 1)} \text{ if } l_2 = l_1 + 1 \text{ and } \Delta m = -1$$

$$= \frac{1}{N_\theta^2(l_1 - 1, m_1 + 1)} \frac{1}{(2l_1 + 1)} \text{ if } l_2 = l_1 - 1 \text{ and } \Delta m = +1 \quad ..(83)$$

$$= 0 \text{ otherwise.}$$

The integral involving v can be evaluated similarly as above and leads to results identical to eqn. (83).

The selection rule on l can, therefore, be summarized as

$$\Delta l = l_2 - l_1 = \pm 1 \qquad ..(84)$$

If these relations are not satisfied, the integral in eqn. (67) vanishes and corresponding transition is forbidden.

3. The r factor of the transition integral

From eqn. (67), the integral depending on r has the value

$$N_{n_1} N_{n_2} \int R_{n_1 \, l_1}(r) R_{n_2 l_2}(r) \ r^3 \, dr \qquad ..(85)$$

There is no formula relating $R_{nl}(r)$ to only a few functions with different values of n as is possible for the Legendre polynomials. Instead, such as expression involves an infinite series of such functions ranging over all values of n. Thus the integral in eqn. (85) is non vanishing for any set of n_1 and n_2 values.

Questions

4.1   Derive an expression for magnetic moment of hydrogen atom. Find the value of Bohr magneton (in MKS units).

4.2   Derive an expression for Larmor frequency.

4.3 Write the values of quantum numbers (l, s, j) for each of the following one electron terms

$$2S_{1/2}, 2P_{1/2}, 2P_{3/2}, 2D_{3/2}$$

$$\left[\textbf{Ans. } \left(0, \tfrac{1}{2}, \tfrac{1}{2}\right); \left(1, \tfrac{1}{2}, \tfrac{1}{2}\right); \left(1, \tfrac{1}{2}, \tfrac{3}{2}\right); \left(2, \tfrac{1}{2}, \tfrac{3}{2}\right)\right]$$

4.4 Enumerate the possible values of j and m_j for states having $l = 3, s = \frac{1}{2}$.

$$\left[\textbf{Ans. } j = \frac{7}{2}; m_j = \frac{7}{2}, \frac{5}{2}, \frac{3}{2}, \frac{1}{2}, -\frac{1}{2}, -\frac{3}{2}, -\frac{5}{2}, -\frac{7}{2}\right.$$

$$\left. \text{and } j = \frac{5}{2}; m_j = \frac{5}{2}, \frac{3}{2}, \frac{1}{2}, -\frac{1}{2}, -\frac{3}{2}, -\frac{5}{2}\right]$$

4.5 Calculate the angle between l and s vectors in the $2P_{3/2}$ state of a one electron atom.   **[Ans. 65.9°]**

4.6 Write the term symbol for ground state of Mn atom.   **[Ans. $^6S_{5/2}$]**

4.7 Explain vector model of the atom.

4.8 Discuss briefly Lande's interval rule.

4.9 Calculate the spin orbit interaction energy for a single non-penetrating valence electron.

4.10 How does the spin orbit interaction when combined with the relativity correction, explain the hydrogen fine structure?

4.11 Explain Lamb shift. How has it modified our knowledge about the spectrum of the hydrogen atom?

4.12 Discuss L-S and j-j coupling for an atom having two non-equivalent electrons.

4.13 Write short notes on:

(*i*) L-S coupling

(*ii*) j-j coupling

(*iii*) Lande's interval rule

(*iv*) Normal and Inverted terms

(*v*)   Lamb shift

5

Spectra of He Atoms and Alkali Atoms

5.1 THE STATES AND SPECTRA OF THE ATOM

He atom is a two electron system ($1s^2$). For this system $l = 0$ for both electrons and the only way the spins can align is antiparallel (according to exclusion principle). Thus, $S = 0$ for the combined electrons and $L = 0 + 0 = 0$ for the orbital angular momentum. Thus, $J = L + S$ is also zero. Hence, the term designation for the ground state of helium is $^{2S+1}L_J = {}^1S_0$.

In case of excited states, the designation becomes a bit complicated, since there being many possible ways of combining the electrons. Suppose that one of the electrons is excited to the $2s$ level, yielding a configuration $1s\,2s$. This state (and energy level of atom) also has $l = 0$ for both the electrons so $L = 0 + 0 = 0$. But the spins can be either parallel or antiparallel (the exclusion principle now doesn't apply because the two electrons belong to different n value). Therefore, we have two possible states for the spin $S = s_1 + s_2 = 1$ or 0.

In the first case, the multiplicity (number of levels) is 1 even though $2S + 1 = 3$ since the multiplicity is determined by whichever is the smaller of $2S + 1$ and $2L + 1$. The second case leads to a multiplicity of $1(S = 0)$. We thus have two possible states $^3S_1$ and $^1S_0$ for the electron configuration $1s\,2s$. (The first is referred to as a *triplet* and the latter as *singlet*).

For the excited state $1s\,2p$, $l = 0$ for one electron and $l = 1$ for the other. This leads to only one possible value for L viz. $L = 0 + 1 = 1$

112

and so we have an angular momentum designation of P. The spins combine with two possibilities, leading to $S = 1$ and $S = 0$. Thus for $S = 1, L = 1$, we have $J = 1 + 1 = 2, J = 1 + 0 = 1$ and $J = 1 - 1 = 0$ (The first one corresponds to spin and orbital angular momentum aligned, the second spin perpendicular to orbital and the third spin antiparallel to orbital).

The possible values of J vary by integral values, i.e. we start with the maximum value of $J (= L + S)$, decrease successively by 1 for either state, until we reach $J = (L - S)$.

$$1s(l_1 = 0, s_1 = 1/2), \; 2p(l_2 = 1, s_2 = 1/2)$$
$$L = l_1 + l_2 = 0 + 1 = 1$$
$$S = s_1 + s_2 = \frac{1}{2} + \frac{1}{2} = 1$$
$$S = s_1 - s_2 = 0$$

$S = 1$: $\qquad J = L + S = 1 + 1 = 2 \rightarrow {}^3P_2$
$$= 1 + 0 = 1 \rightarrow {}^3P_1$$
$$= 1 - 1 = 0 \rightarrow {}^3P_0$$

$S = 0$: $\qquad J = L + S = 1 + 0 = 1 \rightarrow {}^1P_1$

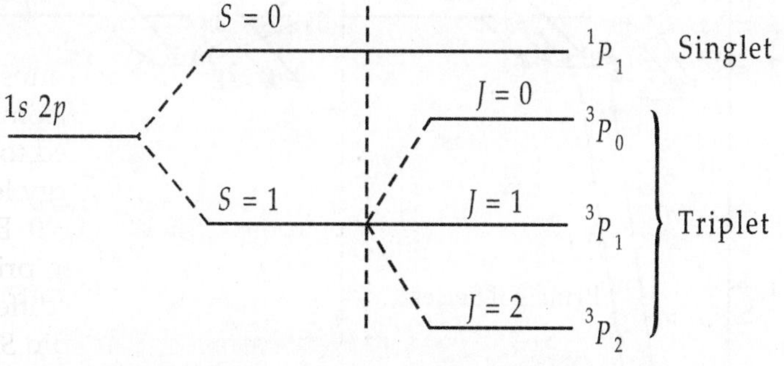

Fig. 5.1: Singlet and triplet energy levels of the excited state $1s\,2p$ of the He atom

The emission spectrum of helium shows a number of series in the visible region, as well as in the near and far ultraviolet regions of spectrum.

The relevant selection rules (in general for many electron systems) are

$$\Delta S = 0, \quad \Delta L = \pm 1, \quad \Delta J = 0, \pm 1.$$

The spectrogram for He shows two systems (i) singlet system, containing four chief series (sharp, principal, diffuse and fundamental like alkali metals) and (*ii*) triplet system which also contains all the four chief series.

Due to the strictness of the rule $\Delta S = 0$ (in the light elements), the atom does not show any intercombination lines, i.e. no transition between states of different multiplicities. This strictness divides the spectrum into two parts—the singlet system belonging to *para-helium* (spins antiparallel) and triplet system belonging to *ortho-helium* (spins parallel).

An ortho-helium atom can lose excitation energy in a collision and become one of para-helium, while a para-helium atom can gain excitation energy in a collision and become one of ortho-helium; ordinary liquid or gaseous helium is, therefore, a mixture of both. The Fig. 5.2 shows energy levels of the electrons in the singlet and triplet states of the helium atom along with some allowed transitions.

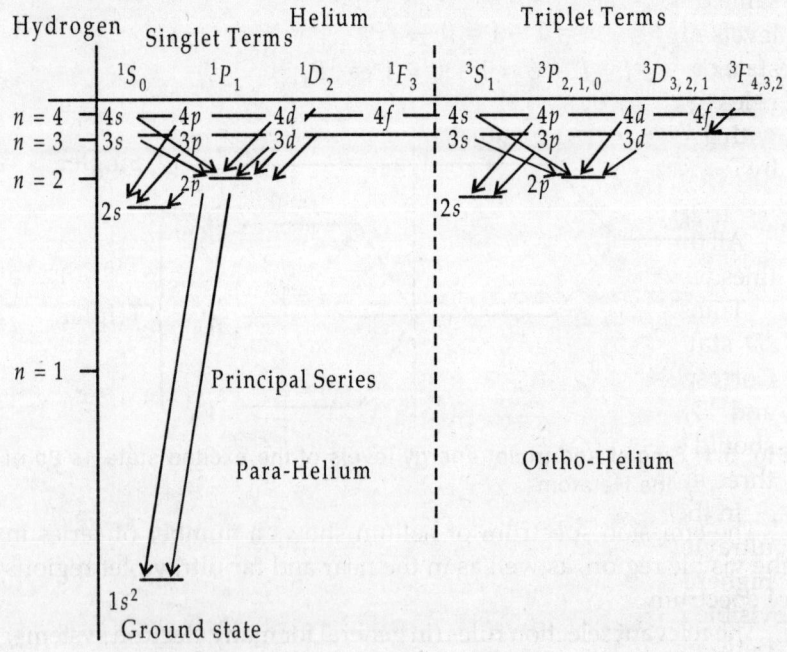

Fig. 5.2: Few allowed transitions in He atom*

---

*For simplicity, we have dropped $1s$ appearing before each of the levels, e.g. $1s$ $2s$, $1s$ $2p$, etc. are designated $2s$ and $2p$ respectively.

There is a further rule that a state with $J = 0$ cannot make a transition to another $J = 0$ state (not important in present case). Also, since S cannot change during a transition, the singlet ground state can undergo transitions only to other singlet states.

In order to investigate possible transitions, it is assumed that only one electron undergoes transition and the other remains in the $1s$ orbital. The left hand side of the Fig. 5.2 shows the energy levels for the singlet states. Initially, the $1s^2\ ^1S_0$ state can undergo a transition to (only) $1s^1\ np^1$ states in the latter $L = 1$, $S = 0$ and hence, $J = 1$ only so the transition is given by

$$1s^2\ ^1S_0 \rightarrow 1s^1\ np^1\ ^1P_1$$

From the $^1P_1$ state, the system would either revert to $^1S_0$ states or undergo transitions to the higher $^1D_2$ states. Now, regarding the triplet states (spins parallel), an outstanding feature is that there is no ground state corresponding to that of the singlet system, due to Pauli's principle, the electrons are forbidden from occupying the same orbital, thus the lowest state is $1s\ 2s$. The triplet state energy levels are shown on the right hand side of the Fig. 5.2. The $1s\ 2s$ state has $S = 1$, $L = 0$ and hence $J = 1$ only, so it is $^3S_1$; by the selection rules, it can undergo transitions only to $1s^1\ np^1$ triplet states: those with $S = 1$, $L = 1$ have $J = 2, 1$ or 0 and so, transitions are represented by

$$^3S_1 \rightarrow\ ^3P_2,\ ^3P_1,\ ^3P_0$$

All three transitions are allowed, since $\Delta J = 0$ or ± 1, so the resulting lines will be triplets.

Transitions from the ^3P states may occur to either ^3S states or to ^3D states. For ^3D, we have $S = 1$, $L = 2$ hence $J = 3, 2$ or 1. Correspondingly, we have in Fig. 5.3, the transitions between ^3P and ^3D states (selection rule $\Delta J = 0, \pm 1$). The complete spectrum should consist of six lines, however, due to lack of resolution, only three lines are seen in the spectrum.

In the spectrum of para-helium, the principal series is in the far ultraviolet region (584 Å–504 Å) and arises due to transitions from higher P terms to the ground state. The principle series lying in the visible region and near ultraviolet region is due to the transitions between higher P terms and 2^1S_0 state (in para-helium). The triplet terms are very narrow. The difference between the terms $1s\ 2p\ ^3P_2$ $- 1s\ 2p\ ^3P_1$ is only $0.077\ \text{cm}^{-1}$ and that between $1s\ 2p\ ^3P_1 - 1s\ 2p\ ^3P_0$ is $0.099\ \text{cm}^{-1}$.

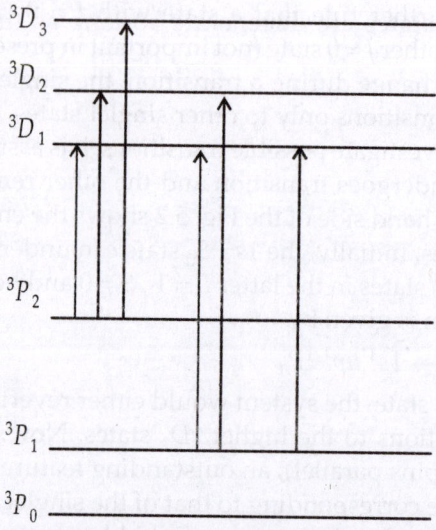

Fig. 5.3: The transitions between ^3P and ^3D levels in the He atom

5.2 QUANTUM MECHANICAL EXPLANATION FOR SPLITTING OF HELIUM LEVELS

The He atom is a two electron system. This requires that the total wave funcion must be antisymmetric with respect to the exchange of the space and spin coordinates of the two electrons. This necessitates a coupling between these two types of coordinates; the symmetric space wave function couples with the (single) antisymmetric spin function and the antisymmetric space wave function couples with the (three) symmetric spin functions. Consequently, the total wave functions of the He atom are of the form:

$$\frac{1}{\sqrt{2}}\left\{\psi_a(1)\psi_b(2)+\psi_b(1)\psi_a(2)\right\}\frac{1}{\sqrt{2}}\left\{\beta_1^+\beta_2^- - \beta_1^-\beta_2^+\right\}$$

:Singlet State   (Antiparallel spins)

$$\frac{1}{\sqrt{2}}\left\{\psi_a(1)\psi_b(2)-\psi_b(1)\psi_a(2)\right\}\beta_1^+\beta_2^+$$

$$\frac{1}{\sqrt{2}}\left\{\psi_a(1)\psi_b(2)-\psi_b(1)\psi_a(2)\right\}\frac{1}{\sqrt{2}}\left\{\beta_1^+\beta_2^- + \beta_1^-\beta_2^+\right\}$$

$$\frac{1}{\sqrt{2}}\left\{\psi_a(1)\psi_b(2)-\psi_b(1)\psi_a(2)\right\}\beta_1^-\beta_2^-$$

Triplet states (Antiparallel spins)

All these four states are degenerate so long as the Coulomb interaction between the two electrons is ignored. This interaction, however, removes the exchange degeneracy. It raises and splits each (ground) state into a singlet state and a 3 fold degenerate triplet state (Fig. 5.4).

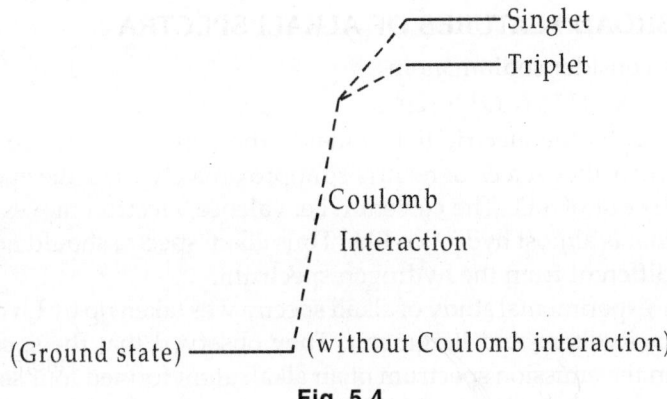

Fig. 5.4

For the ground state, both the electrons (however) have the same space quantum numbers ($a = b = 1s$) so there is a single space wave function $\psi_{1s}(1)\,\psi_{1s}(2)$ which is symmetric with respect to exchange of electrons. This can only be coupled with the antisymmetric spin function $\left(\dfrac{1}{\sqrt{2}}\right)\left\{\beta_1^+\,\beta_2^- - \beta_1^-\,\beta_2^+\right\}$, which can give rise to only a singlet state. Hence the triplet state corresponding to the singlet ground state is missing. This absence of a triplet ground state is a consequence of Pauli's principle. For all other states, triplet states exist along with the singlet states.

Since, in the triplet state, there is a tendency for the two electrons to remain apart, consequently, the Coulomb interaction energy is lesser in the triplet state than in the singlet state. This causes the triplet energy states to be lower than the corresponding singlet states.

Another important feature of the He energy level diagram is that the ground state level lies very much below the other energy levels. This is so, because in the ground state, both electrons are in the innermost orbit and strongly attracted by the nucleus. For other states, one electron exists in a higher approximately hydrogen like orbit. The (other) inner electron moves in the field of doubly charged

He nucleus (as compared to H atom) but the outer electron moves in the field of a singly charged nucleus ($\because$ the inner electron shields the nuclear charge by one unit). Consequently, the ground He level lies much deeper than the H ground level, whereas the excited He levels are akin to the corresponding hydrogen levels.

5.3 BROAD FEATURES OF ALKALI SPECTRA

Let us consider sodium atom

$$Na\ (11) = 1s^2 2s^2 2p^6 3s^1$$

As far as the electric field outside the core of 10 electrons is concerned, they screen or neutralise approximately ten of the nuclear charges out of +11. The eleventh, i.e. valence, electron moves in a field that is almost hydrogen like. Thus alkali spectra should not be very different from the hydrogen spectrum.

The experimental study of alkali spectra was taken up by Liveing, Dewar, Rydberg and Bergmann. They observed that the spectral lines in the emission spectrum of an alkali atom formed four series: a *principal series* of bright and most persistent lines, a *sharp series* of fine lines, a *diffuse series* of comparative broader lines and a *fundamental series*. The strongest is the principal series which (except its first member) lies in the ultraviolet, the sharp and diffuse series lies in the visible part while the fundamental series lies in the infrared region.

Similar to Balmer formula for hydrogen atom $\left(viz.\ \nu = \nu_\infty - \dfrac{R}{n^2} \right)$

Rydberg, pointed out that the alkali series could be represented by

Principal Series $\nu_m^p = \nu_\infty^p - \dfrac{R}{(m+p)^2}$; $m = 2,3\ldots\ldots\infty$

Sharp Series $\nu_m^s = \nu_\infty^s - \dfrac{R}{(m+s)^2}$; $m = 2,3\ldots\ldots\infty$

Diffuse Series $\nu_m^d = \nu_\infty^d - \dfrac{R}{(m+d)^2}$; $m = 3,4\ldots\ldots\infty$

Fundamental Series $\nu_m^f = \nu_\infty^f - \dfrac{R}{(m+f)^2}$; $m = 4,5\ldots\ldots\infty$ ..(1)

where $\nu_\infty's$ are the wave numbers of the *convergence limits* of the corresponding series and are called *fixed terms*. In the so called *running terms, s, p, d* and *f* represent Rydberg corrections (< 1) for the corresponding series.

Rydberg noticed following relations among different series of the same atom:

(*i*) The sharp and diffuse series have a common convergence limit $\left(v_\infty^s = v_\infty^d \right)$.

(*ii*) The common convergence limit is equal to the first running term (with $m = 2$) of the principal series.

$$v_\infty^s = v_\infty^d = \frac{R}{\left(2+p\right)^2} \quad ..(2)$$

(*iii*) The convergence limit of the principal series is equal to a running term (with $m = 1$) of the sharp series.

$$v_\infty^p = \frac{R}{\left(1+s\right)^2} \quad ..(3)$$

(*iv*) The convergence limit of the fundamental series is equal to the first running term (with $m = 3$) of the diffuse series

$$v_\infty^f = \frac{R}{\left(3+d\right)^2} \quad ..(4)$$

In view of these observations, we have

Principal Series $\qquad v_m^p = \dfrac{R}{\left(1+s\right)^2} - \dfrac{R}{\left(m+p\right)^2} ; m = 2,3 \ldots\ldots\infty$

Sharp Series $\qquad v_m^s = \dfrac{R}{\left(2+p\right)^2} - \dfrac{R}{\left(m+s\right)^2} ; m = 2,3 \ldots\ldots\infty$

Diffuse Series $\qquad v_m^d = \dfrac{R}{\left(2+p\right)^2} - \dfrac{R}{\left(m+d\right)^2} ; m = 3,4 \ldots\ldots\infty$

Fundamental Series $\quad v_m^f = \dfrac{R}{\left(3+d\right)^2} - \dfrac{R}{\left(m+f\right)^2} ; m = 4,5 \ldots\ldots\infty \quad ..(5)$

The following two laws were found to hold in the alkali series:

(*i*) **Rydberg-Schuster Law:** The wave number difference between the principal series limit and the sharp (or diffuse) series limit is equal to the wave number of the first line of the principal series

$$v_\infty^p - v_\infty^{s\,(\text{or }d)} = \frac{R}{\left(1+s\right)^2} - \frac{R}{\left(2+p\right)^2} = v_2^p \quad ..(6)$$

(*ii*) **Runge's Law:** The wave number difference between the diffuse series limit and the fundamental series limit is equal to the wave number of the first line of the diffuse series

$$\nu_\infty^d - \nu_\infty^f = \frac{R}{(2+p)^2} - \frac{R}{(3+d)^2} = \nu_3^d \qquad ..(7)$$

5.4 EXPLANATION FOR BROAD FEATURES OF ALKALI SPECTRA

An alkali atom (atomic number Z) consists of an inert gas core composed of the nucleus and a few completely filled subshells having $(Z - 1)$ electrons, plus a single valence electron in the (partially filled) outermost subshell. For example,

$$^3Li = 1s^2\, 2s^1$$
$$^{11}Na = 1s^2\, 2s^2\, 2p^6\, 3s^1$$
$$^{19}K = 1s^2\, 2s^2\, 2p^6\, 3s^2\, 3p^6\, 4s^1$$

In the optical (low-energy) excitation processes, only the outermost, i.e. valence, electron gets excited and is thus responsible for optical spectra. The inner, inert gas core is a spherically symmetrical distribution playing no part in optical excitation.

Consider the case of sodium atom. Its normal outer configuration is 3s. The various possible excited configurations and the corresponding terms are

$$4s,\ 5s,\ 6s\ .....S\text{-terms}$$
$$3p,\ 4p,\ 5p\ .....P\text{-terms}$$
$$3d,\ 4d,\ 5d\ .....D\text{-terms}$$
$$4f,\ 5f,\ 6f\ .....F\text{-terms}$$

In an excitation, the energy of excitation is that of the optically active, i.e. valence, electron and the total energy of the core does not change which may be taken to be zero.

We are now interested in comparing the various energies of the levels of the sodium atom with the corresponding ones of the hydrogen atom. For hydrogen atom

$$E_n = -\frac{2\pi^2\mu(Ze)^2 e^2}{n^2 h^2} = -\frac{2\pi^2\mu(e)^2 e^2}{n^2 h^2} \qquad (\because Z = 1) \quad ..(8)$$

i.e. total energy depends upon n only. In case of an alkali atom, the optical electron moves not only in the central field of nucleus but also the core electrons. Consequently, its energy

$$E = -\frac{2\pi^2\mu\left(Z_ne\right)^2 e^2}{n^2h^2} \qquad ..(9)$$

Since nuclear charge Ze is screened by the charge $(Z-1)e$ of the core electrons, and since the screening is not perfect, the effective nuclear charge Z_ne for the nth orbit is always greater than charge $+e$. Hence comparing eqns (8) and (9), the energy of the alkali atom (for a given n) is more negative than the energy of the H atom for the same n. Consequently, the energy levels of the alkali atom are lower than the corresponding levels of the H atom. However, with increasing n values, the alkali levels approach the corres-ponding hydrogen levels since screening effect also increases.

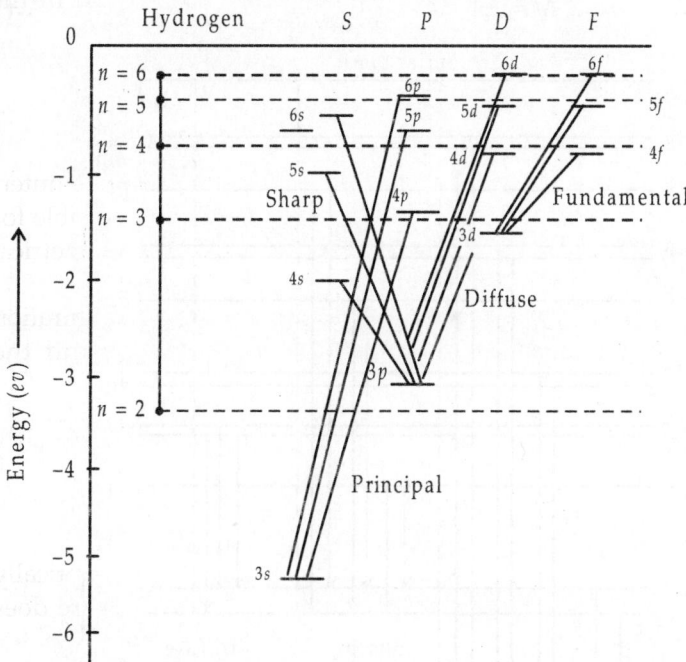

Fig. 5.5: The energy levels and transitions in sodium atom-spectrum as compared to levels of H atom

Further, in case of an alkali atom, unlike hydrogen, the energy in a certain state depends not only upon n but also the l value of the valence electron. This is so because the probability of finding the electron near the nucleus is given by

$$|\psi|^2 \propto r^{2l} \qquad ..(10)$$

and is largest for the s-electron ($l = 0$), decreasing very rapidly with increasing l. That is, for a given n, the energy is most negative for the s-electron (which spends most of the time near the nucleus), becoming less and less negative for the p, d, f, ...electrons respectively. Hence all alkali levels (for a given n) are shifted lower than that of hydrogen, with shift becoming gradually smaller for the p, d, f, ...levels.

When the sodium atom is excited by some means, it is raised to any of the excited levels from where it then jumps back through any selected path as per the selection rules

$$\Delta n = \text{any integer}$$
$$\Delta l = \pm 1 \qquad\qquad ..(11)$$

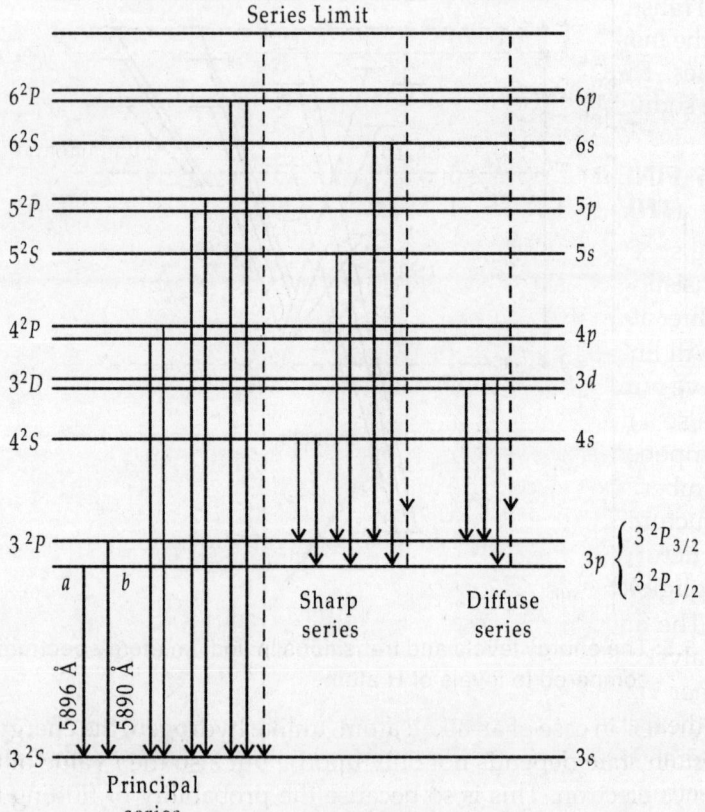

Fig. 5.6: Transitions showing first members of the sharp, principal and diffuse series of the sodium spectrum

The observed series in the emission spectrum arise from the following transitions:

$np \rightarrow 3s$, $n > 2$: Principal Series

$ns \rightarrow 3p$, $n > 3$: Sharp Series

$nd \rightarrow 3p$, $n > 2$: Diffuse Series

$nf \rightarrow 3d$, $n > 3$: Fundamental Series   ..(12)

The Fig. 5.6 shows the transitions for first members of the sharp, principal and diffuse series for the sodium atom. The figure shows the normal state or ground state as $3^2 S$ and the succeeding excited states as $3^2 P, 4^2 S, 3^2 D, 4^2 P$, etc. These level designations correspond to the orbit designations (3s), 3p, 4s, 3d, 4p, etc. The superscript 2 indicates that all levels (except S states) are doublets.

Transitions from the $3^2 P$ levels to the ground state $3^2 S$ give rise to the most prominent lines, the yellow D_1 and D_2 lines of principal series. These two particular lines account for the yellow colour of the sodium light and are called the *resonance lines*.

5.5 FINE STRUCTURE IN ALKALI SPECTRA (THE SPIN ORBIT INTERACTION)

The spectral lines emitted by alkali atoms show a fine structure consisting of doublets whose splitting (separation) is small for the lighter atoms but increases rapidly with increasing atomic number.

All lines of the sharp series are close doublets having the same wave number separation. Similarly each line of the principal series is also a doublet but the wave number separation between the two components decreases rapidly towards the lines of increasing wave number. The lines of diffuse series show a three component fine structure and are called compound doublet (not triplets). The fine structure in the lines of the fundamental series is, however, negligible.

The fine structure (splitting) is traced back to level splitting. An analysis of alkali spectra shows that the S-levels are single but all other viz. P, D, F...are doublet levels. This level splitting is result of the fact that spin magnetic moment of the optically active electron interacts with the internal magnetic field created motion of this electron through the nuclear electric field. This is known as *spin-orbit interaction*. As a result, the orbital angular momentum *l* of the optical electron is coupled to the spin angular momentum *s* to form a resultant *j* about which both *l* and *s* precess. *j* is also the resultant angular momentum of the alkali atom because the core contributes zero. J can take the following two values:

$$j = l \pm s = l \pm \frac{1}{2} \qquad ..(13)$$

according as l and s are parallel or antiparallel. This causes splitting of each of the l-levels into two, except the S-levels (for which $l = 0$). The resulting levels are shown in Table 5.1.

TABLE 5.1

Level	l	s	Multiplicity $(2s + 1)$	j	Resulting Levels
S	0	$\frac{1}{2}$	2	$\frac{1}{2}$	$2S_{\frac{1}{2}}$
P	1	$\frac{1}{2}$	2	$\frac{3}{2}, \frac{1}{2}$	$2P_{\frac{3}{2}}, 2P_{\frac{1}{2}}$
D	2	$\frac{1}{2}$	2	$\frac{5}{2}, \frac{3}{2}$	$2D_{\frac{5}{2}}, 2D_{\frac{3}{2}}$
F	3	$\frac{1}{2}$	2	$\frac{7}{2}, \frac{5}{2}$	$2F_{\frac{7}{2}}, 2F_{\frac{5}{2}}$

5.5.1 Spin Orbit Interaction (and Term Values)

An electron behaves as though it possesses an intrinsic angular momentum $(h/2\pi)$ and magnetic moment of one Bohr magneton. This electron moves in the electric field E of the nucleus with velocity v. Thus magnetic field experienced by the electron is

$$B = \frac{E \times v}{c} \qquad \text{(CGS Units)} \qquad ..(14)$$

where c is the velocity of light.

But $\quad E = \dfrac{Ze}{r^3} r \qquad\qquad\qquad\qquad\qquad ..(15)$

where Ze is the nuclear charge and r, the radial distance of optical electron from the nucleus.

$$\therefore \qquad B = \frac{Ze}{cr^3}(r \times v) \quad \text{(CGS Units)} \qquad ..(16)$$

According to Bohr's quantum condition

$$m_e \; r \times v = l^* \frac{h}{2\pi} \qquad\qquad\qquad ..(17)$$

where $l^* = \sqrt{l(l+1)}$ and m_e is the rest mass of the electron. Hence,

$$B = l^* \frac{h}{2\pi} \frac{Ze}{m_e c} \left(\frac{1}{r^3} \right) \qquad \text{(CGS Units)} \qquad ..(18)$$

In this field, the spinning electron undergoes Larmor precession about the field direction. The angular velocity of precession is given by Larmor's theorem as the product of field strength and the ratio between the magnetic and mechanical moments of the spinning electron.*

$$\omega_L = |B| \cdot 2\frac{e}{2m_e c} = l^* \frac{h}{2\pi} \cdot \frac{Ze^2}{m_e^2 c^2 r^3} \qquad ..(19)$$

A relativistic treatment of the problem by Thomas has revealed that, in addition to the Larmor precession ω_L, there is a relativistic precession ω_R, one half as great and in the opposite direction (due to spin). The resulting precession of the spinning electron is just equal to ordinary Larmor precession:

$$\omega' = |\omega_L + \omega_R| = \frac{1}{2}l^* \frac{h}{2\pi} \frac{Ze^2}{m_e^2 c^2}\left(\frac{1}{r^3}\right) \qquad ..(20)$$

The interaction energy is just the kinetic energy of electron precession around the field B. In absence of the field, energy

$$E(\omega) = \frac{1}{2}I\omega^2 \qquad ..(21)$$

and in the presence of the field, energy

$$E = \frac{1}{2}I(\omega + \omega') \cdot (\omega + \omega')$$

$$\simeq \frac{1}{2}I\omega^2 + I\omega\omega' \qquad \left(\text{neglecting } \frac{1}{2}I\omega'^2\right)$$

$$..(22)$$

We have

$$\Delta\bar{E}_{l,s} = (I\omega)\bar{\omega}'$$

$$\equiv \bar{\omega}\left[s^* \frac{h}{2\pi}\cos(l^*s^*)\right] \qquad ..(23)$$

(because angular momentum $(I\omega)$ is due to spin and orientation is due to l, s interaction)

$$= \frac{Ze^2}{2m_e^2 c^2}\frac{h^2}{4\pi^2}\left\langle\frac{1}{r^3}\right\rangle l^* s^* \cos(l^*s^*) \qquad ..(24)$$

---

*This ratio is just twice the corresponding ratio for the electron's orbital motion.

We will now try to evaluate the last two factors viz. $\left\langle \dfrac{1}{r^3} \right\rangle$ and $\cos(l^* s^*)$. To evaluate $\left\langle \dfrac{1}{r^3} \right\rangle$ we will have to take the time average over a complete cycle. Assuming the orbit to be elliptical

$$\frac{\text{Area of the ellipse}}{\text{Time}} = \frac{\pi ab}{T} = \frac{1}{2} r^2 \frac{d\psi}{dt} \qquad ..(25)$$

(a and b are semimajor and semiminor axes respectively).

$$\therefore \quad \left\langle \frac{1}{r^3} \right\rangle = \frac{1}{T} \int \frac{1}{r^3} dt = \frac{1}{2\pi ab} \int_0^{2\pi} \frac{1}{r} d\psi \quad \text{(substituting for } 1/r^3 \text{)} \quad ..(26)$$

Now, from Sommerfeld's theory,

$$\frac{1}{r} = \frac{1 + \varepsilon \cos \psi}{a(1 - \varepsilon^2)} \qquad ..(27)$$

and $b = a(1 - \varepsilon^2)^{1/2}$ ..(28)

$$\therefore \quad \left\langle \frac{1}{r^3} \right\rangle = \frac{1}{2\pi ab} \int \frac{1 + \varepsilon \cos \psi}{a(1 - \varepsilon^2)} d\psi$$

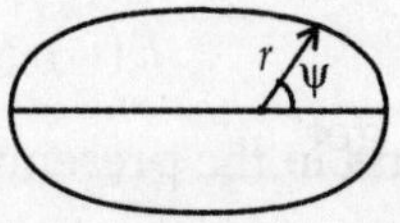

Fig. 5.7: Elliptical orbit

$$= \frac{1}{2\pi b^3} \int_0^{2\pi} (1 + \varepsilon \cos \psi) d\psi$$

$$= \frac{1}{b^3} \qquad ..(29)$$

Further, $b = \left(\dfrac{nk}{Z} \right) a_0$; where $a_0 = \dfrac{b^2}{4\pi^2 me^2}$ is Bohr's first orbit radius and azimuthal quantum number k is related to l by

$$k^3 = l\left(l + \frac{1}{2}\right)(l + 1) \qquad ..(30)$$

Hence $\left\langle \dfrac{1}{r^3} \right\rangle = \dfrac{Z^3}{n^3 k^3 a_0^3} = \dfrac{Z^3}{n^3 l\left(l + \dfrac{1}{2}\right)(l + 1) a_0^3}$ ..(31)

To evaluate $\cos(l^* s^*)$, we have to consider vector model of the atom. In calculating the precession frequency of s^* around the field produced by the orbital motion, the vector l^* was assumed fixed in space. In field free space, both orbit and spin are free to move so that l^* and s^* will precess around their mechanical resultant j^*. By law of conservation of angular momentum, the resultant j^* and hence the angle between l^* and s^* must remain invariant, so that with angle fixed, the cosine is not to be averaged.

$\therefore \qquad\qquad j^{*2} = l^{*2} + s^{*2} - 2l^* s^* \cos(l^*s^*)$

$\therefore \qquad\qquad l^* s^* \cos(l^* s^*) = \dfrac{j^{*2} - l^{*2} - s^{*2}}{2} \qquad\qquad ..(32)$

$\therefore \Delta E_{l,\,s} = \dfrac{Ze^2 h^2}{2m_e^2 c^2 \times 4\pi^2} \times \dfrac{Z^3}{n^3 a_0^3 l\left(l + \dfrac{1}{2}\right)(l+1)}\left[\dfrac{j^{*2} - l^{*2} - s^{*2}}{2}\right] \quad ..(33)$

On using the Rydberg constant

$$R = \frac{2\pi^2 m e^4}{ch^3}$$

and the square of the fine structure constant

$$\alpha^2 = \frac{4\pi^2 e^4}{c^2 h^2}$$

we get $\quad \Delta E_{l,\,s} = \dfrac{R\alpha^2 chZ^4}{n^3 l\left(l + \dfrac{1}{2}\right)(l+1)}\left[\dfrac{j^{*2} - l^{*2} - s^{*2}}{2}\right] \qquad ..(34)$

Total energy (taking into consideration the relativistic correction and spin orbit interaction) is

$$E_{nlj} = E_n - E_{nl} + E_{ls} \qquad\qquad ..(35)$$

or $\qquad T_{nlj} = \dfrac{E_n}{ch} - \dfrac{E_{nl}}{ch} + \dfrac{E_{ls}}{ch} \qquad\qquad ..(36)$

$$= -\frac{RZ^2}{n^2} - \frac{R\alpha^2 Z^4}{n^3}\left[\frac{1}{l + \dfrac{1}{2}} - \frac{3}{4n}\right] + \frac{R\alpha^2 Z^4}{n^3 l\left(l + \dfrac{1}{2}\right)(l+1)}$$

$$\times \left[\frac{j(j+1) - l(l+1) - s(s+1)}{2}\right] \qquad ..(37)$$

Since $\qquad s = \dfrac{1}{2} \quad$ so $\quad j = l \pm \dfrac{1}{2}$

Case (i) $\qquad j = \left(l + \dfrac{1}{2}\right)$

$$T_{nlj} = -\frac{RZ^2}{n^2} - \frac{R\alpha^2 Z^4}{n^3}\left(\frac{1}{l+\frac{1}{2}} - \frac{3}{4n}\right) + \frac{R\alpha^2 Z^4}{n^3 l\left(l+\frac{1}{2}\right)(l+1)}$$

$$\left[\frac{\frac{2l+1}{2} \times \frac{2l+3}{2} - l^2 - l - \frac{3}{4}}{2}\right]$$

$$= -\frac{RZ^2}{n^2} - \frac{R\alpha^2 Z^4}{n^3}\left[\frac{1}{l+\frac{1}{2}} - \frac{3}{4n}\right] + \frac{R\alpha^2 Z^4 l}{2ln^3\left(l+\frac{1}{2}\right)(l+1)}$$

$$= -\frac{RZ^2}{n^2} - \frac{R\alpha^2 Z^4}{n^3}\left[\frac{1}{l+\frac{1}{2}} - \frac{1}{2\left(l+\frac{1}{2}\right)(l+1)} - \frac{3}{4n}\right]$$

$$= -\frac{RZ^2}{n^2} - \frac{R\alpha^2 Z^4}{n^3}\left[\frac{2l+2-1}{2\left(l+\frac{1}{2}\right)(l+1)} - \frac{3}{4n}\right] \quad ..(38)$$

or $$\frac{E_{nlj}}{ch} = -\frac{RZ^2}{n^2} - \frac{R\alpha^2 Z^4}{n^3}\left[\frac{1}{l+1} - \frac{3}{4n}\right] \quad ..(39)$$

Case (ii) $j = \left(l - \frac{1}{2}\right)$

$$T_{n,l,j=l-\frac{1}{2}} = -\frac{RZ^2}{n^2} - \frac{R\alpha^2 Z^4}{n^3}\left\{\frac{1}{l+\frac{1}{2}} - \frac{3}{4n}\right\}$$

$$+ \frac{R\alpha^2 Z^4}{n^3 l\left(l+\frac{1}{2}\right)(l+1)}\left[\frac{\left(l-\frac{1}{2}\right)\left(l-\frac{1}{2}+1\right) - l(l+1) - 3/4}{2}\right]$$

$$= -\frac{RZ^2}{n^2} - \frac{R\alpha^2 Z^4}{n^3}\left\{\frac{1}{l+\frac{1}{2}} - \frac{3}{4n}\right\}$$

$$+ \frac{R\alpha^2 Z^4}{2n^3 l\left(l+\frac{1}{2}\right)(l+1)}\left[l^2 - \frac{1}{4} - l^2 - l - \frac{3}{4}\right]$$

$$= -\frac{RZ^2}{n^2} - \frac{R\alpha^2 Z^4}{n^3} \left\{ \frac{1}{\left(l+\frac{1}{2}\right)} - \frac{3}{4n} \right\} + \frac{R\alpha^2 Z^4}{2n^3 l \left(l+\frac{1}{2}\right)}$$

$$= -\frac{RZ^2}{n^2} - \frac{R\alpha^2 Z^4}{n^3} \left\{ \frac{1}{l+\frac{1}{2}} + \frac{1}{2l\left(l+\frac{1}{2}\right)} - \frac{3}{4n} \right\}$$

$$= -\frac{RZ^2}{n^2} - \frac{R\alpha^2 Z^4}{n^3} \left[\frac{2l+1}{2l\left(l+\frac{1}{2}\right)} - \frac{3}{4n} \right]$$

$$= -\frac{RZ^2}{n^2} - \frac{R\alpha^2 Z^4}{n^3} \left(\frac{1}{l} - \frac{3}{4n} \right) \qquad ..(40)$$

We therefore see from eqns (39) and (40) that wave numbers are different. Putting $l = j - 1/2$ in eqn. (39) and $l = j + 1/2$ in eqn. (40) we get

$$T_{nlj} = -\frac{RZ^2}{n^2} - \frac{R\alpha^2 Z^4}{n^3} \left[\frac{1}{j+\frac{1}{2}} - \frac{3}{4n} \right] \qquad ..(41)$$

5.5.2 Splitting up of the Levels in Alkali Spectra

Wave number separation or term splitting

$$\Delta T = T_n \left(l + \frac{1}{2} \right) - T_n \left(l - \frac{1}{2} \right)$$

$$= -\frac{R\alpha^2 Z^4}{n^3} \left\{ \left[\frac{1}{l+1} - \frac{3}{4n} \right] - \left[\frac{1}{l} - \frac{3}{4n} \right] \right\}$$

$$= -\frac{R\alpha^2 Z^4}{n^3} \left(\frac{1}{l+1} - \frac{1}{l} \right)$$

$$= -\frac{R\alpha^2 Z^4}{n^3} \frac{l-l-1}{l(l+1)}$$

or $\qquad \Delta T = \frac{R\alpha^2 Z^4_{eff}}{n^3 l(l+1)}$ cm^{-1} $\qquad ..(42)$

where $j = \left(l + \dfrac{1}{2}\right)$ for the upper level and $j = \left(l - \dfrac{1}{2}\right)$ for the lower level.

Substituting the value of Rydberg's constant
$$R = 109737 \text{ cm}^{-1}$$

and the fine structure constant $\left(\alpha = \dfrac{1}{137}\right)$ with
$$\alpha^2 = 5.3 \times 10^{-5}$$

we get

$$\Delta T = 5.82 \; \frac{Z_{eff}^4}{n^3 \, l \, (l+1)} \; \text{cm}^{-1} \qquad ..(43i)$$

where $Z_{eff} = Z - \sigma$, σ being the screening constant. It should be noted here that measured from the series limit down, the term value of any fine structure level will be
$$T = T_0 - \Gamma$$

where $T_0 \equiv \left(\dfrac{E_n}{ch} \cdot \dfrac{E_{nl}}{ch}\right)$ is a hypothetical term value for the centre of gravity of the doublet and T factor is the spin orbit interaction energy given by

$$\Gamma = a \left(\frac{j^{*2} - l^{*2} - s^{*2}}{2}\right)$$

with
$$a = \frac{R\alpha^2 Z^4}{n^3 l \left(l + \dfrac{1}{2}\right)(l+1)} \; \text{cm}^{-1}$$

Γ factor gives the shift of each fine structure level from T_0.

We can write ΔT from eqn. (41) in terms of j as

$$\Delta T = T_n(j + 1) - T_n(j)$$

$$= -\frac{R\alpha^2 Z_{eff}^4}{n^3 (j + 3/2)} + \frac{R\alpha^2 Z_{eff}^4}{n^3 (j + 1/2)}$$

$$= R\alpha^2 Z_{eff}^4 \left(\frac{1}{j + \dfrac{1}{2}} - \frac{1}{j + \dfrac{3}{2}}\right)$$

$$= \frac{R\alpha^2 Z_{eff}^4}{n^3 \left(j + \dfrac{1}{2}\right)\left(j + \dfrac{3}{2}\right)} = \frac{5.82 \, Z_{eff}^4}{n^3 \left(j + \dfrac{1}{2}\right)\left(j + \dfrac{3}{2}\right)} \qquad ..(43ii)$$

Formulae (43i) or (43ii) give doublet separation in remarkably good agreement with experimental results. The relative splitting of levels for given n but varying j (i.e. ^2P, ^2D and ^2F terms) are depicted in Fig. 5.8. ΔT gives the separation between two lines of a doublet. Note that Z occurs in the numerator and n and l occur in the denominator. This is in agreement with experimental observations:

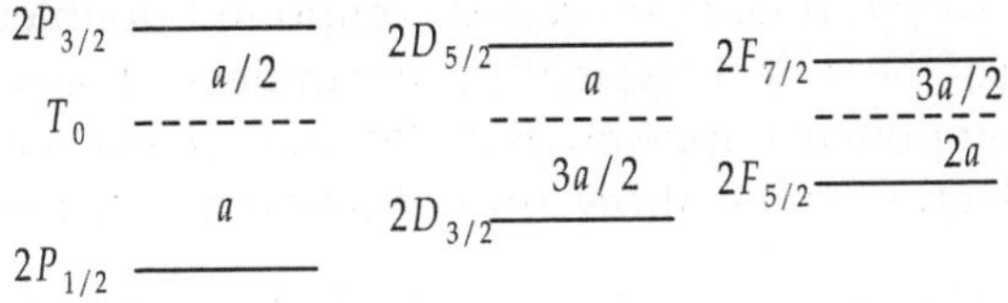

Fig. 5.8: Doublet separation due to the spin-orbit interaction

Doublet term separations (1) increase with increasing atomic number e.g. in going from NaI to KI, (2) decrease with increasing n i.e. in going to higher members of a given series; (3) decrease with increasing l i.e. in going to different series p, d, f etc. (Note here that $T \equiv v/c = \bar{v}$, the wave number in cm^{-1}).

Notations: In the doublet spectra of atomic systems containing but one valence electron, the small letters s, p, d, f.. for different electron orbits are replaced by the corresponding capitals S, P, D, F ...for the terms. The small superscript 2 in front of each term indicates that the level in question has doublet properties, and belongs to a doublet system. Although S levels are single, their doublet nature reveals itself when atom is placed in a magnetic field. Inorder to distinguish between two fine structure levels having the same n and l values, half integral subscripts are used. This subscript to each term (earlier called the *inner quantum number* by Sommerfeld) gives the total angular momentum of all the extranuclear electrons. This inner quantum number is also referred to as the electron quantum number j or the term quantum number J.

The selection rules which explain the production of doublet structure of spectral lines are

$$\Delta n = 0, 1, 2, ...$$
$$\Delta l = \pm 1$$
$$\Delta j = 0, \pm 1.$$

For instance, the principal resonance doublet of sodium consists of D_1 and D_2 lines which are due to the following transitions:

$$3^2P_{1/2} \to 3^2S_{1/2} \qquad\qquad (D_1 \leftrightarrow 5896 \text{ Å})$$
$$3^2P_{3/2} \to 3^2S_{1/2} \qquad\qquad (D_2 \leftrightarrow 5890 \text{ Å})$$

The wave number separation between these two lines is equal to the separation between $3^2P_{1/2}$ and $3^2P_{3/2}$ levels. Since the splitting of the levels decreases with increasing n (see Fig. 5.8), the wave number separation between the doublet components decreases in the lines of higher and higher wave numbers.

The lines of the diffuse and fundamental series show a three component structure and are called compound doublets, because the splitting of the ^2D level is small so that under ordinary resolution only two components appear. (Both $3d$ and $3p$ levels are splitted and the selection rules allow three transitions) (as depicted in Fig. 5.9).

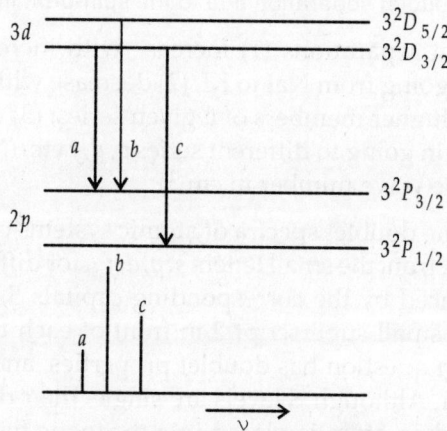

Fig. 5.9: The lines of the diffuse series

Although the lines have three components they are called doublets and not triplets because they arise from transitions between doublet levels.

5.6 INTENSITY RATIO FOR DOUBLETS

Quantitative rules for the relative intensities (discovered by Burger, Dorgelo and Ornstein) for doublets and all multiplets are as follows:

(*i*) The sum of the intensities of those lines of a multiplet which come from a common initial level is proportional to the quantum weight ($2j + 1$) of that level.

(*ii*) The sum of the intensities of those lines of a multiplet which end on a common level is proportional to the quantum weight of that level.

As an example, let us consider the case of a principal series doublet. There are two component lines starting from $^2P_{3/2}$ and $^2S_{1/2}$. The ratio of the intensities of the two lines would be the same as the ratio of the quantum weights of the doublet levels:

$$\frac{I_b}{I_a} = \frac{2 \times \frac{3}{2} + 1}{2 \times \frac{1}{2} + 1} = 2 : 1 \qquad ..(44)$$

Let us now consider the diffuse series compound doublet. The line b starts from $^2D_{5/2}$ while a and c start from $^2D_{3/2}$.

$$\therefore \quad \frac{I_b}{I_a + I_c} = \frac{2 \times \frac{5}{2} + 1}{2 \times \frac{3}{2} + 1} = 3 : 2 \qquad ..(45)$$

Again, the line c ends on the level $^2P_{1/2}$ while a and b end on the level $^2P_{3/2}$. Therefore,

$$\frac{I_c}{I_a + I_b} = \frac{2 \times \frac{1}{2} + 1}{2 \times \frac{3}{2} + 1} = 1 : 2 \qquad ..(46)$$

Solving eqns (45) and (46), we have

$$c = \frac{5}{a} b, \quad a = \frac{1}{9} b$$

$$\Rightarrow \qquad a : b : c = \frac{1}{9} : 1 : \frac{5}{9}$$

$$= 1 : 9 : 5$$

If a is not resolved from b, we will see two lines having an intensity ratio

$$(1 + 9) : 5 = 2 : 1.$$

Example 5.1: If the doublet splitting of the first excited state $2\,^2P_{3/2} - 2\,^2P_{1/2}$ of He$^+$ is 5.84 cm^{-1} calculate the corresponding separation for H.

Solution: The doublet splitting of a one-electron atomic state, due to spin-orbit interaction is given by

$$\Delta T = \frac{R_\infty \alpha^2 Z^4}{n^3 \, l(l+1)} \quad \text{cm}^{-1}$$

where R_∞ is Rydberg's constant, α is fine structure constant and Z is the atomic number, for a given state (n, l constant), we have

$$\Delta T \propto Z^4$$

For He$^+$ and H, $Z = 2$ and 1 respectively

$$\therefore \frac{\Delta T_{He^+}}{\Delta T_H} = \frac{2^4}{1^4} = 16$$

$$\therefore \Delta T_H = \frac{1}{16} \Delta T_{He^+} = \frac{1}{16} \times 5.84 \text{ cm}^{-1}$$

$$= 0.365 \text{ cm}^{-1}.$$

5.7 INTERACTION ENERGY IN L-S AND J-J COUPLINGS

As a result of spin orbit interaction the atomic terms consist of multiplet components of slightly different energy, each corresponding to a different value of J. The interaction, and hence the multiplet splitting, increases rapidly with atomic number Z, and is specially large in excited states of heavier atoms.

We know (from Chapter 4, eqn. 47) that for a single electron atom, the interaction energy, i.e. the shift of each fine structure level from the hypothetical centre is given by

$$-\Delta T_{ls} = \frac{R_\infty \alpha^2 Z^4}{2n^3 l \left(l + \frac{1}{2}\right)(l + 1)}$$

$$\times [j(j + 1) - l(l + 1) - s(s + 1)] \text{ cm}^{-1}$$

$$\equiv a\left[\frac{j^{*2} - l^{*2} - s^{*2}}{2}\right] \qquad ..(47)$$

where $\qquad a = \dfrac{R_\infty \alpha^2 Z^4}{n^3 l \left(l + \dfrac{1}{l}\right)(l + 1)} \text{ cm}^{-1} \qquad ..(48)$

$$j^* = \sqrt{j(j + 1)}, \, l^* = \sqrt{l(l + 1)}, \, s^* = \sqrt{s(s + 1)} \quad ..(49)$$

Since $\qquad j^{*2} = l^{*2} + s^{*2} + 2l^*s^* \cos(l^*, s^*)$, we may write

$$-\Delta T_{ls} = al^*s^* \cos(l^*, s^*) \qquad ..(50)$$

In the case of two optical electrons, there are four angular momenta, l_1^*, l_2^*, s_1^* and s_2^* with six possible interactions:

(1) s_1^* with s_2^*,   (2) l_1^* with l_2^*,   (3) l_1^* with s_1^*

(4) l_2^* with s_2^*,   (5) l_1^* with s_2^*,   (6) l_2^* with s_1^*

L-S Coupling

In *L-S* coupling, the interactions (1) and (2) predominate over (3) and (4) while (5) and (6) are negligibly small.

Applying eqn. (47), the energies corresponding to the interactions (1), (2), (3) and (4) are

$$\Delta T_1 = a_1\, s_1^*\, s_2^*\, \cos\,(s_1^*, s_2^*)$$
$$\Delta T_2 = a_2\, l_1^*\, l_2^*\, \cos\,(l_1^*, l_2^*)$$
$$\Delta T_3 = a_3\, l_1^*\, s_1^*\, \cos\,(l_1^*, s_1^*)$$
$$\Delta T_4 = a_4\, l_2^*\, s_2^*\, \cos\,(l_2^*, s_2^*)$$

..(51)

Now, in *L-S* coupling, s_1^* and s_2^* precess rapidly with fixed angles around their resultant S^*, which remains invariant in magnitude.

Therefore,

$$S^{*2} = s_1^{*2} + s_2^{*2} + 2s_1^*\, s_2^*\, \cos\,(s_1^*, s_2^*)$$

..(52)

This gives

$$\Delta T_1 = \frac{1}{2}\, a_1 (S^{*2} - s_1^{*2} - s_2^{*2})$$

..(53)

Similarly l_1^* and l_2^* precess rapidly with fixed angles around their resultant L^* so that

$$\Delta T_2 = \frac{1}{2}\, a_2 (L^{*2} - l_1^{*2} - l_2^{*2})$$

..(54)

Now L^* and S^* precess around their resultant J^* in the same way as l^* and s^* of a single electron precess around their resultant j^*. The interaction energy corresponding to this precession is due to coupling between l_1^* and s_1^* and between l_2^* and s_2^*, that is, ΔT_3 and ΔT_4. Here, the average values of the cosines must be evaluated, since the angles between the vectors are continually changing.

The average values are given by

$$\overline{\cos\,(l_1^*\, s_1^*)} = \cos\,(l_1^*, L^*)\, \cos\,(L^*, S^*)\, \cos\,(S^*, s_1^*)$$
$$\overline{\cos\,(l_2^*\, s_2^*)} = \cos\,(l_2^*, L^*)\, \cos\,(L^*, S^*)\, \cos\,(S^*, s_2^*)$$

..(55)

Using these average values of the cosines in eqn.—, we get

$$\Delta T_3 + \Delta T_4 = [a_3\, l_1^*\, s_1^*\, \cos\,(l_1^*, L^*)\, \cos\,(S^*, s_1^*)$$
$$+ a_4\, l_2^*\, s_2^*\, \cos\,(l_2^*, L^*)\, \cos\,(S^*, s_2^*)]\, \cos\,(L^*, S^*) \quad ..(56)$$

Applying cosine law for various terms, we get

$$\Delta T_3 + \Delta T_4 = \left[a_3 l_1^* s_1^* \left(\frac{l_1^{*2} + L^{*2} - l_2^{*2}}{2 l_1^* L^*} \right) \left(\frac{S^{*2} + s_1^{*2} - s_2^{*2}}{2 S^* s_1^*} \right) \right.$$

$$\left. + a_4 l_2^* s_2^* \left(\frac{l_2^{*2} + L^{*2} - l_1^{*2}}{2 l_2^* L^*} \right) \left(\frac{S^{*2} + s_2^{*2} - s_1^{*2}}{2 S^* s_2^*} \right) \right] \times \frac{J^{*2} - L^{*2} - S^{*2}}{2 L^* S^*}$$

or $$\Delta T_3 + \Delta T_4 = \left[a_3 \frac{l_1^{*2} + L^{*2} - l_2^{*2}}{2 L^{*2}} \frac{S^{*2} + s_1^{*2} - s_2^{*2}}{2 S^{*2}} \right.$$

$$\left. + a_4 \frac{l_2^{*2} + L^{*2} - l_1^{*2}}{2 L^{*2}} \frac{S^{*2} + s_2^{*2} - s_1^{*2}}{2 S^{*2}} \right] \frac{J^{*2} - L^{*2} - S^{*2}}{2}$$

This may be written as

$$\Delta T_3 + \Delta T_4 = \frac{1}{2} (a_3 \alpha_3 + a_4 \alpha_4) (J^{*2} - L^2 - S^{*2}) \qquad ..(57)$$

where

$$\alpha_3 = \left(\frac{l_1^{*2} - l_2^{*2} + L^{*2}}{2 L^{*2}} \right) \left(\frac{s_1^{*2} - s_2^{*2} + S^{*2}}{2 S^{*2}} \right) \qquad ..(58)$$

and

$$\alpha_4 = \left(\frac{l_2^{*2} - l_1^{*2} + L^{*2}}{2 L^{*2}} \right) \left(\frac{s_2^{*2} - s_1^{*2} + S^{*2}}{2 S^{*2}} \right)$$

For any given triplet $l_1^*, l_2^*, s_1^*, s_2^*, L^*, S^*$, etc. are fixed in magnitude, so, a_3, a_4, α_3 and α_4 are constants. Writing

$$a_3 \alpha_3 + a_4 \alpha_4 = A \qquad ..(59)$$

eqn. (57) becomes

$$\Delta T_3 + \Delta T_4 = \frac{1}{2} A (J^{*2} - L^{*2} - S^{*2}) \qquad ..(60)$$

We may now write any fine structure term by the formula

$$T = T_0 - \Delta T_1 - \Delta T_2 - \Delta T_3 - \Delta T_4 \qquad ..(61)$$

where T_0 is a hypothetical term value for the centre of gravity of the entire electron configuration.

As an example, consider a *ps* configuration:

$$l_1 = 1, \qquad l_2 = 0, \qquad \therefore \ L = 1 \ (P \text{ term})$$
$$s_1 = 1/2, \qquad s_2 = 1/2 \qquad S = 0, 1$$

For $S = 0$ (singlet state)

$$J = |L - S|, \ |L - S| + 1, ...(L + S) = 1$$

and for $S = 1$ (triplet state)

$$J = 0, 1, 2$$

The configuration gives a singlet term $^1P_1$ and a triplet term $^3P_{0,1,2}$. The shift of each term from the centre of gravity is $\Delta T_1 + \Delta T_2$. Now

$$\Delta T_1 + \Delta T_2 = \frac{1}{2} a_1 (S^{*2} - s_1^{*2} - s_2^{*2}) + \frac{1}{2} a_2 (L^{*2} - l_1^{*2} - l_2^{*2}) \quad ..(62)$$

Substituting $S = 0$, $s_1 = \frac{1}{2}$, $s_2 = \frac{1}{2}$, $L = 1$, $l_1 = 1$, $l_2 = 0$, we get

$$\Delta T_1 + \Delta T_2 = -\frac{3a_1}{4} \quad ...(63)$$

Thus, the singlet term is shifted up the hypothetical centre by $3a_1/4$, as shown in Fig. 5.10. Again, substituting $S = 1$, $s_1 = \frac{1}{2}$, $s_2 = \frac{1}{2}$, $L = 1$, $l_1 = 1$, $l_2 = 0$, we get

$$\Delta T_1 + \Delta T_2 = \frac{a_1}{4} \quad ..(64)$$

The triplet term is shifted down the hypothetical centre by $a_1/4$ as shown in Fig. 5.10. (This has been so chosen because the singlet levels lie above the corresponding triplet levels of the same electron configuration).

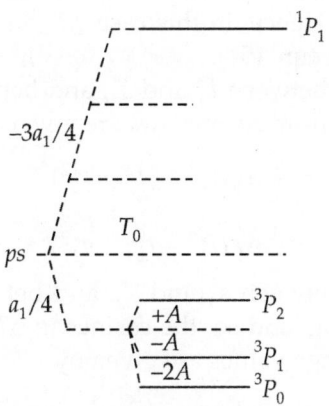

Fig. 5.10

Now, the shift of each fine structure level from the hypothetical centre of the triplet term is $\Delta T_3 + \Delta T_4$. Now,

$$\Delta T_3 + \Delta T_4 = \frac{1}{2} A \ (J^{*2} - L^{*2} - S^{*2}) \quad ..(65)$$

Putting $L = 1$, $S = 1$ and $J = 0, 1, 2$, we get

$$\Delta T_3 + \Delta T_4 = -2A, -A, +A \quad ..(66)$$

Taking it a regular triplet, the levels $^3P_0$ and $^3P_1$ are lowered by $2A$ and A from the centre, while the level $^3P_2$ is raised up by A.

The total ^3P separation, i.e. $^3P_0 - {}^3P_2$ is $3A$. By eqn. (58) and (59), we have

$$A = a_3\alpha_3 + a_4\alpha_4$$

or $$A = a_3 \frac{l_1^{*2} - l_2^{*2} + L^{*2}}{2L^{*2}} \frac{s_1^{*2} - s_2^{*2} + S^{*2}}{2S^{*2}}$$

$$+ a_4 \frac{l_2^{*2} - l_1^{*2} + L^{*2}}{2L^{*2}} \frac{s_2^{*2} - s_1^{*2} + S^{*2}}{2S^{*2}} \qquad ..(67)$$

Substituting $l_1 = 1, l_2 = 0, L = 1, s_1 = \frac{1}{2}, s_2 = \frac{1}{2}, S = 1$, we get

$$A = \frac{a_3}{2} \qquad ..(68)$$

Thus, in ps configuration, the total ^3P separation is $3A = 3a_3/2$.

jj Coupling

In jj coupling, the interaction beteen l_1^* and s_1^*, and that between l_2^* and s_2^* are stronger than the interaction between l_1^* and l_2^* and that between s_1^* and s_2^*. Hence, in this case ΔT_3 and ΔT_4 predominate over ΔT_1 and ΔT_2 in eqn. (51).

Since the angles between l_1^* and s_1^*, and between l_2^* and s_2^* are fixed, ΔT_3 and ΔT_4 (from cosine law) are given by

$$\Delta T_3 = \frac{1}{2} a_3 \left(j_1^{*2} - l_1^{*2} - s_1^{*2}\right)$$
$$\Delta T_4 = \frac{1}{2} a_4 \left(j_2^{*2} - l_2^{*2} - s_2^{*2}\right) \qquad ..(69)$$

But the angles between s_1^* and s_2^*, and between l_1^* and l_2^* are continually changing, and so, the cosines in ΔT_1 and ΔT_2 must be averaged. The average values are given by

$$\overline{\cos(s_1^* s_2^*)} = \cos(s_1^*, j_1^*) \cos(j_1^*, j_2^*) \cos(j_2^*, s_2^*)$$

$$\overline{\cos(l_1^* l_2^*)} = \cos(l_1^*, j_1^*) \cos(j_1^*, j_2^*) \cos(j_2^*, l_2^*)$$

Using these average values in eqn. (51), we get

$$\Delta T_1 + \Delta T_2 = [a_1 s_1^* s_2^* \cos(s_1^*, j_1^*) \cos(j_2^*, s_2^*)$$
$$+ a_2 l_1^* l_2^* \cos(l_1^*, j_1^*) \cos(j_2^*, l_2^*)] \cos(j_1^*, j_2^*)$$

or $$\Delta T_1 + \Delta T_2 = \frac{1}{2} (a_1\beta_1 + a_2\beta_2)(J^{*2} - j_1^{*2} - j_2^{*2}) \qquad ..(70)$$

where
$$\beta_1 = \frac{s_1^{*2} + j_1^{*2} - l_1^{*2}}{2 j_1^{*2}} \frac{s_2^{*2} + j_2^{*2} - l_2^{*2}}{2 j_2^{*2}}$$
..(71)
$$\beta_2 = \frac{l_1^{*2} + j_1^{*2} - s_1^{*2}}{2 j_1^{*2}} \frac{l_2^{*2} + j_2^{*2} - s_2^{*2}}{2 j_2^{*2}}$$

Writing
$$a_1\beta_1 + a_2\beta_2 = B$$

eqn. (70) becomes
$$\Delta T_1 + \Delta T_2 = \frac{1}{2} B (J^{*2} - j_1^{*2} - j_2^{*2}) \qquad ..(72)$$

We may write any fine-structure term by the formula:
$$T = T_0 - \Delta T_1 - \Delta T_2 - \Delta T_3 - \Delta T_4$$

A good example of a *ps* configuration having *jj* coupling is found in tin. We have

$$l_1 = 1, \qquad s_1 = \frac{1}{2} \qquad \therefore j_1 = \frac{1}{2}, \frac{3}{2}$$

$$l_2 = 0, \qquad s_2 = \frac{1}{2} \qquad \therefore j_2 = \frac{1}{2}$$

This gives rise to two (j_1, j_2) combinations, namely $\left(\frac{1}{2}, \frac{1}{2}\right)$ and $\left(\frac{3}{2}, \frac{1}{2}\right)$. We know that

$$J = |j_1 - j_2|, \ |j_1 - j_2| + 1, \ ... |j_1 + j_2|$$

Thus the J-values corresponding to the above combinations are 0, 1 and 1, 2 respectively. Thus, we have the terms

$$\left(\frac{1}{2}, \frac{1}{2}\right)_{0,1} \text{ and } \left(\frac{3}{2}, \frac{1}{2}\right)_{1,2}$$

The shift of each term from the centre of gravity is

$$\Delta T_3 + \Delta T_4 = \frac{1}{2} a_3(j_1^{*2} - l_1^{*2} - s_1^{*2}) + \frac{1}{2} a_4(j_2^{*2} - l_2^{*2} - s_2^{*2}) \quad ..(73)$$

Substituting $j_1 = \frac{1}{2}, j_2 = \frac{1}{2}$ and the values of l_1, s_1, l_2, s_2 we get

$$\Delta T_3 + \Delta T_4 = - a_3 \qquad ..(74)$$

Thus, the term $\left(\frac{1}{2}, \frac{1}{2}\right)_{0,1}$ is shifted down the hypothetical centre by a_3.

Again, substituting $j_1 = \dfrac{3}{2}$ and $j_2 = \dfrac{1}{2}$ and the values of l_1, s_1, l_2 and s_2, we get

$$\Delta T_3 + \Delta T_4 = \frac{a_3}{2} \qquad ..(75)$$

Thus, the term $\left(\dfrac{3}{2}, \dfrac{1}{2}\right)_{1,2}$ is shifted up by $a_3/2$. (Fig. 5.11).

For the separation of fine structure levels, we calculate $\Delta T_1 + \Delta T_2$, which is

$$\Delta T_1 + \Delta T_2 = \frac{1}{2} B \left(J^{*2} - j_1^{*2} - j_2^{*2}\right)$$

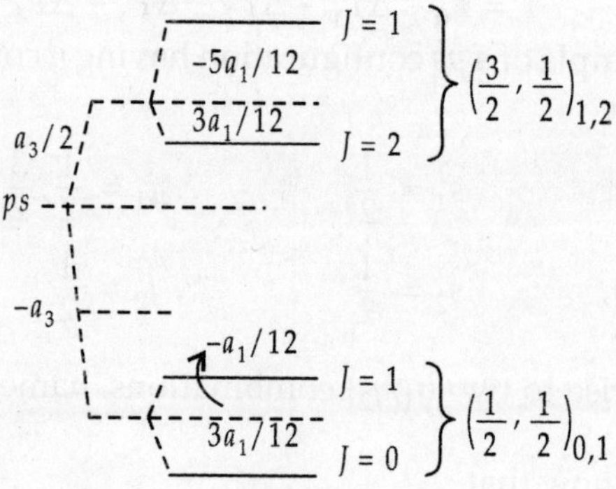

Fig. 5.11

For the term $\left(\dfrac{3}{2}, \dfrac{1}{2}\right)_{1,2}$, we have $B = \dfrac{a_1}{3}$

and $\qquad \Delta T_1 + \Delta T_2 = -\dfrac{5}{4} B, \dfrac{3}{4} B$

$$= -\frac{5}{12} a_1, \frac{3}{12} a_1$$

For the term $\left(\dfrac{1}{2}, \dfrac{1}{2}\right)_{0,1}$, we have

$$B = -a_1/3$$

and $\qquad \Delta T_1 + \Delta T_2 = -\dfrac{3}{4} B, \dfrac{1}{4} B = \dfrac{3}{12} a_1, -\dfrac{1}{12} a_1$

These are shown in Fig. 5.11. It can be seen that the total ^3P separation, i.e. between $J = 0$ level and $J = 2$ level is $3a_3/2$ which is the same as in *L-S* coupling.

Questions

5.1 Explain the occurrence of ortho and para states of He atom.

5.2 Use quantum mechanical considerations to explain the ground state of the He atom.

5.3 State Pauli's principle and hence show that the He atom in its ground state can exist only in singlet state.

5.4 Explain with reason the transition from ^1P to ^3S state in He is forbidden.

5.5 Discuss the salient features of the spectrum of the He atom. How is the spectrum explained quantum mechanically?

5.6 Discuss broad features of alkali spectra.

5.7 Briefly explain broad features of alkali spectra.

5.8 Calculate the spin-orbit interaction energy for a single non-penetrating valence electron. How will you explain the separation of ^2P and ^2D terms of alkali spectra?

5.9 Show how does the concept of spinning electron accounts for the doubling of levels in the alkali spectra.

Interaction with Electric and Magnetic Fields

6.1 INTRODUCTION

The quantum states of an atom are modified when the atom is subjected to external magnetic or electric fields. States which are formerly degenerate (i.e. having the same energy) become separated and spectral lines arising from transitions between these states may split into several components.

The effect of magnetic field on the spectrum was first discovered by Zeeman in the year 1896 and is called Zeeman effect after his name.

When an atom (spectral source) is placed in an external magnetic field, the spectral lines, it emits, are split into several polarised components. For fields less than several tenths of 1 tesla, the splitting is proportional to the strength of field. This is known as the *Zeeman effect*. A singlet spectral line viewed normal to the field is split into three plane-polarized components—a central unshifted line (with the electric field vector parallel to the field and called π-component) and two other lines equally displaced, (one on either side of π), with electric field vector perpendicular to the field (called σ-components). This is termed as *normal Zeeman effect*.

The fine structure components of a multiplet spectral line, however, in a magnetic field, show a complex Zeeman pattern. For instance, the D_1 and D_2 components of sodium doublet give four and six lines respectively. This is *anomalous Zeeman effect*. The Zeeman splitting is smaller than fine structure splitting.

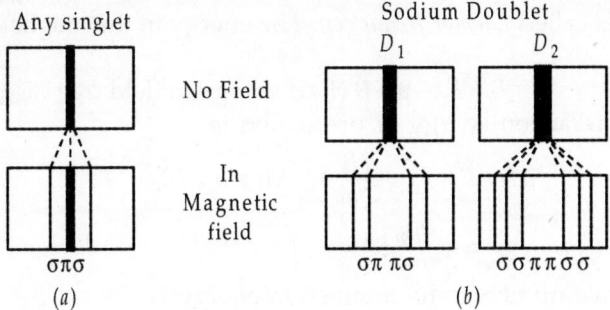

Fig. 6.1: (*a*) Normal and (*b*) Anomalous Zeeman effect

6.2 NORMAL ZEEMAN EFFECT

For a one electron atom, e.g. H atom, the orbital angular momentum L and orbital magnetic moment μ are related by

$$\frac{|\mu|}{|L|} = \frac{e}{2mc} \text{ (This is known as } gyromagnetic\ ratio).}$$

or more exactly $\mu = -\dfrac{e}{2mc}L$

Because of the negative sign of charge (viz. electron), the magnetic moment has direction opposite to the angular momentum.

For a multi-electron atom, the angular momentum is $J\ (h/2\pi)$ and the magnitude of the magnetic moment is

$$|\mu| = -\frac{e}{2mc}\frac{h}{2\pi}|J| \qquad\qquad ..(1)$$

Suppose the atom is placed in a magnetic field B along z-axis. The component of μ in the field direction is then *neglecting spin*

$$|\mu_B| = -\frac{e}{2mc}\frac{h}{2\pi}M_J \qquad\qquad ..(2)$$

where $\qquad M_J = -J, ...0, .., J - 1, J \qquad\qquad ..(3)$

i.e. a total of $(2J + 1)$ values.

The vector J precesses round the field direction with its z-component given by

$$J_Z = M_J\frac{h}{2\pi} \qquad\qquad ..(4)$$

and angular velocity of percession

$$\omega = \frac{eB}{2mc} \text{ or } \nu = \frac{1}{2\pi}\frac{eB}{2mc} \qquad\qquad ..(5)$$

This is called *Larmor frequency*. The energy in the magnetic field is

$$E = E_0 - \boldsymbol{\mu}_B \cdot \boldsymbol{B} \quad (E_0 \text{ is energy in field free case)} \quad ..(6)$$

So, interaction energy of precession is

$$\Delta E = E - E_0 = \frac{eB}{2mc}\frac{h}{2\pi}M_J \equiv h \, \nu \, M_J$$

$$= \frac{eh}{4\pi mc}BM_J \qquad ..(7)$$

In wave numbers, the interaction energy is

$$-\Delta T = \frac{\Delta E}{hc} = \frac{eB}{4\pi mc^2}M_J \text{ cm}^{-1} \qquad ..(8)$$

The terms with different J values will have different numbers of components $(2J + 1)$ in a magnetic field, but the separation (ΔT) of consecutive components would be the same for all terms of an atom for a given field strength. This separation is $h\nu$.

Figure 6.2 illustrates normal Zeeman effect for a transition $J = 2$ to $J = 1$.

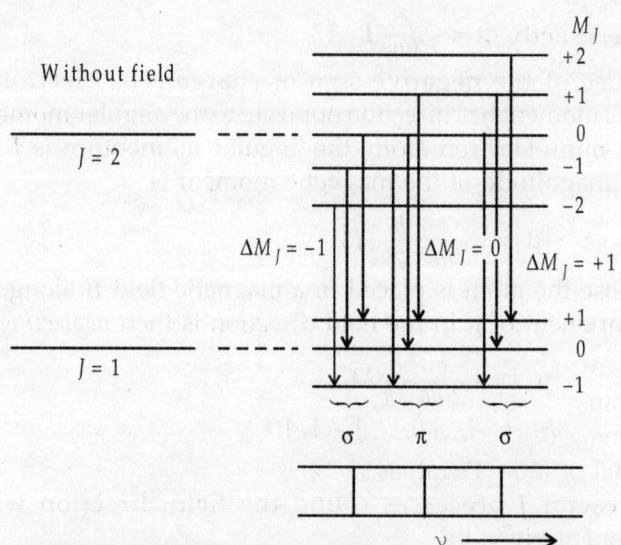

Fig. 6.2: Normal Zeeman effect for $J = 2 \rightarrow J = 1$.

The selection rule for M_J is

$$\Delta M_J = 0, \pm 1 \qquad ..(9)$$

Because there is equal splitting for all terms, the number of line components is always three, since the arrows representing the transition form three groups and arrows in each group have equal length, so they give rise to one and the same line in the splitting pattern.

Line with $\Delta M_J = 0$ falls in the position of the original field free line, whereas lines with $\Delta M_J = \pm 1$ lie to the right and left. This is normal Zeeman effect and is observed only for singlet lines ($S = 0$).

6.3 ANOMALOUS ZEEMAN EFFECT

The spectral lines arising from transition between components of multiplet levels produce a complex Zeeman pattern and is known as anomalous Zeeman effect. Its origin lies in the spin of the electron. It is observed in case of a weak magnetic field.

According to vector model of the atom, vectors L and S precess rapidly around the total angular momentum J. In presence of a weak magnetic field (along the z-axis), the magnetic moment of the atom μ_J associated with vector J causes the vector J to precess slowly around the field direction. Permissible values of J_2 are $M_J = h/2\pi$

where $\qquad M_J = J, J - 1, J - 2, .., 0, ..J(2J + 1$ values) ..(10)

Thus J level is split into $2J + 1$ levels giving rise to a fine structure.

Let us consider a single valence electron atom. Classically,

$$\frac{|\mu_L|}{|L|} = \frac{e}{2mc} \ (\mu_L \text{ is antiparallel to } L) \qquad ..(11)$$

From quantum mechanics,

$$\frac{|\mu_S|}{|S|} = 2 \times \frac{e}{2mc} \qquad\qquad ..(12)$$

(More exactly $\mu_S = -2.0023 \dfrac{e}{2mc} S$)

μ_S is antiparallel to S

Because of the inequality of the two ratios (11) and (12) the total magnetic moment

$$\mu = \mu_L + \mu_S$$

is not exactly antiparallel to J since J is a constant of the motion, vectors L, S, μ_L, μ_S and μ precess around J. The component of μ perpendicular to J averages out to zero. Only the component parallel to J contributes to the magnetic moment of the atom and is given by

$$\mu_J = |\mu_L| \cos(L, J) + |\mu_S| \cos(S, J) \qquad ..(13)$$

μ_J precesses around the z-axis.

or $$\mu_J = \frac{e}{2mc}|L|\cos(L, J) + \frac{2e}{2mc}|S|\cos(S, J)$$

$$= \frac{e}{2mc}\left\{\sqrt{L(L+1)}\cos(L, J) + 2\sqrt{S(S+1)}\cos(S, J)\right\}\frac{h}{2\pi} \quad ..(14)$$

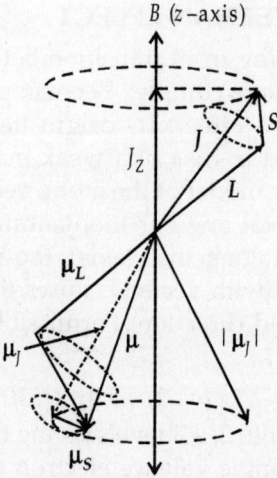

Fig. 6.3: Illustrating the origin of weak field, i.e. anomalous Zeeman effect

Applying the cosine law to the triangle formed by vectors S, L and J, we get

$$S^2 = L^2 + J^2 - 2LJ\cos(L, J)$$

or $$\cos(L, J) = \frac{J^2 + L^2 - S^2}{2JL}$$

or $$\cos(L, J) = \frac{J(J+1) + L(L+1) - S(S+1)}{2\sqrt{J(J+1)}\sqrt{L(L+1)}} \quad ..(15)$$

Similarly,

$$\cos(S, J) = \frac{J(J+1) + S(S+1) - L(L+1)}{2\sqrt{J(J+1)}\sqrt{S(S+1)}} \quad ..(16)$$

Hence $$\mu_J = \frac{e}{2mc}\left[\frac{J(J+1) + L(L+1) - S(S+1)}{2J(J+1)} + \right.$$

$$\left.\frac{J(J+1) + S(S+1) - L(L+1)}{2J(J+1)}\right]\sqrt{J(J+1)}\frac{h}{2\pi} \quad ..(17)$$

or
$$\mu_J = \frac{e}{2mc} \times g \times \sqrt{J(J+1)} \frac{h}{2\pi} = g \frac{e}{2mc} |J| \qquad ..(18)$$

where
$$g \equiv 1 + \frac{J(J+1) + S(S+1) - L(L+1)}{2J(J+1)} \qquad ..(19)$$

is termed as the *Lande's g-factor*. It gives the relative separation of the Zeeman levels for the different terms

$$\frac{\mu_J}{|J|} = g \frac{e}{2mc} \qquad ..(20)$$

By Larmor's theorem, the angular velocity of precession

$$\omega = \frac{\mu_J}{|J|} B = g \frac{e}{2mc} B \qquad ..(21)$$

The energy of precession

$$\Delta E = \omega J_z = g \frac{e}{2mc} B J_z$$

$$= g \frac{e}{2mc} B M_J \frac{h}{2\pi}$$

$$= g M_J \frac{eh}{4\pi mc} B \qquad ..(22)$$

In wave numbers, energy

$$-\Delta T = \frac{\Delta E}{hc} = g M_J \frac{eB}{4\pi mc^2} \text{ cm}^{-1} \qquad ..(23)$$

Thus, in a weak magnetic field, each J level is split into $(2J + 1)$ sublevels which are equispaced (spacing $= \Delta E$) and spacing depends on the value of g for that level.

$$\frac{eB}{4\pi mc^2} = L' \qquad ..(24)$$

is termed as the *Lorentz unit*.

Let us consider the Zeeman splitting of sodium D_1 and D_2 lines which arise from the transitions

$$^2P_{1/2} \rightarrow {}^2S_{1/2} \ (D_1 \leftrightarrow 5896 \text{ Å})$$

$$^2P_{3/2} \rightarrow {}^2S_{1/2} \ (D_2 \leftrightarrow 5890 \text{ Å})$$

The Table 6.1 depicts g, M_J and gM_J values for the various terms. The splittings are shown in Fig. 6.4.

TABLE 6.1

Terms	No. of Levels (2J + 1)	g	M_J (+J...–J)	Shift gM_J in Lorentz Units
$^2S_{1/2}$ ($L = 0, S = 1/2, J = 1/2$)	2	2	$\pm\dfrac{1}{2}$	± 1
$^2P_{1/2}$ ($L = 1, S = 1/2, J = 1/2$)	2	2/3	$\pm\dfrac{1}{2}$	$\pm\dfrac{1}{3}$
$^2P_{3/2}$ ($L = 1, S = 1/2, J = 3/2$)	4	4/3	$\pm\dfrac{3}{2}, \pm\dfrac{1}{2}$	$\pm 2, \pm\dfrac{2}{3}$

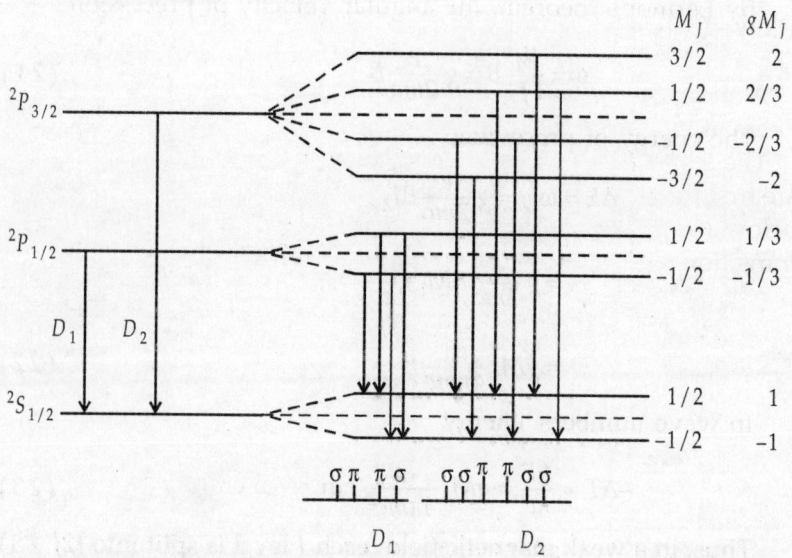

Fig. 6.4: Anomalous Zeeman effect for sodium resonance lines D_1 and D_2

6.4 SELECTION RULES

The selection rules are

$$\Delta M_J = 0, \pm 1 \qquad \text{(but } M_J = 0 \leftrightarrow M_J = 0 \text{ if } \Delta J = 0\text{)}$$

This gives four components for the D_1 line and six components for the D_2 lines. The components corresponding to $\Delta M_J = 0$ have their electric vector polarised parallel to the magnetic field (π-components) whereas those corresponding to $\Delta M_J = \pm 1$ have electric vector perpendicular to the field (σ-components). In terms of direction of view, polarization rules may be stated as follows:

$$\text{View} \perp \text{to field} \begin{cases} \Delta M_J = 0; \text{ plane polarised } \parallel \text{ to } B; \\ \quad \pi\text{-components} \\ \Delta M_J = \pm1; \text{ plane polarized } \perp \text{ to } B; \\ \quad \sigma\text{-components} \end{cases} \quad ...(25)$$

$$\text{View} \parallel \text{to field} \begin{cases} \Delta M_J = 0; \text{ forbidden}; \pi\text{-components} \\ \Delta M_J = \pm1; \text{ Circularly polarized}; \\ \quad \sigma\text{-components} \end{cases} \quad ...(26)$$

Thus, for the stronger of the two field-free lines, there are six allowed transitions and two forbidden transitions (i.e. for D_2). For the other line (viz. D_1), there are four allowed transitions.

A method frequently employed for a rapid calculation of Zeeman patterns is as follows:

The separation factors gM_J for both the initial and final states are first written down in two rows with equal values of M_J directly below (or above) each other. For example, for a $^2D_{5/2} \rightarrow {}^2P_{3/2}$ transition, we have

$$M_J = \frac{5}{2} \quad \frac{3}{2} \quad \frac{1}{2} \quad -\frac{1}{2} \quad -\frac{3}{2} \quad -\frac{5}{2}$$

gM_J initial state $\quad \dfrac{15}{5} \quad \dfrac{9}{5} \quad \dfrac{3}{5} \quad -\dfrac{3}{5} \quad -\dfrac{9}{5} \quad -\dfrac{15}{5}$

gM_J final state $\quad\quad \dfrac{6}{3} \quad \dfrac{2}{3} \quad -\dfrac{2}{3} \quad -\dfrac{6}{3}$

In the above array, the vertical arrows indicate the π-components, $\Delta M_J = 0$ and the diagonal arrows the σ-components, $\Delta M_J = \pm 1$. The differences expressed with a least common denominator are as follows:

Vertical differences **π-components**	**Diagonal differences** **σ-components**
$+\dfrac{3}{15}, \; +\dfrac{1}{15}, \; -\dfrac{1}{15}, \; -\dfrac{3}{15}$	$\pm\dfrac{15}{15}, \; \pm\dfrac{17}{15}, \; \pm\dfrac{19}{15}, \; \pm\dfrac{21}{15}$

In brief, we may write these as

$$\Delta\bar{\nu} = \frac{(\pm1), (\pm3), \pm15, \pm17, \pm19, \pm21}{15} L' \text{ cm}^{-1}$$

The four π-components are given in parentheses followed by the eight σ-components.

6.5 PASCHEN-BACK EFFECT

Normal Zeeman effect is observed when the external magnetic field is weak as compared to fields due to spin and orbital motion of valence electrons. When the external field B becomes stronger as compared with the internal field, the coupling between J and B exceeds that between L and S and precession of J about B becomes faster than that of L and S about J. Then, coupling between L and S is partially broken. With further increase in B, L and S start precessing independently about B with quantized components L_Z and S_Z along the field direction (B).

Their magnitudes are $M_L \hbar$ and $M_S \hbar$ respectively.

$$M_L = \left| -L, \ -L+1, \ ...0...L-1, L \right.$$
$$M_S = \left| -S, \ -S+1, \ ...0...S-1, S \right| \qquad ..(27)$$

The angular velocities of precession of L and S about the field B are given by

$$\omega_L = \left| \frac{e}{2mc} B \right.$$
$$\omega_S = \left| 2\frac{e}{2mc} B \right| \qquad ..(28)$$

The energies of the two precessions are given by

$$\Delta E_L = \omega_L L_Z = \left| \frac{e\hbar}{2mc} B M_L \right.$$
$$\Delta E_S = \omega_S S_Z = \left| 2\frac{e\hbar}{2mc} B M_S \right| \qquad ..(29)$$

The energy shift from the unperturbed level is
$$\Delta E = \Delta E_L + \Delta E_S$$

$$= \left(M_L + 2M_S \right) \frac{e\hbar}{2mc} B \qquad ..(30)$$

The shift in the wave number is

$$-\Delta T = \frac{\Delta E}{hc} = \left(M_L + 2M_S \right) \frac{eB}{4\pi mc^2} \ \text{cm}^{-1} \qquad ..(31)$$

or, in Lorentz units,

$$-\Delta T = (M_L + 2M_S)L' \ \text{cm}^{-1} \qquad ..(32)$$

This is the strong field interaction energy ignoring spin orbit interaction. Thus each of the field free levels is split into $(2L + 1)$ $(2S + 1)$ levels ($\because M_L$ can take $2L + 1$ values and M_S can take $2S + 1$ values).

As an example, we consider the D_1 and D_2 lines of sodium
$$^2P_{1/2,\,3/2} \rightarrow {}^2S_{1/2}$$
The strong field levels are as follows:

TABLE 6.2

Term Field Levels $(2L + 1)(2S + 1)$	Number of Strong	$\mathbf{M}_L$	M_S $M_L + 2M_S$	Shift
$^2P\left(L = 1, S = \dfrac{1}{2}\right)$	6	1	$\dfrac{1}{2}, -\dfrac{1}{2}$	2, 0
		0	$\dfrac{1}{2}, -\dfrac{1}{2}$	1, –1
		–1	$\dfrac{1}{2}, -\dfrac{1}{2}$	0, –2
$^2S\left(L = 0, S = \dfrac{1}{2}\right)$	2	0	$\dfrac{1}{2}, -\dfrac{1}{2}$	1, –1

The splittings are shown in Fig. 6.5.

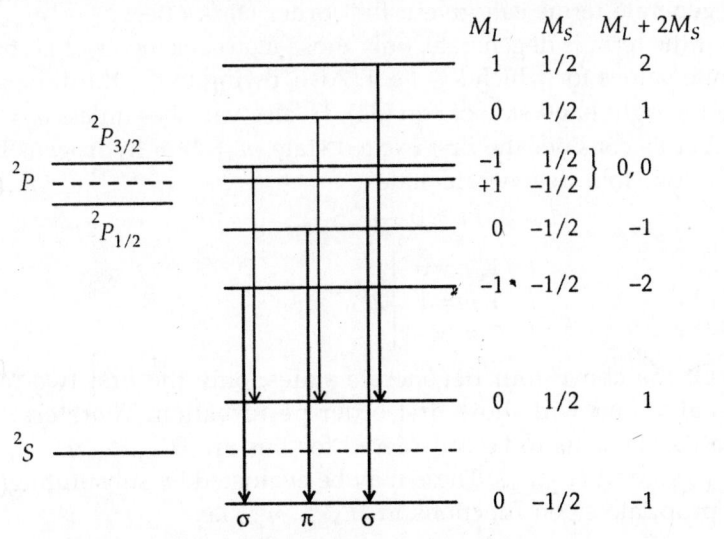

Fig. 6.5: Strong field pattern

The selection rules in this case are

$$\Delta M_L = 0 \text{ (Components polarized } || \text{ to the field)}$$
$$= \pm 1 \text{ (Components polarized } \perp \text{ to the field)}$$
$$\Delta M_S = 0 \qquad \qquad \qquad ..(33)$$

These selection rules lead to a pattern same as normal Zeeman triplet.

6.6 STARK EFFECT

Perturbation theory of quantum mechanics permits us to calculate the first order energy correction for an atom placed in a uniform electric field F directed along the z-axis. The perturbation, i.e. extra energy of the nucleus and the electron is $H' = eFz = eFr \cos\theta$

The matrix element is

$$H'_{nlm,n'l'm'} = eF \int_0^\infty R_{nl} R_{n'l'} \, r^2 dr \times \int_0^\pi P_l^m P_{l'}^{m'} \cos\theta \sin\theta d\theta \int_0^{2\pi} e^{i(m-m')} d\phi \quad ..(34)$$

As $\cos\theta$ is an odd function, the above integral vanishes unless the spherical harmonics $Y_{lm}(\theta, \phi)$ and $Y_{l'm'}(\theta, \phi)$ have opposite parity. The first order perturbation energy, therefore, vanishes unless the term is degenerate. For a non-degenerate state, e.g. ground state ($n = 1$, $l = 0$, $m = 0$) of hydrogen atom, which has even parity, the first order perturbation is zero, i.e. only degenerate terms can give a first order stark effect.

If the term is degenerate, only those matrix elements of H' have finite values for which $l' = l \pm 1$. Also, owing to the third integral on the right hand side of eqn (34), H' also vanishes unless $m = m'$.

Let us consider the first excited state ($n = 2$) of hydrogen. This is a four-fold degenerate state

$$l = 0, \ m = 0 \leftrightarrow 2S$$

$$\left. \begin{array}{l} l = 1, \ m = 0 \\ l = 1, \ m = 1 \\ l = 1, \ m = -1 \end{array} \right\} 2P$$

Of the above four degenerate states, only the first two have equal m and will show first order perturbation. Therefore, the matrix elements to be considered for this are $H'_{200, 210}$, $H'_{210, 200}$, $H'_{210, 210}$ and $H'_{200, 200}$. These may be evaluated by substituting the appropriate eigen functions in $H'_{nlm, n'l'm'}$, i.e.

$$<\Psi_{nlm} | H' | \Psi_{n'l'm'}>$$

For hydrogen $Z = 1$, and

$$\psi_{200} = \frac{1}{4\sqrt{2\pi}}\left(\frac{1}{a_0}\right)^{3/2}\left(2 - \frac{r}{a_0}\right)e^{-\frac{r}{2a_0}} \qquad ..(35)$$

$$\psi_{210} = \frac{1}{4\sqrt{2\pi}}\left(\frac{1}{a_0}\right)^{3/2}\frac{r}{a_0}e^{-\frac{r}{2a_0}}\cos\theta \qquad ..(36)$$

$$H'_{200,210} = H'_{210,200} = \frac{eF}{32\pi a_0^3}\int_0^\infty\left(2 - \frac{r}{a_0}\right)\frac{r}{a_0}e^{-\frac{r}{a_0}}r^3 dr$$

$$\times\int_0^\pi\cos^2\theta\sin\theta d\theta\times 2\pi$$

$$= 3eFa_0 \qquad ..(37)$$

and $\qquad H'_{210,210} = H'_{200,200} = 0 \qquad ..(38)$

The eigen values E' of the matrix H' are given by

$$\begin{vmatrix} H'_{210,200} - E' & H'_{200,210} \\ H'_{210,200} & H'_{210,210,} - E' \end{vmatrix} = 0$$

or $\qquad \begin{vmatrix} -E' & 3eFa_0 \\ 3eFa_0 & -E' \end{vmatrix} = 0 \qquad ..(39)$

The roots of this equation are

$$E' = \pm 3efa_0 \qquad ..(40)$$

This result agrees with the old quantum theory results for the states $n = 2$, $n_2 - n_1 = 1$.

Questions

6.1 Derive an expression for Lande's splitting factor and explain with its help the Zeeman effect of the sodium doublet components D_1 and D_2.

6.2 Illustrate with the help of diagrams the splitting of energy levels of sodium D_1 and D_2 lines and the allowed transitions giving rise to (a) anomalous Zeeman effect, (b) Paschen-Back effect.

6.3 Discuss with necessary theory the splitting of sodium lines when (a) a weak magnetic field, (b) a strong magnetic field, is applied.

6.4 Write a short note on Stark effect.

The Nuclear Spin
and its Consequences:
Hyperfine Structure

7.1 INTRODUCTION

It has been observed that many fine structure components of spectral lines, when viewed by instruments of highest resolution, are further split up into components having separations ~1 cm⁻¹ which is very much smaller than those of the ordinary multiplet structure.

This is known as the *hyper fine structure* and is caused by properties of atomic nuclei.

There are two types of nuclear effects that give rise to hyperfine structure. The first is due to presence of isotopes in the specimen under investigation.

The second type of effect is because the charged nucleus can possess spin angular momentum I and the associated magnetic moment. As nuclear magnetic moments are smaller than electronic moments by ~10^{-3}, the hyperfine splitting (because of interaction between nuclear magnetic moment and internal field of surrounding electrons is smaller than the spin orbit splitting by the same factor.

7.2 ISOTOPE EFFECT IN HYPERFINE STRUCTURE

Since the nuclear mass enters in the expansion for Rydberg constant, (through the reduced mass of the atom), different isotopes have slightly different values of Rydberg constant.

It has been observed in hydrogen spectrum, each of the first four members of Balmer series (H_α, H_β, H_γ and H_δ) has a very weak companion (on the short wavelength side) at distances of 1.79, 1.33,

1.19 and 1.12 Å respectively. These are attributed to presence of an isotope of mass 2 viz. deuterium.

7.3 NUCLEAR SPIN AND HYPERFINE SPLITTING

In most cases, isotope effect fails to explain the hyperfine structure. The number of hyperfine components is more than the number of isotopes. Certain elements, e.g. Bismuth) have only one isotope but still have hyperfine structure, e.g. six components for Bi at wavelength 4722 Å.

It was pointed out by Pauli (in 1924) that hyperfine structure could be explained assuming that the atomic nucleus possesses an intrinsic angular momentum I having an associated magnetic moment μ_I.

$$|I| = \sqrt{I(I+1)}\,\hbar$$

where I is the nuclear spin quantum number. I has integral values for nuclei of even mass number and half integral values for odd mass nuclei.

Similar to L, S, J, I has z-component given by

$$I_Z = M_I \hbar \qquad \qquad ..(1)$$

where nuclear magnetic quantum number M_I has $2I + 1$ values, given by

$$M_I = I, I - 1, .., -I \qquad \qquad ..(2)$$

The motions of the nuclear protons produce a magnetic moment μ_I, given by

$$|\mu_I| = g_I \left(\frac{e}{2m_p c} \right) I \qquad \qquad ..(3)$$

where e and m_p are the charge and mass of the proton respectively. The term g_I is called the "nuclear g-factor".

$$\therefore \qquad \mu_I = g_I \left(\frac{e}{2m_p c} \right) \sqrt{I(I+1)}\,\hbar \qquad \qquad ..(4i)$$

$$\text{or} \qquad \mu_I = g_I \sqrt{I(I+1)} \left(\frac{eh}{4\pi m_p c} \right) \qquad \qquad ..(4ii)$$

The quantity $\left(\dfrac{eh}{4\pi m_p c} \right)$ forms a natural unit for the measurement of nuclear magnetic moments and is called the *nuclear magneton*, denoted by μ_N. It is 1/1836 times the Bohr magneton.

$$\therefore \qquad \mu_I = g_I \sqrt{I(I+1)}\, \mu_N \qquad\qquad ..(5)$$

and z-component of μ_I is

$$\mu_{IZ} = g_I M_I \mu_N \qquad\qquad ..(6)$$

Since the maximum value of M_I is I, the maximum possible value of M_I is $g_I I \mu_N$ and is commonly termed as the "nuclear magnetic moment". It is roughly 1000 times smaller than electron magnetic moment.

7.4 THE VECTOR MODEL WITH NUCLEAR SPIN INTO ACCOUNT

The total angular momentum of the whole atom is the sum

$$F = L + S + I \qquad\qquad ..(7)$$

where L is the total electron angular momentum, S is total electron spin angular momentum and I is the nuclear spin angular momentum.

Due to spin orbit interaction, L and S precess rapidly around their resultant J. Further, I and J couple and they precess around their resultant F (Fig. 7.1). This precession s, however, about 1000 times slower than that of L and S about J (because nuclear magnetic moment is much smaller).

The quantised values of the total angular momentum F are $\sqrt{F(F+1)}\,\hbar$ where the hyperfine splitting quantum number F can take the possible values

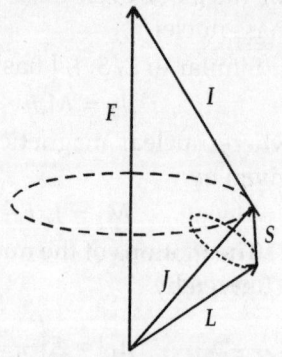

Fig. 7.1: Vector model with nuclear spin I into consideration

$$F = J + I, J + I - 1, J + I - 2, \ldots |J - I| \qquad ..(8)$$

These are $2J + 1$ values if $I \geq J$ and $2I + 1$ values if $I \leq J$. Thus each fine structure J level splits into $2J + 1$ levels (for $I \geq J$) or $2I + 1$ levels (for $I \leq J$). The I-J interaction can be shown to be given by

$$E'_{IJ} = \frac{1}{2} A\left[F(F+1) - I(I+1) - J(J+1)\right] \qquad ..(9)$$

where A' is a constant. The various *hfs* levels of a given term of an atom have the same I and J but different F values. Hence the separation between two *hfs* levels is obtained as

$$\Delta E' = E'(F + 1) - E'(F)$$

$$= \frac{1}{2} A' \left[(F + 1)(F + 2) - F(F + 1) \right]$$

$$= A'(F + 1) \qquad ..(10)$$

i.e. the energy interval between consecutive *hfs* levels F and $F + 1$ is proportional to the larger of the F values (Lande's interval rule). The order of *hfs* levels in some hyper multiplets is normal (smallest F level deepest) while in others it is inverted (largest F level deepest).

The selection rule for F is similar to that for J

$$\Delta F = 0, \pm 1, \text{ but } F = 0 \nleftrightarrow F = 0 \qquad ..(11)$$

As an example, let us consider the hyperfine structure of the fine structure component D_2 of sodium $\left(I = \frac{3}{2} \right)$. The transition is $^2P_{3/2} - ^2S_{1/2}$. For the $^2P_{3/2}$ term, we have

$$J = 3/2, I = 3/2 \therefore F = 3, 2, 1, 0 \qquad ..(12)$$

This term is thus split into four *hfs* components. For the $^2S_{1/2}$ term,

$$J = 1/2, I = 3/2 \therefore F = 2, 1 \qquad ..(13)$$

This term is split into two *hfs* components. The selection rule ($\Delta F = 0, \pm 1$) allows six transitions. However, since the factor A' for the $^2P_{3/2}$ term is much smaller than that for the $^2S_{1/2}$ term, the splittings of levels in the former are comparatively much smaller and remain unresolved. Hence, practically, only two *hfs* components (corresponding to splitting of $^2S_{1/2}$) are seen (for $F = 2$ and 1).

7.5 NORMAL AND INVERTED HYPERFINE STRUCTURE

The energy eigenvalues for the field produced H_z by orbital motion of electrons (characterized by $\vec{L}$) and spin of the electrons (characterized by $\vec{S}$), at the nucleus (of spin I) are given by

$$E_F = \frac{A}{2} [F(F + 1) - I(I + 1) - J(J + 1)]$$

where $\vec{J}$ is the resultant of $\vec{L}$ and $\vec{S}$ and

$$\vec{F} = \vec{L} + \vec{S} + \vec{I}$$

is the total angular momentum of the atom and F is the hyperfine splitting quantum number. (A is a constant). The various *hfs* levels of a given term of an atom have the same I and J but different F values.

The hyperfine splitting of a particular component of a multiplet level is said to be normal if the level with highest value of F has the largest energy and level with the lowest value of F has the least energy. A *hfs* with the reverse ordering is said to be inverted.

Let us consider two fine structure levels with $J_1 = L + S$ and $J_2 = L - S$ assuming $L > S$. If the nuclear spin is I, the hyperfine components arising from these levels can be summarized as follows, assuming J_1 and $J_2 > I$:

$$F_1 = J_1 + I = L + S + I$$
$$F_1' = J_1 - I = L + S - I$$
$$F_2 = J_2 + I = L - S + I$$
$$F_2' = J_2 - I = L - S - I$$

Here the unprimed F represents the highest value and F' represents the lowest value for the hyperfine structure level in question.

Classically, the orbital motion of the electron produces a magnetic field which is opposite in direction to its mechanical moment $L(L + 1)\hbar$. The most stable state for the nucleus in this field, (therefore the state of least energy) will be the one for which the nuclear magnetic moment is most nearly parallel to the field. The nuclear magnetic moment has the same direction as the mechanical moment I (since it is positively charged). Hence the state of least energy would be the one for which I and L are oppositely directed. This means that the levels F_1' and F_2' correspond to the least energy and F_1 and F_2, to the largest energy in their respective groups if the nuclear-electron spin interaction is neglected.

Now, the nucleus and the electron may be regarded as tiny magnets acting on each other at some distance from one another as far as their spin-spin interactions are concerned. The most stable configuration of the two dipoles in an external field is that for which their magnetic moments are oppositely directed. Since the nucleus and the electron are oppositely charged, the corresponding mechanical moments ($\bar{I}$ and $\bar{S}$) must be in the same direction.

If we apply the above condition to the hyperfine levels under consideration, we find that it tends to decrease the energy of F_1 and increase that of F_1' i.e. there is tendency to invert the hyperfine components of the fine structure level J_1. On the other hand, the energy of F_2 tends to increase on account of this spin-spin interaction and that of F_2' tends to decrease. This treatment is strictly true only for an atom with single valence electron.

7.6 BACK-GOUDSMIT EFFECT

If the magnetic field strengths used commonly in studying the Zeeman effect are employed in a study of the spectrum of an element having a nuclear magnetic moment also in addition to mechanical moment, a phenomenon analogous to Paschen-Back effect in fine structure is observed. This effect was first observed by Back and Goudsmit, therefore, it has been referred to as the Back-Goudsmit effect.

In this effect, a magnetic field which is a weak field for fine structure is, in reality, a strong field for hyperfine structure. In such a weak field, the coupling between the nuclear moment I^* and the electrons' resultant J^* is broken down and each is quantized separately with the field direction.

Questions

7.1 Discuss hyperfine structure of spectral lines. What light does this study throw on the spin and magnetic moment of atomic nuclei?

7.2 How does the nuclear spin affect the hyperfine structure of the emission spectrum of an atom?

7.3 What is Back-Goudsmit effect?

Basic Aspects of Molecular Spectroscopy

8.1 TYPES OF MOLECULAR SPECTROSCOPY

When we compare the spectra obtained by molecular species with the atomic spectra, e.g. that of hydrogen atom we observe that in both the cases, regular series of lines or bands are present. However, qualitatively, a fundamental difference exists. It is found that while for the atoms, the line separation in a series decreases rapidly, e.g. Rydberg series for (say infrared molecular spectra) it is almost constant, so the molecular spectra cannot be explained with the help of stationary states of an electron revolving about a core as it could be in the case of atomic spectra.

It has been observed that the spectrum emitted by any kind of molecule belongs to any of the following three categories, depending on the respective spectral ranges to which they belong:

(*i*) Far infrared region

(*ii*) Infrared spectrum

(*iii*) Visible and ultraviolet spectrum

These could be explained by transitions in molecular quantum states due to the following three physical situations respectively:

(*i*) **Rotation of Molecules:** A molecule may be considered as a rigid structure which may have electric charges so disposed that the molecule possesses an electric dipole moment. If such a molecule rotates, then according to classical electromagnetic theory it would emit radiation—conversely, when radiation falls on such a molecule, it would set it into rotation, due to absorption of radiation.

Spectra lines corresponding to this physical situation give rise to emission and absorption spectra in the far infrared and are termed rotation spectra. Typical photon energies involved in these are 10^{-5} to 10^{-3} eV.

(ii) **Combined Vibration and Rotation:** If the molecules were not rigid but possess atoms capable of vibrating about their equilibrium positions, then radiation would be emitted (according to classical theory) due to vibration of atoms.

If, however, the amplitude of vibrations becomes large, the vibrations still remain periodic but are no longer simple harmonic. Then the emitted radiations could be resolved by Fourier analysis into waves corresponding to fundamental frequency and in addition harmonic overtones of separate frequencies, of rotations (as described above in (i)), are present, would be split up further due to molecular rotations.

Spectra corresponding to this situation are observed in the infrared region and are called vibration rotation spectra. Photon energies are in the range of 0.2 to 2.0 eV.

(iii) **Electronic Spectra:** Finally, (classically) an electron in a molecule may vibrate by itself and therefore, radiate. The emitted radiation would be affected by vibration of atoms and rotation of the molecule as a whole. Principally, a valence electron, i.e. lying outside core, known as the optical electron would radiate with its frequency affected basically by instantaneous position and motion of nuclei. The rotation of the molecule would tend to split up the emitted lines as in vibration rotation spectra. This physical situation corresponds to molecular spectra in the visible and ultraviolet.

The study of interaction of electromagnetic radiation with matter (generally in molecular form) is termed as *molecular spectroscopy*. In atomic spectra, the emitting substance is in the atomic state. In molecular spectra, the emitting substance is in molecular state containing two or more atoms.

The molecular spectroscopy serves as an important tool to investigate the structure of matter, e.g. about size and shape of molecules and bond lengths, etc. therefore, it becomes important to discuss first the nature of electromagnetic radiation in relation to molecules. *Similar to atomic energy levels, there are molecular energy levels.*

When a molecule jumps from a higher energy state E_2 to a lower level E_1, the excess energy is emitted as a photon. On the other hand, when a molecule jumps from a lower energy level to a higher one, energy is absorbed in the form of a photon. In either of these two cases, the energy difference between states E_2 and E_1 is given by

$$E_2 - E_1 = h\nu = \frac{hc}{\lambda} = hc\bar{\nu}$$

where $\nu\lambda \equiv c$ is the velocity of light, ν is the frequency, $\bar{\nu}$, the wave number and λ, the wavelength of the emitted or absorbed radiation. The energy difference $E_2 - E_1 \equiv \Delta E$ is expressed in joules per mole, frequency in per second (Hz), wavelength in cm and wave number in cm^{-1}. Since ΔE is directly proportional to the wave number, therefore, it is a common practice in spectroscopy to express ΔE in terms of wave number. ΔE in ergs can be obtained by multiplying wave number by $h \times c$. For example, an energy difference of 20 cm^{-1} corresponds to an energy state-separation equivalent to a radiation of wave number value 20 cm^{-1}. A spectrum can be obtained in three ways: (*i*) by emission spectroscopy (*ii*) by absorption spectroscopy and (*iii*) Raman spectroscopy.

Emission Spectroscopy: Atoms or molecules may be excited to higher energy state by subjecting them to high temperatures or electric discharge. On returning to the lower energy state, which is usually the ground state, the atom or the molecule emits a photon and the corresponding frequency emitted is recorded as the emission spectrum.

Absorption Spectroscopy: In this, the absorbing sample is placed between the source of light and the spectrometer. The spectro-meter records the percentage of light absorbed against the range of incident frequencies absorbed by the sample and an absorption spectrum is obtained.

Raman Spectroscopy: It is a technique which explores energy levels of molecules by the scattering of light due to Raman effect. Photons of frequency ν_{in} are scattered by collision with the molecules of the sample, when new frequencies are added, because photons can lose or acquire energy during collision. If the molecules are excited by light during the collision, molecules withdraw some energy from the photons and so scattered light emerges with a lower frequency $(\nu_{in} - \nu)$. If the molecules are already excited before the photons collide with them, they may give up energy to photons, so emerging photons will have higher frequency $(\nu_{in} + \nu)$.

With increasing frequency, the various spectroscopic regions encountered are:

1. **Radiofrequency Region:** $(3 \times 10^6$–3×10^{10} Hz) Nuclear magnetic resonance (NMR) and electron spin resonance (ESR) spectroscopy. The energy change involved is that arising from the reversal of spin of a nucleus or electron and is of the order of 0.001–10 joules/mole.

2. **Microwave Region:** $(3 \times 10^{10}$–3×10^{12} Hz) Rotational spectroscopy, separation between the rotational levels of molecules are of the order of hundreds of joules/mole.

3. **Infrared Region:** $(3 \times 10^{12}$–3×10^{14} Hz) Vibrational spectroscopy, one of the most valuable spectroscopic regions. Energy separations are of the order of 10^4 joules/mole.

4. **Visible and Ultraviolet Regions:** $(3 \times 10^{14}$–3×10^{16} Hz) Electronic spectroscopy, the separation between the energies of valence electrons are some hundreds of kilojoules/mole.

5. **X-ray Region:** $(3 \times 10^{16}$–3×10^{18} Hz) Energy changes involving the "inner" electrons of an atom or a molecule may be of the order of 10,000 kilo joules.

6. **γ-ray Region:** $(3 \times 10^{18}$–3×10^{20} Hz) Energy changes involve the rearrangement of nuclear particles, having energies of 10^9–10^{11} joules per gram atom.

The Table 8.1 depicts common spectroscopic techniques based on various wavelength ranges of electromagnetic radiation.

TABLE 8.1: Common spectroscopic techniques

Spectroscopy Type	Usual Wavelength Range	Type of Transition
1. Gamma-ray emission	0.005–1.4 Å	Nuclear
2. X ray absorption emission, fluorescence and diffraction	0.1–100 Å	Inner electron
3. Ultraviolet, visible absorption, emission and fluorescence	180–780 nm	Bonding electrons
4. Infrared absorption and Raman scattering	0.8–300 μm	Rotation/vibration of molecules
5. Microwave absorption	0.8–3.8 mm	Rotation of molecules
6. Electron spin resonance	3 cm	Spin of electrons in an external magnetic field
7. Nuclear magnetic resonance	0.6–10 m	Spin of nuclei in an external magnetic field

Though Table 8.1 lists the types of spectroscopies, there are yet other forms or types than those listed in this Table and it would be impossible to be all inclusive. Probably, the biggest omission from this list is Raman spectroscopy, which is a differential type of spectroscopy (i.e. the spectrum is measured by differences in photon energy rather than the energy of the photons themselves). This Table also omits hybrid forms of spectroscopy like photo acoustic spectroscopy.

8.2 INSTRUMENTAL METHODS IN SPECTROSCOPY

8.2.1 Terms Employed in Absorption Spectroscopy

Table 8.2 lists some common terms employed in absorption spectroscopy.

TABLE 8.2: Important terms employed in absorption spectroscopy

Term and Symbol	Definition
1. Radiant power P, P_0	Energy of radiation (in joules) striking a 1 m^2 area of detector per second
2. Absorbance, A	$\log \dfrac{P_0}{P}$
3. Transmittance, T	$\dfrac{P}{P_0}$
4. Path length of radiation (b) in cm	—
5. Absorptivity*, a	$\dfrac{A}{bc} L g^{-1} cm^{-1}$ (c in gm/litre)
6. Molar absorptivity(*), $\in$	$\dfrac{A}{bc} L mol^{-1} cm^{-1}$ (c in mol/litre)

*(c is concentration of absorbing solution.)

Figure 8.1 depicts a beam of parallel radiation passing through a layer of solution having a thickness b cm and a concentration c of an absorbing species.

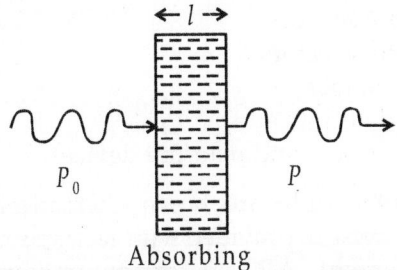

Absorbing
solution having concentration c.

Fig. 8.1: Attenuation of a radiation beam by an absorbing solution

Due to interaction with the absorbing particles, the power of the beam is attenuated from P_0 to P. The transmittance T of the solution is given by

$$T = \frac{P}{P_0} \quad \text{or} \quad \% T = \frac{P}{P_0} \times 100$$

The absorbance A of a solution is defined by

$$A = -\log_{10} T = \log \frac{P_0}{P}$$

i.e. the absorbance of a solution increases as the attenuation of the beam increases.

Absorbance is directly proportional to the path length b through the solution and the concentration c of the absorbing species in the solution. Hence

$$A = abc$$

where a is a proportionality constant called the *absorptivity*. (The magnitude of a will depend upon the units of b and c, as indicated in Table 8.2).

For measurement of transmittance or absorbance, the power of the beam transmitted by the analyte solution is compared with the power of the beam transmitted by an identical cell containing only solvent. Then absorbance is obtained as

$$A = \log \frac{P_{\text{solvent}}}{P_{\text{solution}}} \cong \log \left(\frac{P_0}{P} \right)$$

8.2.2 Instrument-Components

Instruments for measuring the absorption of ultraviolet, visible and near infrared radiation consist of

1. one or more sources
2. wavelength selectors
3. sample containers
4. radiation detectors and
5. signal processors and read-out devices.

The instrument modules are shown schematically in Fig. 8.2. The radiation source must be provided with each spectral region having its own requirement. All spectrophotometers include some arrangement to discriminate between different radiation frequencies either through use of filters, prisms or gratings. The sample absorbs a portion of the incident radiation, the remainder is transmitted to a detector where it is changed into an electrical signal and displayed, usually after amplification, on a meter, chart recorder, or any other type of read-out device.

A *photometer* is defined as an instrument that furnishes the ratio or some function of the ratio of radiant power of two electromagnetic beams.

An *optical spectrometer* is an instrument, with an entrance slit, a dispersing device and one or more exit slits, with which measurements are made at selected wavelengths within the spectral range. The quantity detected is a function of radiant power.

A *spectrophotometer* is a spectrometer plus photometer so that it furnishes the ratio (or a function of the ratio) of the radiant power of the two beams as a function of spectral wavelength. These two beams may be separated in time or space or both.

For the purpose of molecular absorption measurements, a continuous source is required whose power does not change sharply over a considerable range of wavelengths. Some common sources are as follows:

Dueterium and Hydrogen Lamps: A truly continuous spectrum in the ultraviolet region is produced by electrical excitation of deuterium or hydrogen at low pressure.

Both deuterium and hydrogen lamps produce a continuous spectrum in the region of 160 to 370 nm. At longer wavelengths (> 400 nm), the lamps produce emission lines, superimposed on the continuous spectrum. These lines, though represent a nuisance, can be used for wavelength calibration of absorption instruments.

Tungsten Filament Lamps: The most common source of visible and near infrared radiation is the tungsten filament lamp. The energy distribution of the source approximates that of a blackbody and is thus temperature dependent. In most absorption instruments, the operating filament temperature is 2870 K; the bulk of the energy is emitted, therefore, in the infrared region. A tungsten filament lamp is useful for the wavelength region between 350 and 2500 nm.

Xenon Arc Lamps: This lamp produces radiation by the passage of current through an atmosphere of xenon. The spectrum is continuous over the range between 250 to 600 nm, with the peak intensity occurring at about 500 nm.

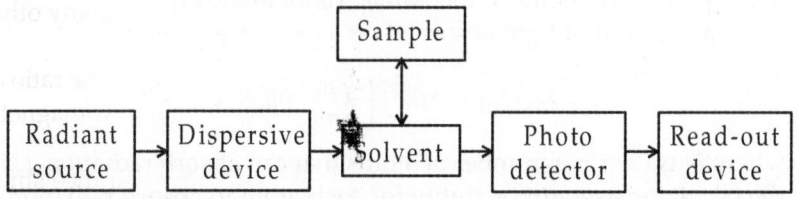

Fig. 8.2: Instrument modules present in absorption spectrophotometry in visible/ultraviolet regions of spectrum

8.3 ATOMIC ABSORPTION SPECTROMETRY

It was first introduced by Alan Walsh in the mid 1950's. Since its introduction, it has proved to be the most powerful technique for the quantitative determination of trace metals in liquids. The method provides a total metal content of the sample and is almost independent of the molecular form of the metal in the liquid. In this technique, it is not necessary to separate the test element from the other elements present in the sample, so it saves a great deal of time. The technique is not only restricted to aqueous solutions but is also applicable to non-aqueous solutions. Another chief advantage is that it does not demand sample, preparation, so it is an ideal tool for non-chemists also, e.g. engineers, biologists or clinicians as they are interested only in the significance of the results.

Principle: It is based on the absorption of energy by ground states of atoms in their gaseous state.

When a solution containing metallic species is introduced into a flame, the vapour of metallic species will be obtained. Some of the metal atoms may be raised to an energy level sufficiently high to emit the characteristics radiation of the metal, but a large percentage of metal atoms will remain in the ground state. These ground state atoms of a particular element can absorb light radiation of their own specific wavelength which they emit (when excited) in their emission spectrum. Thus, when a light of this wavelength is allowed to pass through a flame having atoms of metallic species, a part of the light will be absorbed and the absorption will be proportional to the density of atoms in the flame. In this technique, i.e. atomic absorption spectroscopy, we determine the amount of light absorbed. Once this is determined, we can know the concentration of the particular metallic element since the absorption is proportional to the density of atoms in the flame. Mathematically,

Total amount of light absorbed (at a given frequency)

$$= \text{constant} \times N \times f \left(= \frac{\pi e^2}{mc} Nf \right)$$

where N is the total number of atoms that can absorb radiation, f is the oscillator strength or ability for each atom to absorb radiation, (e, the electronic charge, m, the mass of the electron and c is the speed of light)

It is to be noted that

(*i*) there is no term involving the wavelength (or frequency of absorption).

(*ii*) there is no temperature dependence.

Thus, absorption by atom is independent of the wavelength of absorption and the temperature of the atoms. The signal in atomic absorption spectroscopy is obtained from the difference between the intensity of the source in the absence of metallic elements (present in the liquid) and the (decreased) intensity obtained when metallic elements are present in the liquid. The absorption depends on the number of unexcited atoms and the absorption intensity does not depend upon the temperature of the flame directly. The relation between absorbance and concentration is fairly linear, i.e. Beer's law is obeyed over a wide concentration range.

Instrumentation: A schematic diagram of the atomic absorption spectrophotometer is shown in Fig. 8.3.

Light of a certain wavelength (produced by a special kind of lamp), which can emit spectral lines corresponding to the energy required for an electronic transition (from ground state to an excited state) is allowed to pass through the flame. At the same time, the sample solution is aspirated into the flame. Before it enters the flame, the solution gets dispersed into a mist of very small droplets which evaporates in the flame to give first the dry salt, and then its vapour. A part of this vapour will be dissociated into atoms of the element whose concentration is to be measured. Thus the flame possesses free, unexcited atoms which are capable of absorbing radiation from an external source when the radiation corresponding exactly to the emission spectrum of the elements of the free atoms is allowed to fall in the flame. Then the remaining unabsorbed radiation is allowed to pass through a monochromator which isolates the exciting spectral lines of the light source. From the monochromator the unabsorbed radiation is then led into the detector which is then registered by a photodetector, whose output is amplified and measured on a recorder. Absorbance is then measured by the difference in transmitted signal in the presence and absence of test element.

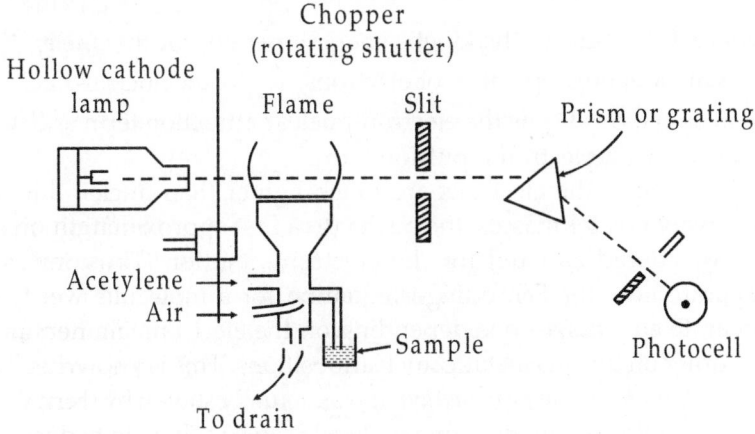

Fig. 8.3: Arrangement for atomic absorption spectroscopy

A rotating wheel (known as Chopper) is used between the hollow cathode lamp and the flame. It breaks the steady light from the lamp into an intermittent (i.e. pulsating) light. This gives a pulsating current in the photocell. There is also a steady current caused by light which is emitted by the flame, but only the pulsating current is amplified and recorded. Thus, the absorption of light is measured without interference from the light emitted by flame itself.

8.4 BORN-OPPENHEIMER APPROXIMATION

As compared to atoms, the energy-level scheme of molecules is relatively complicated. We cannot classify molecular energy states according to electronic angular momentum since it is not conserved in molecules. However, in diatomic molecules, in addition to electronic motion, the two nuclei vibrate along the internuclear axis and the whole system may rotate about the centre of mass. The energy of these vibrational and rotational motions is also quantized, hence there are many more energy levels in a molecule. Consequently, the Hamiltonian for a molecule is rather complex due to electronic and nuclear motions and their interactions. Therefore, the exact solution of Schrodinger's equation in case of molecules is not possible. However, we may write the Hamiltonian operation for a molecule.

$$\hat{H} = \hat{T}_N + \hat{T}_e + \hat{V}_{NN} + \hat{V}_{Ne} + \hat{V}_{ee}$$

where $\hat{T}_N$ stands for the kinetic energy operator for the nuclei, $\hat{T}_e$, the kinetic energy operator for electrons, $\hat{V}_{NN}$ is the nuclear-nuclear repulsion term, $\hat{V}_{Ne}$ is the electron-nuclear attraction term and $\hat{V}_{ee}$ is the electron-electron repulsion term.

Now, since the electrons are much lighter than nuclei, due to relatively greater masses, the nuclei (to a first approximation) may be considered as fixed for the electronic motion. Thus we can separate now the Schrodinger equation for a molecule into two separate equations—one depending on the electronic motion and the other on the quasi-static nuclear positions. This is known as the *Born-Oppenheimer approximation*. It was actually shown by them that the Schrodinger equation for a molecule can be solved in two steps:

(i) first solving the wave equation for the electrons alone in which positions of nuclei occur only as parameters and then (*ii*) solving a wave equation for the nuclei alone in which characteristics eigen value of the electronic equation occurs as part of the potential energy function of the nuclear motion. We now proceed to formulate this approximation.

For a molecule consisting of α nuclei and i electrons, the Hamiltonian is

$$H = -\sum_{\alpha} \frac{h^2}{8\pi^2 M_{\alpha}} \nabla_{\alpha}^2 - \sum_{i} \frac{h^2 \nabla_i^2}{8\pi^2 m_e} + V_{NN} + V_{Ne} + V_{ee}$$

$$..(1)$$

where M_{α} is mass of the nucleus α and m_e is the electronic mass. The wave equation is

$$H\psi = E\psi \qquad \qquad ...(2)$$

Under the assumption of fixed nuclei, the first term in eqn (1) is zero and V_{NN} is a constant, then, for electrons,

$$H_e = -\sum_{i} \frac{h^2}{8\pi^2 m_e} \nabla_i^2 + V_{Ne} + V_{ee} \qquad \qquad ..(3)$$

and $\qquad H_e \psi_e = E_e' \psi_e$ $\qquad \qquad ..(4)$

where ψ_e and E_e' are respectively the eigen functions and the corresponding eigen values for electrons. We write, for nuclei,

$$H_N = -\sum_{\alpha} \frac{h^2}{8\pi^2 M_{\alpha}} \nabla_{\alpha}^2 + V_{NN} \qquad \qquad ..(5)$$

According to Born and Oppenheimer, the total wave function is given by

$$\psi = \psi_e \psi_N \qquad \qquad ..(6i)$$

where ψ_e is a function of the coordinates of both the electrons and the nuclei and ψ_N is a function of coordinates of nuclei alone. Thus

$$H \psi_e \psi_N = E \psi_e \psi_N \qquad \qquad ..(6ii)$$

or $\quad -\sum_{\alpha} \frac{h^2}{8\pi^2 M_{\alpha}} \nabla_{\alpha}^2 \psi_e \psi_N - \sum_{i} \frac{h^2}{8\pi^2 m_e} \nabla_i^2 \psi_e \psi_N + (V_{NN} + V_{Ne} + V_{ee})\psi_e \psi_N$

$$= E \psi_e \psi_N \quad ..(7)$$

Now, $\quad \nabla_{\alpha}^2 \psi_e \psi_N = \psi_N \nabla_{\alpha}^2 \psi_e + \psi_e \nabla_{\alpha}^2 \psi_N + 2\nabla_{\alpha} \psi_e \nabla_{\alpha} \psi_N \qquad ..(8i)$

and $\qquad \nabla_i^2 \psi_e \psi_N = \psi_N \nabla_i^2 \psi_e \qquad \qquad ..(8ii)$

(because ∇_i is independent of nuclear coordinates).

Further, ψ_e is a slowly varying function

i.e. $\nabla_\alpha \psi_e \ll \nabla_\alpha \psi_N$

and $\nabla_\alpha^2 \psi_e \ll \nabla_\alpha^2 \psi_N$

Hence (8i) becomes

$$\nabla_\alpha^2 \psi_e \psi_N = \psi_e \nabla_\alpha^2 \psi_N \qquad\qquad ..(8iii)$$

Substituting eqns (8ii) and (8iii) in eqn (7),

$$-\psi_e \sum_\alpha \frac{h^2}{8\pi^2 M_\alpha} \nabla_\alpha^2 \psi_N - \psi_N \sum_i \frac{h^2}{8\pi^2 m_e} \nabla_i^2 \psi_e + (V_{NN} + V_{Ne} + V_{ee})\psi_e \psi_N$$

$$= E \psi_e \psi_N$$

$$-\frac{\psi_e}{\psi_N} \sum_\alpha \frac{h^2}{8\pi^2 M_\alpha} \nabla_\alpha^2 \psi_N + \left\{ -\sum_i \frac{h^2}{8\pi^2 m_e} \nabla_i^2 \psi_e + (V_{Ne} + V_{ee})\psi_e \right\}$$

$$+ V_{NN} \psi_e = E \psi_e$$

$$-\frac{\psi_e}{\psi_N} \sum_\alpha \frac{h^2}{8\pi^2 M_\alpha} \nabla_\alpha^2 \psi_N + E_e' \psi_e + V_{NN}\psi_e = E\psi_e$$

(using eqns (3) and (4))

$$-\sum_\alpha \frac{h^2}{8\pi^2 M_\alpha} \nabla_\alpha^2 \psi_N + E_e' \psi_N + V_{NN}\psi_N = E\psi_N$$

$$\left[-\sum_\alpha \frac{h^2}{8\pi^2 M_\alpha} \nabla_\alpha^2 + E_e' + V_{NN} \right] \psi_N = E\psi_N$$

or $(H_N + E_e') \psi_N = E\psi_N$ (using eqn (5))   ..(9)

This is the wave equation for the nuclear motion. The effective Hamiltonian is the term within the bracket on the left hand side. It contains the eigen value E_e' as a part of the potential energy function of nuclear motion. (The other part is V_{NN}). Thus with Born-Oppenheimer approximation we have been able to solve approximately the molecular wave equation. The eigen values E are the characteristic energies for the whole molecule in a given electronic state.

Apart from translational motion, molecules possess following three types of energies:

(*i*) The electronic energy E_e

(*ii*) The vibrational energy E_v due to periodic displacement of its atoms from their equilibrium position and (*iii*) The rotational energy E_r due to bodily rotation of its atoms about some axis within the molecule.

Thus, the structure of electronic spectra involves the changes of at least three quantum numbers simultaneously namely electronic, vibrational and rotational quantum numbers. Under Born-Oppenheimer approximation, the rotational (E_r), vibrational (E_v) and electronic energy levels are independent of one another. Hence we may write

$$E = E_e + E_v + E_r \qquad\qquad ..(10)$$

A change in total energy as a result of electronic transition in a molecule is

$$\Delta E = \Delta E_e + \Delta E_v + \Delta E_r \text{ Joules}$$

or $\qquad\qquad \Delta \epsilon = \Delta \epsilon_e + \Delta \epsilon_v + \Delta \epsilon_r \text{ cm}^{-1} \qquad\qquad ..(11)$

The approximate orders of magnitude of these changes are

$$\Delta \epsilon_e = \Delta \epsilon_v \times 10^3 \approx \Delta \epsilon_r \times 10^6 \text{ cm}^{-1} \qquad\qquad ..(12)$$

Therefore vibrational changes give a *coarse structure* and *rotational changes* give a *fine structure* to the electronic spectra. It must be noted here that whereas pure rotation spectra are shown only by molecules possessing a permanent electric dipole moment, and vibrational spectra require a change of dipole moment during the motion, electronic spectra are given by all molecules. Since the changes in electron distribution in a molecule are always accompanied by a dipole change, therefore, homonuclear diatomic molecules which show no pure rotation or vibration rotation spectra, do produce an electronic spectrum having vibrational and rotational structure in their spectra. This is depicted schematically in Fig. 8.4.

The rotational, vibrational and electronic energy separations for a diatomic molecule are of the order of 1–300 cm^{-1}, 300–4000 cm^{-1} and 10^6 cm^{-1} respectively.

8.5 STABILITY OF MOLECULAR STATES

We say that a molecule is stable if its ground state is stable, i.e. the ground state has an appreciable potential minimum such that the energy of the lowest vibrational level ($n = 0$) is lower than the energy of the separated atoms in their ground state.

The physical stability is to be distinguished from the chemical stability which is possessed by a given molecule only if even on collision with like molecules at low temperatures, it is stable for an appreciable length of time. On the other hand, molecules such as CH, CN, OH, etc. are physically but not chemically stable, since although they appear in chemical reactions, and in electric discharges they do not form a stable gas at ordinary temperatures.

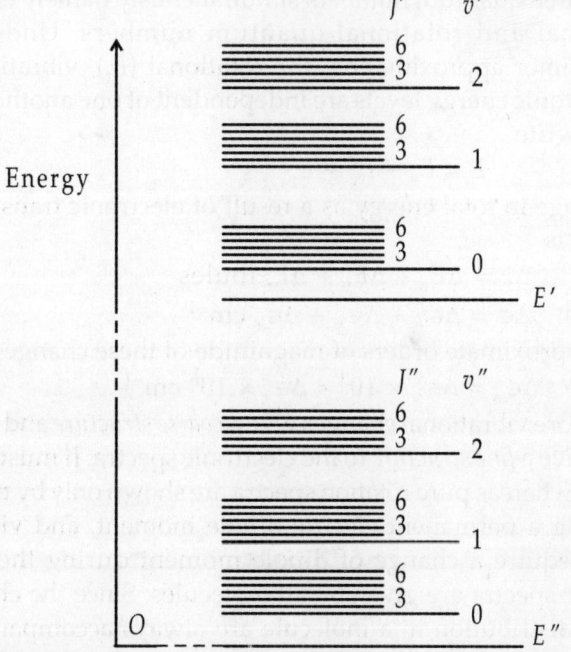

Fig. 8.4: Energy levels of a diatomic molecule. E'', E' are electronic levels; J'' and J' denote rotational levels and v'', v' the vibrational levels

The He_2 molecule is an example of a molecule that is physically and chemically unstable while it has many stable excited states and its lowest state has no potential minimum.

The fact that for many molecules, only a few electronic states are observed appears at first to be in contradiction to the large manifold of terms demanded by theory. Although this has its basis partly in the inadequacy of the observational material obtained upto the present time, the main reason is that a large number of states theoretically predicted are unstable and thereby escape observation. The unstable states are no less real than the stable states and under conditions may be themselves known by continuous or diffuse spectrum.

In discussing the stability of the molecules, three types of binding are generally distinguished:

    (*i*) Homopolar binding.

    (*ii*) Ionic (Electronic) binding.

    (*iii*) Polarization or van der Waal's binding.

We will now consider homopolar binding because this is important in relation to molecular spectra.

8.6 HEITLER AND LONDON THEORY OF H_2 MOLECULE

We picture the H_2 molecule as two separate atoms. When these two atoms are brought close together, then the electrons of the two atoms become indistinguishable and we have to consider the possibility of the electrons belonging to any of the two atoms.

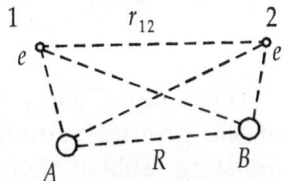

Fig. 8.5: The H_2 molecule

We now consider the problem of H_2 molecule by neglecting the spins of the electrons. The wave equation for the H_2 molecule is solved by starting out from the state of the separated atoms as zero approximation and then introducing the interaction of the two atoms as perturbation.

The system comprises two hydrogen nuclei A and B and the two electrons whose coordinates we designate by 1 and 2.

The potential energy of the two electrons in the field of the two photons is

$$V = \frac{e^2}{R} - \frac{e^2}{r_{1A}} - \frac{e^2}{r_{2A}} - \frac{e^2}{r_{1B}} - \frac{e^2}{r_{2B}} + \frac{e^2}{r_{12}} \qquad ..(13)$$

The Schrodinger equation of H_2 molecule, with the nuclei considered fixed at a distance R is

$$\nabla_1^2 \psi + \nabla_2^2 \psi + \frac{8\pi^2 m}{h^2}(E - V)\psi = 0 \qquad ..(14)$$

From this equation for various values of R, we may find the variation of the electronic energy with R (i.e. the potential energy under which the nucleus moves).

For very large values of R, the potential V is $-\dfrac{e^2}{r_{1A}} - \dfrac{e^2}{r_{2B}}$, i.e. then the wave equation is that of two independent H atoms, at distance R.

Let $\psi_A(1)$ = Hydrogen wave function for electron 1 moving about A.

$\psi_B(2)$ = Hydrogen wave function of electron 2 moving about B.

Both of these eigen functions are referred to the same coordinate system. The eigen function for the system is $\psi_A(1)\,\psi_B(2)$. In the ground state

$$\psi_A(1) = \frac{1}{\sqrt{\pi a_0^3}}\,e^{-r_{1A}/a_0}\,; \quad a_0 = \frac{h^2}{4\pi m e^2} \qquad ..(15)$$

(The first Bohr radius)

The energy of the system for large R is $2E_H$ where E_H is the energy of the ground state of the H atom. Since the electrons are indistinguishable, they may be exchanged without affecting the energy of the system, i.e. for large R, we can use

$$V = \frac{e^2}{r_{2A}} - \frac{e^2}{r_{1B}} \qquad ..(16)$$

Instead of the above, then we can use the eigen function $\psi_A(2)$ $\psi_B(1)$.

Thus we have a *double degeneracy* or *exchange degeneracy* and any linear combination of the two functions $\psi_A(1)\,\psi_B(2)$ and $\psi_A(2)\,\psi_B(1)$ will be a solution.

If the two atoms are brought close together, a splitting of the degeneracy will arise. Since potential energy remains unchanged for an exchange of the two electrons, the eigen function must either also remain unchanged or only change sign for such an exchange. In zero approximation, the eigen functions are, therefore,

$$\psi_s = N_s[\psi_A(1)\,\psi_B(2) + \psi_A(2)\,\psi_B(1)]$$

and $\qquad\qquad \psi_a = N_a[\psi_A(1)\,\psi_B(2) - \psi_A(2)\,\psi_B(1)] \qquad ..(17)$

The first of these is symmetric, remains unchanged for an exchange of the electrons. The second is antisymmetric, changes sign for such an exchange. The normalization constants N_s and N_a are given by

$$N_S = \frac{1}{\sqrt{2+2S}}\,; \quad N_a = \frac{1}{\sqrt{2-2S}} \qquad ..(18)$$

where $\quad S = \int \psi_A(1)\psi_B(1)\psi_A(2)\psi_B(2)d\tau_1 d\tau_2; \qquad ..(19)$

If the calculation of the first order perturbation is carried out using as perturbation function W:

$$W = \frac{e^2}{R} + \frac{e^2}{r_{12}} - \frac{e^2}{r_{1B}} - \frac{e^2}{r_{2A}} \qquad ..(20)$$

then $\qquad E_s = 2E_H + \dfrac{K+J}{1+S}$

$$E_a = 2E_H + \frac{K-J}{1-S} \qquad ..(21)$$

where $\qquad K = \int \psi_A(1)\psi_B(2)W\psi_A(1)\psi_B(2)d\tau_1 d\tau_2 \qquad ..(22)$

and $\qquad J = \int \psi_A(1)\psi_B(2)W\psi_A(2)\psi_B(1)d\tau_1 d\tau_2 \qquad ..(23)$

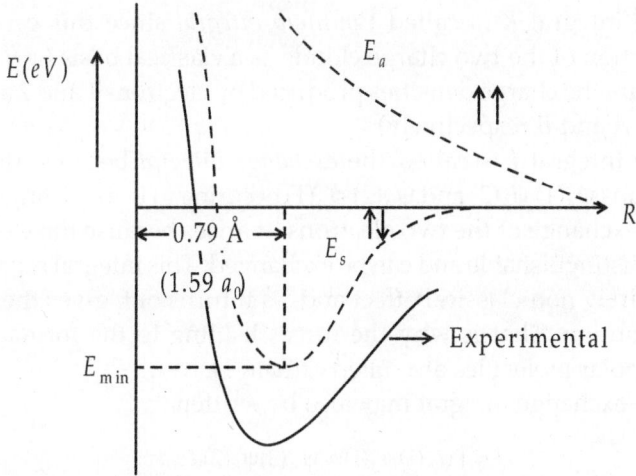

Fig. 8.6: The curve of E vs R (The constant term E_H has been omitted). a_0 is the first Bohr radius.

In the state characterized by ψ_s a potential minimum arises leading to molecule formation. While for the state characterized by ψ_a, there is a repulsion for all values of R. Assuming that the integrals J and K can be evaluated, we obtain the dissociation energy:

TABLE 8.3

	R	Dissociation Energy
Theoretical	0.79 Å	–3.2 eV
Experimental	0.75 Å	–4.4 eV

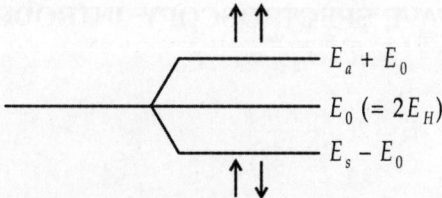

Fig. 8.7: The lifting off of degeneracy, (E_s constitutes singlet state and E_a constitutes triplet state)

The reason for discrepancy between theoretical and experimental values is that we are assuming the bond to be purely covalent but it is actually partly ionic also.

The integral K is called *Coulumb integral* since this gives the interaction of the two charge clouds as a classical basis ($\psi_A^2(1)$ and $\psi_B^2(2)$ are the charge densities produced by electrons 1 and 2 around nuclei A and B respectively).

The integral J is called the *exchange integral* because the two functions $\psi_A(1)\,\psi_B(2)$ and $\psi_A(2)\,\psi_A(1)$ occuring in it are distinguished by an exchange of the two electrons. It arises because the electrons are indistinguishable and can be exchanged. This integral represents an entirely non-classical effect and, as it turns out, gives the major contribution. That is why the forces leading to the formation of homopolar molecules are called *exchange forces*.

The exchange integral may also be written as

$$J = \int \psi_A(1)\psi_B(1)W\psi_A(2)\psi_B(2)d\tau_1 d\tau_2 \qquad ..(24)$$

Here $\psi_A(1)\,\psi_B(1)$ (Similarly $\psi_A(2)\,\psi_B(2)$) is a function depending on the overlapping of the two hydrogen eigen functions belonging to the two nuclei. For intermediate R this function is largest near the two nuclei and consequently the terms $-\dfrac{e^2}{r_{1B}} - \dfrac{e^2}{r_{2A}}$ give contribution to J which is, therefore, negative. Thus, since, we always have $S < 1$, we see that E_s is below the value of $2E_H$ while E_a is above it, consequently, we have attraction and repulsion respectively. For small values of R, the largest contribution to both K and J is that due to $\dfrac{e^2}{R}$ which is positive, i.e. we have repulsion in both the states.

8.7 MICROWAVE SPECTROSCOPY: INTRODUCTION

Microwave spectroscopy is proving in recent years a great asset in the study of various problems in physics, chemistry, electronics and even astronomy. The ability to measure frequencies more precisely in the microwave region allow very accurate calculations then in the infrared, visible, ultraviolet regions. This technique has been found to be very useful in the determination of the structures of those molecules which do not give useful results by using Raman and IR spectroscopy.

The microwave spectroscopy explores that part of e.m. spectrum which extends from 100 μm to 1 cm (i.e. microwave region). This lies between far infrared and conventional r.f. regions. Mircowave spectroscopy utilizes exclusively absorption works rather than emission. In most of the cases, absorption of microwave energy represents changes of the absorbing molecule from one rotational level to another.

Consequently, the microwave spectroscopy deals with the pure rotational levels of molecules and is also known as rotational spectroscopy. The condition for observation is that a molecule must possess permanent dipole moment. When such a molecule rotates, it generates an electric field which can interact with the electric component of the microwave radiation. During such interaction, energy can be absorbed or emitted and thus rotation of molecule gives rise to a spectrum. If molecule lacks permanent dipole moment, interaction is not possible and these molecules are said to be *microwave inactive*, e.g. H_2, Cl_2, etc. On the other hand, the molecules like HCl, CH_3Cl, etc. are having dipole moments and are therefore *microwave active*.

8.7.1 Experimental Arrangement for Studying Microwave Spectra

Microwave spectra can only be obtained in case of gaseous molecules (under very low pressure), because in the condensed phase, no well defined rotational levels exist, so microwave spectra cannot be observed in case of liquids and solids. Further, the molecule must possess a dipole moment. Consequently, HCl, HBr, CO, etc. are microwave active as they possess dipole moments. H_2, Cl_2, Br_2, etc. are microwave inactive. Further, CH_4, C_2H_6, C_2H_4, etc. are also microwave inactive as they do not possess dipole moments.

In microwave spectroscopy, the source is monochromatic, having a well defined single wavelength whose intensity can be rapidly varied. Such radiations are generated using Klystron or magnetron tube. Monochromatic radiations of various known wavelengths are allowed to pass through an absorption cell consisting of a metal tube containing the gaseous sample of the substance under examination. After suitable amplification, the intensity of the transmitted beam is measured with the help of a cathode ray oscillograph acting as detector (or recorder). The resolving power of this arrangement is 10^5 times that of the best infrared grating spectrometer so the wavelength measurements can be made to seven significant figures.

By varying the frequency of the Klystron oscillator, and observing the intensity of transmitted beam, data can be obtained from which moment of inertia and internuclear distances can be calculated. A schematic microwave spectrometer is shown in Fig. 8.8.

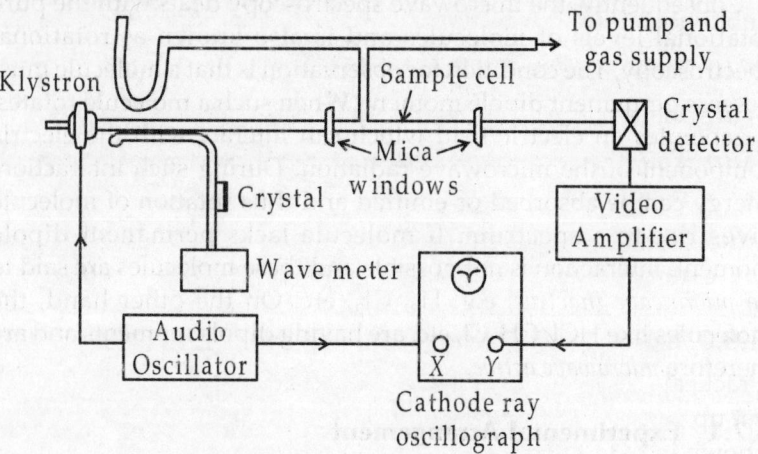

Fig. 8.8: A microwave spectrometer

8.8 INFRARED SPECTROSCOPY

8.8.1 Introduction

Infrared spectroscopy is one of the most powerful analytical techniques which offer the possibility of chemical identification. This technique when coupled with intensity measurements may be used for quantitative analysis. One of the most important advantages of this technique over other usual methods of structural analysis (viz. *XRD*, *ESR*) is that it provides useful information about the structure of molecule quickly (without tiresome evaluation methods).

The technique is based upon the simple fact that a chemical substance shows marked selective absorption in the IR region. After absorption of IR radiations, the molecules of a chemical substance vibrate at many rates of vibration, giving rise to close-packed absorption bands called a IR absorption spectrum, which may extend over a wide wavelength range. There will be present many bands which will correspond to the characteristics functional groups and bonds present in a chemical substance. Thus an IR spectrum of a chemical substance is a fingerprint of its identification.

8.8.2 The Range of Infrared Radiation

IR is that region of electromagnetic spectrum which lies between the visible and microwave regions. However, this region may be subdivided into four sections:

(*i*) **The Photographic Region:** This ranges from visible to 1.2 μ.

(*ii*) **The very near IR Region:** This ranges from 1.2 to 2.5 μ.

(*iii*) **The near IR Region:** This is also known as vibration-rotation region and ranges from 2.5 to 25 μ.

(*iv*) **The far IR Region:** This is known as the rotation region and ranges from 25 to 300–400 μ.

8.8.3 Experimental Arrangement for Studying IR Spectra

Practical IR spectra lie in the range 2.5 μ to 25 μ. The spectroscopic set up consists of a source, a dispersion element and a detector, as shown in the Fig. 8.9.

For absorption spectra, a source capable of emitting all wavelengths over a suitable interval of wavelengths is used and radiation is then focussed onto a monochromator through a convenient thickness of an absorbing sample. The recorder then shows the graph that predicts the energy reaching the detector at each wavelength. To determine absorption, wavelength interval is covered again without absorbing sample. Source and detector should remain stable for the period of these two scans (i.e. with and without sample).

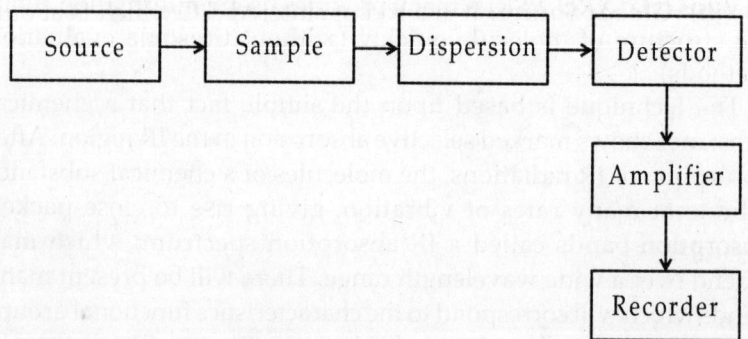

Fig. 8.9: Block diagram of a spectroscopic set up (single-beam) for observation of infrared absorption spectra

8.8.4 Instrumentation for Infrared Spectroscopy

The usual optical materials (glass or quartz) absorb strongly in the infrared region. As a consequence, the apparatus for taking IR spectra is appreciably different from that used for the visible and ultraviolet regions. The main parts of an IR spectrometer are:

1. The IR radiation sources.
2. The monochromators.
3. The sample cells.
4. Detectors.

We now consider these one by one.

1. The IR Radiation Sources: The various popular sources of IR radiations are:

(*i*) **Incandescent Lamp:** An ordinary incandescent lamp generally gives radiation in the near IR region. It has a low spectral emissivity and fails in the far infrared.

(*ii*) **Nernst-Glower:** It consists of a hollow rod which is about 2 mm in diameter and 30 mm in length. The glower is made up of rare earth oxides such as zirconia, yttria and thoria. Nernst glower needs external heating. When heated to a temperature between 1000 to 1800° C, it provides maximum radiation at about 7100 cm^{-1}. Its main disadvantage is that it emits IR radiation over a wide wavelength range and another disadvantage is its frequent mechanical failure.

(*iii*) **Globar Source:** It is a rod of sintered SiC which is about 50 mm long and 4 mm in diameter. When heated to a temperature between 1300 and 1700° C, it starts emitting strongly in the IR region. It emits maximum radiation at 5200 cm^{-1}. Unlike Nernst glower, it is self-starting and more reliable. The main disadvantage is that it is a less intense source than Nernst glower.

(*iv*) **Mercury Arc:** A high pressure mercury arc provides far IR radiation. At shorter wavelengths, the heated quartz envelope emits the radiation, whereas at longer wavelengths the mercury plasma provides the radiation through the quartz envelope.

2. **Monochromators:** The radiation sources emit various frequencies. Since sample in IR spectroscopy absorbs at only a certain frequency, therefore, it becomes necessary to select desired frequency and reject the others. This necessitates the use of monochromators. These are mainly of two types:

   (*a*) Prism monochromator.

   (*b*) Grating monochromator.

   (*a*) **Prism Monochromator:** A single pass monochromator is illustrated in Fig. 8.10. The sample is kept at the focus of the beam, just before the entrance slit *A* to the mono-chromator. The radiation from the source after passing through the sample and the entrance slit, strikes the off axis Littrow mirror *B* which renders the radiation parallel and sends it to the prism *C*. The dispersed radiation after reflection from a plane mirror *D* returns through the prism a second time and focusses onto the exit slit of the monochromator, through which it finally passes to the detector section.

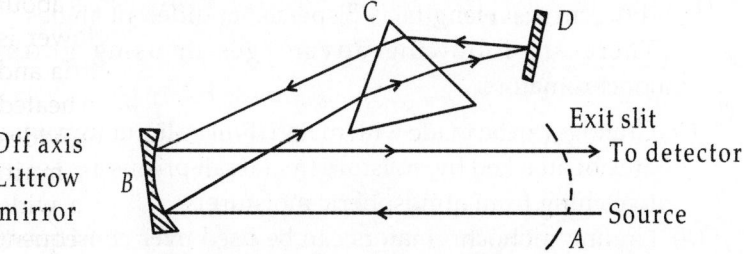

Fig. 8.10: Single pass monochromator

The double pass monochromater is illustrated in Fig. 8.11.

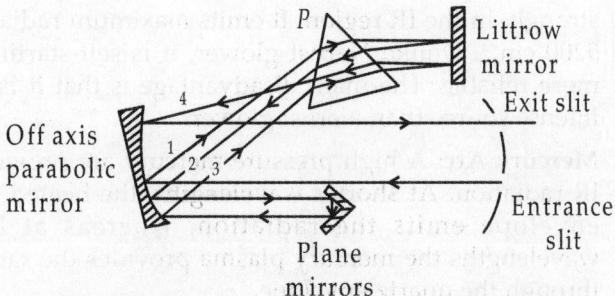

Fig. 8.11: Double pass monochromator

In this there occur a total of four passes of radiation through the prism as indicated by 1, 2, 3, 4. The double pass produces more resolution (than the single pass monochromator) before it finally passes on to the detector.

In both single pass and double pass monochromator, sodium chloride (rock salt) prism is employed for the entire region (2.5 to 15.4 μ).

(b) **Grating Monochromator:** If a prism in a prism-monochromator is replaced by a grating, higher dispersion can be achieved. Dispersion by a grating follows the law:

$$n\lambda = d(\sin i \pm \sin \theta)$$

where n is the order (a whole number), λ, the wavelength of the radiation, d, the distance between two consecutive lines of the grating, i, the angle of incidence of IR radiation and θ, the angle of dispersion of light (of wavelength λ). At a grating, separation of light occurs because light of different wavelengths is dispersed at different angles.

There are following advantages in using grating monochromators:

(i) Gratings can be made with materials like aluminium which are not attacked by moisture (metal salt prisms are subject to etching from atmospheric moisture).

(ii) Grating monochromators can be used over considerable wavelength ranges.

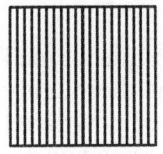 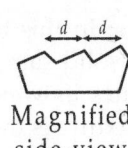 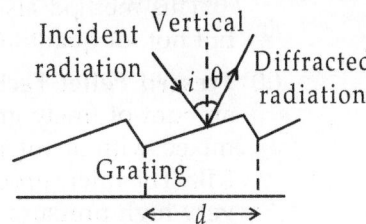

Top view

Magnified side-view

Fig. 8.12: A grating monochromator

Fig. 8.13: Path of IR radiation diffracted by a grating monochromator

A grating is generally used in conjunction with a small prism which acts as an order sorter.

3. **Sample Cells:** The material containing the sample must be transparent to IR radiation. This condition restricts us to select only certain salts like NaCl or KBr. A final choice of salt will depend on the wavelength range to be studied.

As IR spectroscopy is used for the characterization of solids, liquids and gases as well, we now discuss the sampling of solids, liquids and gases.

(A) **Sampling of Solids:** The following techniques are employed:

   (a) **Solids Run in Solution:** If the solution of solid can be prepared in a suitable solvent then the solution is run in one of the cells for liquids. This method cannot be used for all solids.

   (b) **Solid Films:** If a solid is amorphous, the sample is deposited on the surface of a KBr or NaCl cell by evaporation of a solution of the solid. This technique is useful for qualitative analysis but becomes useless for quantitative analysis.

   (c) **Mull Technique:** In this technique, the finely ground solid sample is mixed with a mineral oil (nujol) to make a thick paste which is then made to spread between IR transmitting windows. This is then mounted in a path of IR beam and then the spectrum is run.

This method also is good for qualitative analysis but not for quantitative analysis.

(*d*) **Pressed Pellet Technique:** In this technique a small amount of finely ground solid sample is thoroughly mixed with about 100 times its weight of powdered KBr. The finely ground mixture is then pressed under very high pressure (~2500 psi) to form a small pellet (about 1–2 mm thick and 1 cm in diameter). The resulting pellet is transparent to IR radiation and is run as such.

This method has following advantages over Mull-method:

(*i*) KBr pellets can be stored for long periods of time.

(*ii*) It can be used for quantitative analysis.

(*iii*) The resolution of spectrum in the KBr is superior to that obtained with nujol.

However, there are following disadvantages also:

(*i*) It always has a band (at 3450 cm^{-1}) from the OH group of moisture present in the sample.

(*ii*) The high pressure during the formation of pellets may bring about polymorphic changes in crystallinity of the samples (especially inorganic complexes) which may cause complications in IR spectrum.

(*iii*) This method cannot be used for some polymers (which are difficult to grind with KBr).

(*B*) **Sampling of Liquids:** Samples that are liquid at room temperatures may be put directly (with no preparation) into rectangular cells made of NaCl or KBr and their IR spectra are obtained directly. The sample thickness should be so selected that the transmittance lies between 15 and 70 per cent. If a cell possesses good quality windows, flat and parallel, its thickness in cm can be calculated from:

$$2t = N / \omega_1 - \omega_2$$

where N is the number of fingers between wave numbers ω_1 and ω_2.

(*C*) **Sampling of Gases:** The gas sample is put into a special cell generally about 10 cm long, which is then kept across the path of the IR beam. The end walls of the cell are generally made up of sodium chloride.

Since the sampling area of most spectrometer is restricted in length, mirrors are used to cause multiple reflections to make the effective path lengths as long as 40 cm.

4. **Detectors:** Thermal detectors are generally used (except in near infrared, where we use a photoconductivity cell). These give responses for all frequencies. As the radiant power is low for the infrared region, the detector signal will also be low. In order to locate these low signals, a preamplifier is attached to the detector and the radiation beam is modulated with a low-frequency light interruptor, 10–25 cps. To detect such signals, thermal detectors must possess a short response time and the absorbed heat must be lost rapidly.

The various types of detectors used in IR spectroscopy are bolometers, thermocouples, thermistors and Golay cells.

8.8.5 Single Beam Spectrophotometer

Figure 8.14 depicts the schematic diagram of a single beam infrared spectrophotometer.

The radiation emitted by the source after passing through the sample passes through a fixed prism and a rotating Littrow mirror. Both prism and Littrow mirror select the desired wavelength and allow it to pass on to the detector.

The detector measures the transmitted intensity (i.e. after passage through the sample). Knowing the original intensity, one can measure the radiation that has been absorbed. By measuring the degree of absorption at various wavelengths, the absorption spectrum of the sample can be obtained.

The main disadvantages of a single beam spectrophotometer are as follows:

(i) The intensity of emission of the radiation source varies from point to point in IR absorption spectrum, therefore the resulting spectrum is considerably deformed.

(ii) When the sample is analyzed in solution form, the bands of solvent appear in the spectrum. The correction is done by subtracting the spectrum of the solvent from the resultant spectrum.

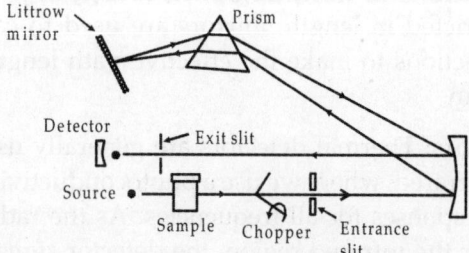

Fig. 8.14: Schematic diagram of a single beam infrared spectrophotometer

In order to overcome the above difficulties, a double beam spectrophotometer is used.

8.8.6 Double Beam Spectrophotometer

In double beam spectrophotometer (Fig. 8.15) the energy emitted by the source is split by the instrument into two beams, which are energetically and optically identical. One of the beams passes through the sample and the other through the reference sample. The sample is placed in the path of the sample beam and the reference material (i.e. solvent) in the reference beam. The two half beams recombine and pass along the optical path to the detector. The signal from the detector passes onto the recording unit through a servomotor.

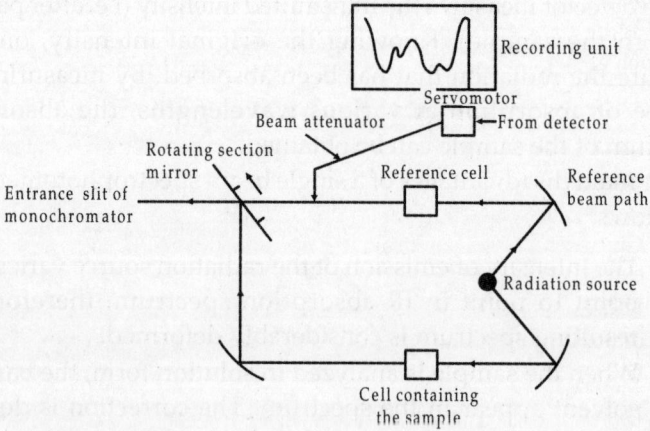

Fig. 8.15: Schematic diagram of a double beam infrared spectrophotometer

8.8.7 Limitations of Infrared Spectroscopy

In spite of being a most valuable method for characterising both qualitatively and quantitatively many inorganic compounds, there are following shortcomings:

1. By infrared spectroscopy, it is generally not possible to know molecular weight of a substance.
2. Generally, the IR spectroscopy does not provide information of the relative positions of different functional groups on a molecule.
3. From the single IR spectrum of an unknown substance, it is not possible to know whether it is a pure compound or a mixture of compounds. For example, a mixture of paraffins and alcohols will give the same IR spectra as by higher molecular weight alcohols.

TABLE 8.4

Microwave Spectroscopy	Infrared Spectroscopy
1. The substance must be in the gaseous state	1. The substance may be in the solid, liquid or gaseous state
2. The spectra observed are nearly always absorption spectra	2. The spectra observed may be absorption or emission spectra
3. The spectrum is a characteristic of the absorbing molecule as a whole	3. The spectrum is characteristic of functional groups present in the molecule
4. Resolution of lines is very much greater	4. Resolution is comparatively smaller

Questions

8.1 Give an outline of quantum theory of H_2 molecule. Comment on the extent of agreement with experimental data on the different parameters.

8.2 Discuss the origin of the various types of spectra obtained from a diatomic molecule.

8.3 Write a short note on Born–Oppenheimer approximation.

8.4 Discuss various types of molecular spectra, referring to quantum of energy involved and the regions in which they fall.

8.5 Write a short note on Heitler and London theory of H_2 molecule.

8.6 Outline the Heitler and London wave-functions for hydrogen molecule. What are singlet and triplet states of hydrogen?

8.7 Explain why the ground state of the hydrogen molecule is a binding state and the first excited state is an unstable state.

8.8 Why are mirrors instead of lenses used in the IR region?

8.9 Why water cannot be used as a solvent for infrared spectroscopy?

8.10 Which crystals are used for optical system in infrared instruments?

9

Pure Rotation Spectra
(Far Infrared Spectroscopy)

The bands which appear in far infrared region at wavelengths of 200×10^4 Å or more are due to transitions involving very small energy changes, about 0.005 eV. With such a small quantum of energy, electronic and vibrational energy states will not be excited or in other words only transitions that are purely rotational in character will appear. Only those molecules which have a permanent electric dipole moment can give rise to pure rotational spectra. This homonuclear diatomic molecules such as H_2, O_2, N_2 do not exhibit pure rotation spectra whereas heteronuclear diatomic molecules such as HF, HCl, HBr, etc. do exhibit pure rotation spectra. These are known only in absorption and only for molecules HCl, HBr, HI, H_2O and NH_3 have been studied in detail.

9.1 RIGID ROTATOR

As a first approximation, rotating diatomic molecule, whose nuclei are considered as being separated by a definite mean distance may be treated as a rigid rotator with a free axis. Suppose masses m_1 and m_2 are joined by a rigid bar (the bond) whose length is $r_0 = r_1 + r_2$ where r_1 and r_2 are the distances of m_1 and m_2 respectively from the centre of mass C of the system.

We shall calculate rotational energy levels, frequency of spectral lines arising due to transition between two energy levels and selection rule. In our analysis, it will be assumed that the bond between the two atoms is stiff and bond length does not change (i.e. a rigid rotator).

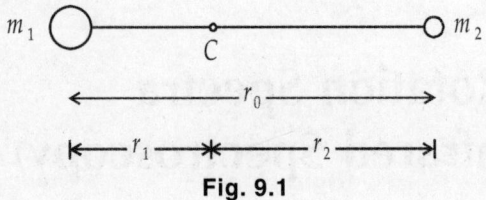

Fig. 9.1

Rotational Energy Levels

Consider a diatomic molecule having internuclear separation r_0. The molecule is free to rotate about an axis passing through its centre of mass C.

$$m_1 r_1 = m_2 r_2 \qquad \qquad ..(1)$$

and

$$r_0 = r_1 + r_2 \qquad \qquad ..(2)$$

$\therefore$

$$r_1 = \frac{m_2}{m_1 + m_2} r_0$$

and

$$r_2 = \frac{m_1}{m_1 + m_2} r_0 \qquad \qquad ..(3)$$

The moment of inertia $I = m_1 r_1^2 + m_2 r_2^2 \qquad ..(4)$

$$= \frac{m_1 m_2^2}{(m_1 + m_2)^2} r_0^2 + \frac{m_2 m_1^2}{(m_1 + m_2)^2} r_0^2$$

$$= \frac{m_1 m_2}{m_1 + m_2} r_0^2 = \mu r_0^2 \qquad \qquad ..(5)$$

where the reduced mass μ is defined as

$$\mu = \frac{m_1 m_2}{m_1 + m_2} \qquad \qquad ..(6)$$

If the molecule is rotating with angular velocity ω then classically, it would have an energy $\frac{1}{2} I\omega^2$. The angular momentum is given by $I\omega$ and it is quantized as

$$I\omega = J\frac{h}{2\pi}, \quad J = 0, 1, 2 \ldots \qquad \qquad ..(7)$$

Therefore, allowed rotational energies will be given by

$$E_r = \frac{1}{2}I\omega^2 = \frac{1}{2I}I^2\omega^2$$

$$= \frac{1}{2I}\frac{J^2h^2}{4\pi^2} = \frac{J^2h^2}{8\pi^2 I} \qquad \qquad ...(8)$$

Here $J = 0, 1, 2 ...$ are termed as rotational quantum numbers.

Energy Levels of the Rigid Rotator on the Basis of Schrodinger's Equation

The Schrodinger equation in spherical polar coordinates is

$$\frac{1}{r^2}\frac{2}{\partial r}\left(r^2\frac{\partial\psi}{\partial r}\right) + \frac{1}{r^2\sin\theta}\frac{\partial}{\partial\theta}\left(\sin\theta\frac{\partial\psi}{\partial\theta}\right) + \frac{1}{r^2\sin^2\theta}\frac{\partial^2\psi}{\partial\phi^2}$$

$$+ \frac{8\pi^2 m}{h^2}(E - V)\psi = 0 \qquad ..(9)$$

In order to apply this equation to the rigid rotator, we have to arrive at some conclusion for m, r and V as discussed below.

In cartesian coordinates, kinetic energy of a single particle of mass m is

$$T = \frac{1}{2}m\left(\dot{x}^2 + \dot{y}^2 + \dot{z}^2\right) \quad \left(\text{where } \dot{x} = \frac{\partial x}{\partial t} \equiv v_x\right)$$

In spherical coordinates

$$T = \frac{1}{2}m(\dot{r}^2 + r^2\dot{\theta}^2 + r^2\dot{\phi}^2\sin^2\theta)$$

If the distance r of the particle from the origin is fixed, $\dot{r} = 0$ and

$$T = \frac{1}{2}mr^2\left(\dot{\theta}^2 + \dot{\phi}^2\sin^2\theta\right) \qquad ..(10)$$

For two particles of a rigid rotator

$$T = T_1 + T_2$$

$$= \frac{1}{2}m_1 r_1^2\left(\dot{\theta}^2 + \dot{\phi}^2\sin^2\theta\right) + \frac{1}{2}m_2 r_2^2\left(\dot{\theta}^2 + \dot{\phi}^2\sin^2\theta\right)$$

Since the interparticle distance is fixed, the potential energy $V = 0$ and so, the total energy

$$E \equiv T = \frac{1}{2}\left(m_1 r_1^2 + m_2 r_2^2\right)\left(\dot{\theta}^2 + \dot{\phi}^2\sin^2\theta\right)$$

$$= \frac{1}{2}I\left(\dot{\theta}^2 + \dot{\phi}^2\sin^2\theta\right) \qquad ..(11)$$

Comparing eqn (11) with eqn (10), we conclude that the rigid rotator behaves like a single particle of mass I placed at a fixed distance, equal to unity $(r = 1)$ from the origin (i.e. centre of mass). Consequently, we put

$$V = 0, r = 1 \text{ and }, \mu \to I$$

in the equation (9), so that for rigid rotator

$$\frac{1}{\sin\theta}\frac{\partial}{\partial\theta}\left(\sin\theta\frac{\partial\psi}{\partial\theta}\right) + \frac{1}{\sin^2\theta}\frac{\partial^2\psi}{\partial\phi^2} + \frac{8\pi^2IE}{h^2}\psi = 0$$

$$..(12)$$

This indicates that ψ is a function of θ and ϕ.

We solve this eqn (12) by the method of separation of variables, we put

$$\psi(\theta, \phi) = \Theta(\theta)\,\Phi(\phi) \qquad ..(13\,i)$$

$$\therefore \qquad \frac{\partial\psi}{\partial\theta} = \Phi\frac{\partial\Theta}{\partial\theta} \qquad ..(13\,ii)$$

and $\qquad \dfrac{\partial^2\psi}{\partial\phi^2} = \Theta\dfrac{\partial^2\Phi}{\partial\phi^2} \qquad ..(13\,iii)$

Putting these values in eqn (12), we get after multiplying the whole by $\dfrac{\sin^2\theta}{\Theta\Phi}$.

$$\frac{\sin\theta}{\Theta}\frac{\partial}{\partial\theta}\left(\sin\theta\frac{\partial\Theta}{\partial\theta}\right) + \frac{1}{\Phi}\frac{\partial^2\Phi}{\partial\phi^2} + \frac{8\pi^2IE}{h^2}\sin^2\theta = 0$$

On separating the variables

$$\frac{\sin\theta}{\Theta}\frac{\partial}{\partial\theta}\left(\sin\theta\frac{\partial\Theta}{\partial\theta}\right) + \beta\sin^2\theta = -\frac{1}{\Phi}\frac{\partial^2\Phi}{\partial\phi^2} \qquad ..(14)$$

where $\quad \beta = \dfrac{8\pi^2IE}{h^2}$

In the above equation, LHS involves the variable θ only whereas RHS involves the variable ϕ only. As the two variables θ and ϕ are independent, the two sides can be equated only when each is equal to a constant say m^2, so that

$$\frac{\sin\theta}{\Theta}\frac{\partial}{\partial\theta}\left(\sin\theta\frac{\partial\Theta}{\partial\theta}\right) + \beta\sin^2\theta = m^2 \qquad ..(15)$$

and $\quad -\dfrac{1}{\Phi}\dfrac{\partial^2\Phi}{\partial\phi^2} = m^2 \qquad\qquad ..(16)$

The solution of eqn (10) is

$$\Phi = Ce^{\pm im\phi}$$

For Φ to be a single valued function, it should have the same value for $\phi = 0$ and $\phi = 2\pi$, i.e.

$$\Phi = Ce^{\pm im\phi} = C$$

and $\Phi = Ce^{\pm i2\pi m}$ should be same

$$\therefore \qquad C = Ce^{\pm i2\pi m}$$

or $\qquad e^{\pm i2\pi m} = 1$ i.e. $\cos 2\pi m \pm i\sin 2\pi m = 1$

$\therefore$ m is to be either 0 or an integer.

Now we consider the solution of eqn (15). We can write eqn (15) on multiplying by $\dfrac{\Theta}{\sin^2\theta}$ as

$$\frac{1}{\sin\theta}\frac{\partial}{\partial\theta}\left(\sin\theta\frac{\partial\Theta}{\partial\theta}\right)+\left(\beta-\frac{m^2}{\sin^2\theta}\right)\Theta=0 \qquad ..(17)$$

Putting $x = \cos\theta$

$$\frac{\partial\Theta}{\partial\theta}=\frac{\partial\Theta}{\partial x}\cdot\frac{\partial x}{\partial\theta}=-\sin\theta\frac{\partial\Theta}{\partial x}$$

or $\qquad \dfrac{\partial}{\partial\theta}=-\sin\theta\dfrac{\partial}{\partial x}$

In above equation, we have

$$\frac{1}{\sin\theta}\times(-\sin\theta)\frac{\partial}{\partial x}\left(-\sin^2\theta\frac{\partial\Theta}{\partial x}\right)+\left(\beta-\frac{m^2}{\sin^2\theta}\right)\Theta=0$$

or $\qquad \dfrac{\partial}{\partial x}\left\{(1-x^2)\dfrac{\partial\Theta}{\partial x}\right\}+\left\{\beta-\dfrac{m^2}{(1-x^2)}\right\}\Theta=0 \qquad ..(18)$

It is Legendre's differential equation, x can assume values from $\cos 0$ (or +1) to $\cos\pi$ (or –1). To solve eqn (18), we write

$$\Theta = (1 - x^2)^{m/2}\cdot G \text{ where } G \text{ is a function of } x \text{ only.}$$

so that

$$\frac{\partial\Theta}{\partial x}=-mx\left(1-x^2\right)^{\frac{m}{2}-1}G.+\left(1-x^2\right)^{\frac{m}{2}}\frac{\partial G}{\partial x}$$

$\therefore$ $\quad \left(1-x^2\right)\dfrac{\partial\Theta}{\partial x}=-mx\left(1-x^2\right)^{\frac{m}{2}}G.+\left(1-x^2\right)^{\frac{m}{2}+1}\dfrac{\partial G}{\partial x}$

Hence from eqn (18)

$$\frac{\partial}{\partial x}\left[(1-x^2)\frac{\partial \Theta}{\partial x}\right] = \left\{-m(1-x^2)^{\frac{m}{2}} - mx\left(\frac{m}{2}\right)(-2x)(1-x^2)^{\frac{m}{2}-1}\right\}G$$

$$-mx(1-x^2)^{\frac{m}{2}}\frac{\partial G}{\partial x} + \left(\frac{m}{2}+1\right)(1-x^2)^{\frac{m}{2}}(-2x)\frac{\partial G}{\partial x} + (1-x^2)^{\frac{m}{2}+1}\frac{\partial^2 G}{\partial x^2}$$

$$= \left\{-m(1-x^2)^{\frac{m}{2}} + m^2x^2(1-x^2)^{\frac{m}{2}-1}\right\}G - \left\{mx(1-x^2)^{\frac{m}{2}} + \right.$$

$$(m+2)(-x)(1-x^2)^{\frac{m}{2}}\left.\right\}\frac{\partial G}{\partial x} + (1-x^2)^{\frac{m}{2}+1}\frac{\partial^2 G}{\partial x^2}$$

$$= \left\{-m(1-x^2)^{\frac{m}{2}} + m^2x^2(1-x^2)^{\frac{m}{2}-1}\right\}G - \left\{2x(m+1)(1-x^2)^{\frac{m}{2}}\right\}\frac{\partial G}{\partial x}$$

$$+ (1-x^2)^{\frac{m}{2}+1}\frac{\partial^2 G}{\partial x^2}$$

Hence from eqn (18)

$$\left\{-m(1-x^2)^{\frac{m}{2}} + m^2x^2(1-x^2)^{\frac{m}{2}-1}\right\}G - \left\{2x(m+1)(1-x^2)^{\frac{m}{2}}\right\}\frac{\partial G}{\partial x}$$

$$+ (1-x^2)^{\frac{m}{2}+1}\frac{\partial^2 G}{\partial x^2} + \left\{\beta - \frac{m^2}{1-x^2}\right\}(1-x^2)^{\frac{m}{2}}G = 0$$

Dividing both sides by $(1-x^2)^{m/2}$, we get

$$(1-x^2)\frac{\partial^2 G}{\partial x^2} - 2(m+1)x\frac{\partial G}{\partial x} + \left\{-m + \frac{m^2x^2}{1-x^2} + \beta - \frac{m^2}{1-x^2}\right\}G = 0$$

or $\quad (1-x^2)G'' - 2(m+1)xG' + \left\{\beta + \frac{-m + mx^2 + m^2x^2 - m^2}{1-x^2}\right\}G = 0$

or $\quad (1-x^2)G'' - 2(m+1)xG' + \left\{\beta + \frac{-m(1-x^2) - m^2(1-x^2)}{1-x^2}\right\}G = 0$

or $\quad (1-x^2)G'' - 2(m+1)xG' + \{\beta - m(1+m)\}G = 0 \qquad ..(19)$

Let $\alpha = (m+1)$ and $\lambda = \beta - m(1+m)$ then

$$(1-x^2)G'' - 2\alpha xG' + \lambda G = 0 \qquad ..(20)$$

Now let us represent G as a power series

$$G = a_0 + a_1x + a_2x^2 + ...$$

$\therefore$ $\qquad G' = a_1 + 2a_2x + 3a_3x^2 + \dots$

and $\qquad G'' = 2a_2 + 6a_3x + 12a_4x^2 + \dots$

Substituting these values in eqn (20), we get

$$(1 - x^2)(2a_2 + 6a_3x + 12a_4x^2 + \dots) - 2\alpha x(a_1 + 2a_2x$$

$+$

$$3a_3x^2 + \dots) + \lambda(a_0 + a_1x + a_2x^2 + \dots) = 0$$

or $\quad (2a_2 + \lambda a_0) + \{6a_3 - (\lambda - 2\alpha)a_1\}x + \{12a_4 + (\lambda - 4\alpha - 2)a_2\}x^2$

$\dots + [(n + 1)(n + 2)a_{n+2} + \{\}\lambda - 2n\alpha - n(n - 1)a_n]x^n + \dots = 0$

This series will be zero for all values of x only if coefficient of each power of x vanishes separately, i.e.

$$(n + 1)(n + 2)a_{n+2}\{\lambda - 2n\alpha - n(n - 1)\}a_n = 0$$

(where $n = 0, 1, 2, 3, \dots$)

or $\qquad a_{n+2} = \dfrac{2n\alpha + n(n-1) - \lambda}{(n+1)(n+2)} a_n$ $\qquad ..(21\,i)$

Substituting the values of α and λ, we get

$$a_{n+2} = \frac{(n+m)(m+m+1) - \beta}{(n+1)(n+2)} a_n \qquad ..(21\,ii)$$

For the wave function to satisfy the boundary condition (i.e. for the wave function to be finite) $G(x)$ should be finite, i.e. it must be a polynomial which vanishes after a definite number of terms. This is possible only when numerator of the above eqn is zero after n terms.

i.e. $\qquad (n + m)(n + m + 1) - \beta = 0$

or $\qquad \beta = (n + m)(n + m + 1)$

Since m is either zero or an integer, and n is also either zero or an integer, therefore, we can put $n + m = l$ (another constant) which is also either zero or an integer, i.e.

$$\beta = l(l + 1), \ l = 0, 1, 2, 3, \dots \qquad ..(22)$$

We know that

$$\beta = \frac{8\pi^2 IE}{h^2} \qquad \therefore \ \frac{8\pi^2 IE}{h^2} = l(l + 1)$$

so that $\quad E = l(l + 1)\dfrac{h^2}{8\pi^2 I}$

Replacing l by J, we get

$$E_r = \frac{h^2}{8\pi^2 I} J(J + 1) \qquad \text{where} \ \ J = 0, 1, 2 \dots$$

$$..(23)$$

This equation relates the allowed rotational energies to a molecular property I and a quantum number J. If this energy E_r is converted to the units of rotational term values, it becomes

$$F(J) = \frac{E_r}{hc} = \frac{h}{8\pi^2 Ic} J(J+1) \qquad ..(24)$$

$F(J)$ is called rotational term (having units m^{-1})
We can rewirte eqn (24) as

$$F(J) = BJ(J+1) \ m^{-1}, \ J = 0, 1, 2, ... \qquad ..(25)$$

where B is called *rotational constant* and is given by

$$B = \frac{h}{8\pi^2 Ic} \ m^{-1} \qquad ..(26)$$

Degeneracy

If N_0 is the Avogadro number, then the number of molecules N_J occupying Jth state of energy E_J, is given by the Boltzmann distribution

$$N_J = N_0 e^{-E_J/kT} \qquad ..(27)$$

at an absolute temperature T kelvin, k is Boltzmann constant. If the degeneracy is taken into account, there are $(2J+1)$ states having the same energy E_J, so that

$$N_J = (2J+1) N_0 e^{-E_J/kT} \qquad ..(28)$$

Since $\qquad \dfrac{E_J}{hc} = BJ(J+1)$

hence $\quad N_J = (2J+1) N_0 e^{-BhcJ(J+1)/kT} \qquad ..(29)$

Frequency of Spectral Lines

If a rotational transition occurs from an upper level (having rotational quantum number J') to a lower level (J'') then frequency of spectral line expressed in wave numbers is

$$\bar{\nu}_r = \frac{E_{r'} - E_{r''}}{hc}$$

$$= \frac{h}{8\pi^2 Ic} \left[J'(J'+1) - J''(J''+1) \right] \qquad \text{using eqn (24)}$$

$$= B[J'(J'+1) - J''(J''+1)] \qquad ..(30)$$

A molecule must have a dipole moment to give rise to rotational spectra otherwise it cannot give energy to or take energy from radiation to produce spectrum.

Spectral Transitions and Selection Rules

Let us denote two energy states by subscripts m and n. Then according to quantum mechanics, the probability of transition between these two states accompanied by the absorption or emission of dipole radiation is given by the matrix element

$$P_{mn}(x) = \int \psi_m^* \left(\Sigma_j \in_j x_j \right) \psi_n d\tau$$

where ψ_m and ψ_n are eigen functions of the two states, $\in_j$ is the electric charge and x_j, the x coordinate of the jth atom in the molecule, $\sum_j \in_j x_j$ (summation over all the atoms in the molecule) gives the component of the electric dipole moment in the direction of x-axis, i.e. μ_x so

$$P_{mn}(x) = \int \psi_m^* \mu_x \psi_n d\tau \qquad ..(31)$$

A molecule which is symmetrical in ground state and so not possessing dipole moment will not interact with the radiation. As a result, molecule will not yield a spectrum unless an electric moment is produced by disturbing the symmetry of electrons or the nuclei.

For a rigid rotator, if we put the appropriate eigen functions of upper and lower states in eqn (31), assuming $\mu_x \neq 0$, it will be found that $P_{mn}(x)$ is zero unless $J' - J'' = \pm 1$. That is only those transitions are permitted which involve an increase or decrease of unity in the rotational quantum number. Thus we have the selection rule.

$$\Delta J = J' - J'' = \pm 1 \qquad ..(32)$$

For the rotator $J' > J''$ (since J' refers to the upper state), and therefore, considering only $\Delta J = +1$, we have for the emitted or absorbed lines of the rotator

$$\overline{\nu}_r = F(J' + 1) - F(J') \qquad ..(33)$$

$$= B(J'' + 1)(J'' + 2) - BJ''(J'' + 1)$$

$$= 2B(J'' + 1); \quad J'' = 0, 1, 2, \dots$$

Writing J in place of J'' for lower state, we get

$$\overline{\nu}_r = 2B(J + 1); \quad J = 0, 1, 2 \dots \qquad ..(34)$$

Putting the values for J, we find that the frequencies for consecutive lines in the pure rotation spectrum of a diatomic molecule are $2B$, $4B$, $6B$, .., etc. Therefore, on the wave number (or frequency) scale, the lines are equidistant. This is depicted in Fig. 9.2.

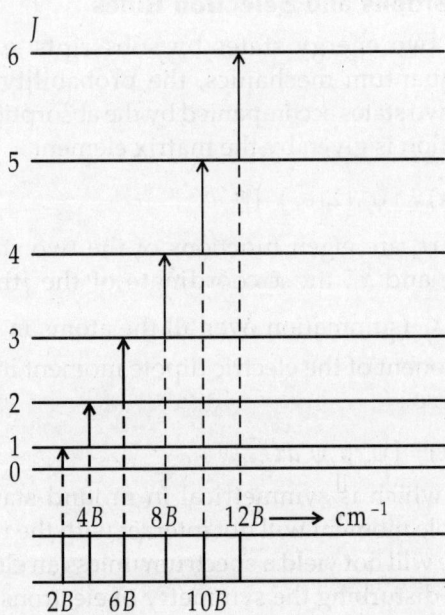

Fig. 9.2: Allowed transitions between rotational energy levels of a diatomic molecule and the resulting absorption spectrum

The Rotational Frequency

The defining equations for the energy and angular momentum of a rigid rotator are

$$E_r = \frac{1}{2}I\omega^2 \quad \text{and} \quad P_r = I\omega \quad \text{respectively}$$

But
$$\frac{E_r}{hc} = BJ(J+1) \qquad \qquad ..(35)$$

or
$$E_r = BhcJ(J+1)$$

∴
$$\frac{1}{2}I\omega^2 = BhcJ(J+1)$$

or
$$\omega = \sqrt{\frac{2Bhc}{I}}\sqrt{J(J+1)} \qquad \qquad ..(36)$$

Using $B = \dfrac{h}{8\pi^2 Ic}$, we have $\dfrac{h}{I} = 8\pi^2 Bc$.

Putting the value of h/I in eqn (36), we get angular frequency

$$\omega = \sqrt{16\pi^2 B^2 c^2} \sqrt{J(J+1)}$$

$$\omega = 4\pi Bc\sqrt{J(J+1)} \qquad ..(37)$$

or rotational frequency

$$\nu_r = \frac{\omega}{2\pi} = 2Bc\sqrt{J(J+1)} \qquad ..(38)$$

or $\qquad \nu_r \simeq c \cdot 2BJ$

i.e. the rotational frequency in any given state of the rotator is approximately equal to the frequency of the spectral line that has this state as upper state.

9.2 ISOTOPE EFFECT IN ROTATIONAL SPECTRA

When a particular atom in a diatomic molecule is replaced by its isotope, the molecule will be chemically identical with the original one and the nature of the chemical bond will remain unchanged. As the internuclear distances in molecules are practically determined by the structure of the chemical bond, the isotopic molecules should have the same internuclear distances. This is also confirmed by various experimental results. However (due to isotopic substitution), the reduced masses in the two cases will be different, consequently the rotational constant

$$B = \frac{h}{8\pi^2 \mu r^2 c}$$

is different. This would reflect a change in the energy levels. The heavier species will show a smaller separation of $2B'e$ between the lines and the lighter ones $2Be$. Thus,

$$\frac{B'}{B} = \frac{\mu}{\mu'} = \frac{\dfrac{mm_1}{(m+m_1)}}{\dfrac{mm_2}{(m+m_2)}} = \frac{m_1}{m_2}\left[\frac{m+m_2}{m+m_1}\right]$$

or $\qquad \dfrac{m_1}{m_2} = \dfrac{m}{m_2} - \dfrac{B'}{B} \bigg/ \left(1 + \dfrac{m}{m_2} - \dfrac{B'}{B}\right)$

where m_1 and m_2 are masses of the two isotopes and m is the mass of the nucleus of the other molecule.

From these measurements, m_1/m_2 can be obtained with great precision. The mass ratio m/m_2 need be known moderately accurately, since it enters into both numerator and denominator. By this procedure, the mass ratio of any two isotopes can be calculated.

Example 10.1: Consider the case of carbon monoxide. We see that on going from ^{12}C^{16}O to ^{13}C^{16}O, there is a mass increase and hence, a decrease in the B value, i.e. $B' < B$ where B' corresponds to ^{13}C^{16}O and B to ^{12}C^{16}O. This change is reflected in the transitions, as depicted in Fig. 9.3.

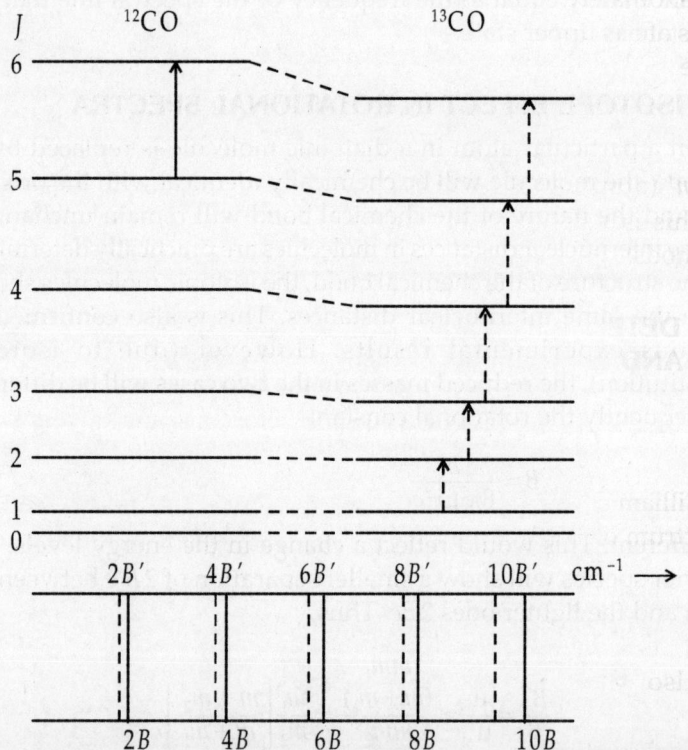

Fig. 9.3: The effect of isotopic substitution on the energy levels and hence rotational level of carbon monoxide (In this figure, the solid lines correspond to 2B, 4B, 6B, 8B, and 10B whereas dashed lines correspond to 2B', 4B', 6B', 8B', and 10B'.)

Evaluation of Atomic Weights

Gilliam *et al.* found the first rotational absorption of $^{12}C\,^{16}O$ to be at 3.84235 cm^{-1} while that of $^{13}C\,^{16}O$ was at 3.67337 cm^{-1}. The values of B determined from these figures are

$$B = 1.92118 \text{ cm}^{-1}$$

and $$B' = 1.83669 \text{ cm}^{-1}$$

where the prime refers to $^{13}C\,^{16}O$.

We have,

$$\frac{B'}{B} = \frac{8\pi^2 Ic}{h} \times \frac{h}{\pi^2 I'C} = \frac{\mu}{\mu'} = \frac{1}{1.046}$$

where μ is reduced mass and it is assumed that the internuclear distance remains unchanged by isotopic substitution. Taking the mass of oxygen to be 15.9994 and that of carbon-12 to be 12.0000 we have

$$\frac{\mu'}{\mu} = 1.046 = \frac{15.9994\, m'}{(15.9994 + m')} \times \frac{(12 + 15.9994)}{12 \times 15.9994}$$

$\Rightarrow$ m' (atomic weight of ^{13}C) = 13.0007

This is within 0.02% of the best value obtained from other methods.

9.3 DETERMINATION OF MOMENT OF INERTIA AND BOND LENGTH

$$\overline{\nu}_{J \to J+1} = 2B(J+1) \text{ cm}^{-1} \qquad ..(39)$$

$\therefore$ $$\overline{\nu}_{0 \to 1} = 2B \text{ cm}^{-1}$$

Gilliam *et al.* have measured the first line ($J = 0$) in the rotation spectrum of CO as 3.84235 cm^{-1}. Hence for CO

$$\overline{\nu}_{0 \to 1} = 3.84235 = 2B \text{ cm}^{-1}$$

$\Rightarrow$ $$B = 1.92118 \text{ cm}^{-1}$$

Also $$B = \frac{h}{8\pi^2 Ic} \text{ cm}^{-1} \qquad ..(40)$$

$\Rightarrow$ $$I_{CO} = \frac{6.626 \times 10^{-34}}{8\pi^2 \times 2.9979 \times 10^{10} \times B}$$

$$= \frac{27.9907 \times 10^{-47}}{B} \text{ kg m}^2$$

$$= 14.5695 \times 10^{-47} \text{ kg m}^2$$

Here we have expressed the velocity of light in cm/s since B is in cm^{-1}. But moment of inertia $I = \mu r^2$.

Knowing the relative atomic weights (H = 1.0080) to be C = 12.0000; O = 15.9994 and the absolute mass of the hydrogen atom to be 1.67343×10^{-27} kg, we can calculate the masses of carbon and oxygen as 19.92168×10^{-27} and 26.56136×10^{-27} kg respectively. Consequently,

$$\mu = \frac{19.92168 \times 26.56136 \times 10^{-54}}{46.48303 \times 10^{-27}}$$

$$= 11.38365 \times 10^{-27} \text{ kg}$$

Hence $\quad r^2 = \dfrac{I}{\mu} = \dfrac{14.5695 \times 10^{-47}}{11.3865 \times 10^{-27}}$

$$= 1.2799 \times 10^{-20} \text{ m}^2$$

$\Rightarrow \qquad r_{CO} = 0.1131$ nm (or 1.131 Å)

From eqn (40) $B = \dfrac{h}{8\pi^2 I c} = \dfrac{h}{8\pi^2 \mu r^2 c}$

$$\Rightarrow \qquad\qquad B \propto \frac{1}{r^2} \qquad\qquad\qquad ..(41)$$

Above example shows that internuclear distances can be calculated from B values obtained from spectroscopic data.

$$\overline{\nu}_{J \to J+1} = 2B(J+1) \text{ cm}^{-1} \qquad\qquad ..(42)$$

9.4 NON-RIGID ROTATOR

Experimental investigations show that the separation between successive lines and consequently the apparent B-value decreases steadily with increasing J. The reason for this is as follows:

At high angular velocities, molecular bonds are not perfectly rigid but are elastic and a centrifugal distortion occurs resulting in execution of harmonic vibrations along with rotation.

In order to detemine the effect of bond elongation on rotational energy levels, we consider a diatomic molecule held together by a bond that (ideally) obeys Hooke's law (it is assumed that vibrations are within the elastic limit). Let k denote the force constant, the restoring force is $k(R - R_0)$, where R is stretched length and R_0 is the equilibrium separation. This restoring force provides the centripetal force for rotation. (The centrifugal force is due to point masses separated by R_0). Let μ denote the effective mass, then centripetal force is $\mu\omega^2 R$ so that

$$k(R - R_0) = \mu\omega^2 R$$

or
$$R = \frac{kR_0}{k - \mu\omega^2} \qquad \qquad ...(43)$$

The total energy of the rotating molecule is its kinetic energy $\frac{1}{2}I\omega^2$ minus the elastic potential energy $\frac{1}{2}k(R - R_0)^2$.

Hence
$$E = (1/2)\,I\omega^2 - (1/2)\,k(R - R_0)^2$$

$$= \frac{1}{2}I\omega^2 - \frac{1}{2}\frac{\mu^2\omega^4 R^2}{k}$$

$$= \frac{1}{2}I\omega^2 - \frac{1}{2}\frac{\left(I\omega^2\right)^2}{kR^2}$$

$$= \frac{L^2}{2I} - \frac{L^4}{2I^2R^2k} \qquad \qquad ...(44)$$

where $L(= I\omega)$ is the magnitude of the angular momentum of the molecule. Since L can have only the values $\sqrt{J(J+1)}\hbar$, the rotational energy levels are specified by

$$E_J = \frac{J(J+1)\hbar^2}{2I} - \frac{J^2(J+1)^2\hbar^4}{2I^2R^2k} \qquad ..(45)$$

In this expression, both the moment of inertia I and the bond length R refer to the rotating molecule, so these two are different for either value of J.

However, we are interested in an expression for E_J in terms of the reduced (effective) mass μ and R_0.

With the help of eqn (43) and the fact that $I\omega = \mu R^2\omega = \sqrt{J(J+1)}\hbar$, we find that

$$E_J \simeq \frac{J(J+1)\hbar^2}{2\mu R_0^2} - \frac{J^2(J+1)^2\hbar^4}{2\mu^2 R_0^6 k} \qquad ..(46)$$

Comparison with equation $\qquad (\because I = \mu R_0^2)$

$$E_r = \frac{\hbar^2 J(J+1)}{2I} \qquad \qquad ..(47)$$

reveals that the rotational energy levels of a non-rigid molecule are lower than the corresponding levels of a rigid molecule with the increasing J.

Due to vibrational energy, if the motion is assumed to be simple harmonic, the force constant is given by

$$k = 4\pi^2 \bar{\omega}^2 c^2 \mu \qquad ..(48)$$

where $\bar{\omega}$ is the vibration frequency expressed in cm^{-1}. Plainly the variation of B with J is determined by the force constant, the weaker the bond, the more readily will it distort under centrifugal forces. This gives the first order effect due to periodic vibration with a certain fundamental frequency.

The second consequence of elasticity is that the quantities r and B vary during a vibration. Then we need to define three different sets of values for B and r. At the equilibrium separation r_e between the nuclei, the rotational constant is Be, in the vibrational ground state, the average internuclear separation is r_0 associated with a rotational constant B_0, while if the molecule has excess vibrational energy, the quantities are r_v and B_v, where v is the vibrational quantum number.

Energy Levels

The Schrodinger wave equation may be set up for a non-rigid molecule and the rotational energy levels are found to be

$$E_J = \frac{h^2}{8\pi^2 I} J(J+1) - \frac{h^4}{32\pi^2 I^2 r^2 k} J^2(J+1)^2 \text{ Joules} \qquad ..(49)$$

or
$$F(J) = \frac{E_J}{hc} = BJ(J+1) - DJ^2(J+1)^2 \text{ cm}^{-1} \qquad ..(50)$$

where B is the rotational constant and D is called the *centrifugal distortion constant* given by

$$D = \frac{h^3}{32\pi^2 I^2 r^2 kc} \text{ cm}^{-1} \qquad ..(51)$$

It is a positive quantity.

Equation (51) is true when force is simple harmonic. If force field is anharmonic, the expression becomes

$$F(J) = BJ(J + 1) - DJ^2(J + 1)^2 + HJ^3(J + 1)^3$$
$$+ KJ^4(J + 1)^4 + ...\text{cm}^{-1} \qquad ..(52)$$

where H, K, etc. are small constants dependent upon the geometry of the molecule.

From the defining expressions for B and D, we find that

$$D = \frac{16B^3 \pi^2 \mu c^2}{k} = \frac{4B^3}{\bar{\omega}^2} \qquad ..(53)$$

where $\bar{\omega}$ is the vibrational frequency of the bond in normal state and k has been expressed according to eqn (48). Vibrational frequencies $\bar{\omega}$ are usually of the order of 10^3 cm^{-1} while B is of the order of 10 cm^{-1}. Thus D is of the order of 10^{-3} cm^{-1} and is very small compared to B. For small J therefore, the correction term $DJ^2(J + 1)^2$ is almost negligible while for J values of 10 or more, it may become appreciable.

Spectrum: The figure shows much exaggerated, the lowering of rotational levels when passing from the rigid to the non-rigid diatomic molecule. The spectra are also compared. It should be noted that the selection rule is still $\Delta J = \pm 1$.

We may easily write an analytical expression for the transitions

$$F(J+1) - F(J) = \bar{v}_J = B[(J+1)(J+1) - J(J+1)]$$
$$-D\left[(J+1)^2(J+2)^2 - J^2(J+1)^2 \right]$$
$$= 2B(J+1) - 4D(J+1)^3 \text{ cm}^{-1} \qquad \ldots(54)$$

where $\bar{v}_J$ represents both upward transition from J to $(J + 1)$ or downward, from $(J + 1)$ to J.

Thus we see analytically and from Fig. 9.4 that the spectrum of the elastic rotor is similar to that of the rigid molecule except that each line is displaced slightly to low frequency, the displacement increasing with $(J + 1)^3$.

A knowledge of D provides two useful pieces of information:

Firstly it allows us to determine the J value of lines in an observed spectrum and secondly, it enables us (although rather not precisely) to determine the vibrational frequency of a diatomic molecule. For hydrogen fluoride, eqn 54 has the value

$$\bar{\omega}^2 = \frac{4B^3}{D} = 16.33 \times 10^6 \left(\text{cm}^{-1} \right)^2$$

i.e. $\bar{\omega} = 4050$ cm^{-1}

The force constant then follows directly

$$k = 4\pi^2 c^2 \bar{\omega}^2 \mu = 960 \text{ N m}^{-1}$$

which indicates, as expected, that H-F is a relatively strong bond.

9.5 ROTATION OF MOLECULES

A molecule is equivalent to a three dimensional rigid body characterised by three principal moments of inertia I_a, I_b and I_c. The

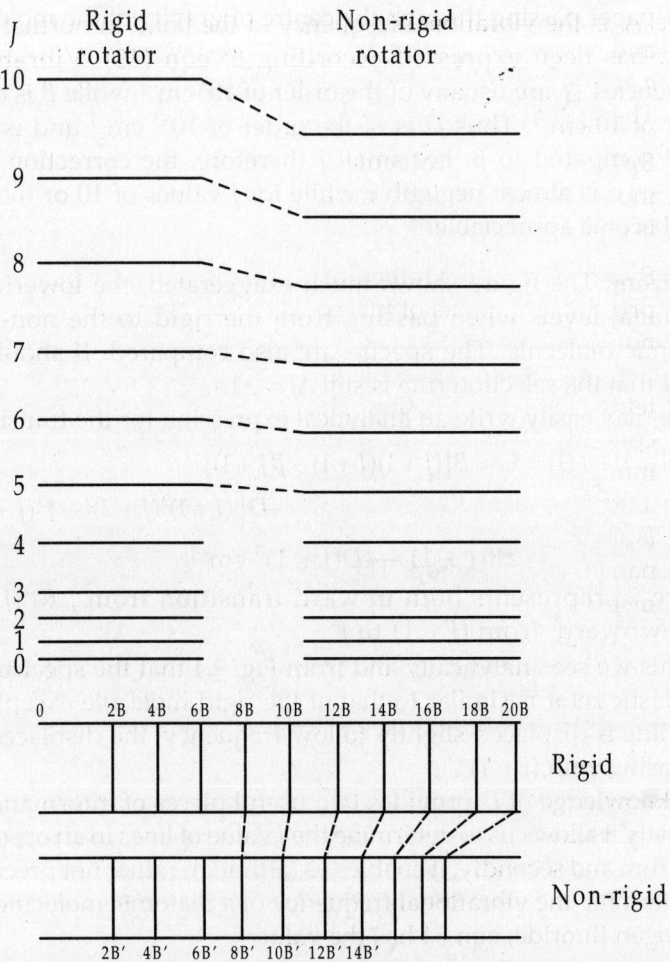

Fig. 9.4: The change in rotational energy levels and rotational spectrum when passing from a rigid to a non-rigid diatomic molecule. Levels on the right are given by $D = 10^{-3}B$

three axes a, b and c, in general are so selected that $I_a < I_b < I_c$. Based on the relative values of I_a, I_b and I_c, the molecules are classified into following four categories:

(i) Linear Molecules: In this case, all the atoms of the molecule are arranged in a straight line, e.g. HCl, CO_2 and C_2H_2. The three mutually perpendicular axes of rotation may be chosen as a—the molecular axis; b—the direction in the plane of the paper passing through the centre of gravity of the molecule and perpendicular to a-axis and c-axis perpendicular to both a and b axes; through the centre of gravity of the molecule. Then $I_a = 0$ and $I_b = I_c$

(*ii*) **Spherical Top Molecules:** When all the three principal moments of inertia of a molecule are equal (i.e. $I_a = I_b = I_c$), it is called a spherical top. Examples are SF_6, CH_4, CCl_4, etc.

(*iii*) **Symmetric Top Molecules:** In a symmetric top, the two principal moments of inertia are equal and all the three are non-zero. Examples are CH_3Cl, NH_3 and BF_3.

Consider the molecule CH_3Cl, in which the carbon has tetrahedral coordination. The C-Cl bond axis is the a-axis (it contains the centre of gravity of the molecule). The two mutually perpendicular axes (viz. b and c axes) lie in a plane $\perp$ to the a-axis. Obviously $I_b = I_c$. A molecule of this type (spinning about the a-axis) is like a spinning top, hence the name symmetric top. These are of two types: prolate symmetric top ($I_a < I_b = I_c$) and oblate symmetric top ($I_a = I_b < I_c$).

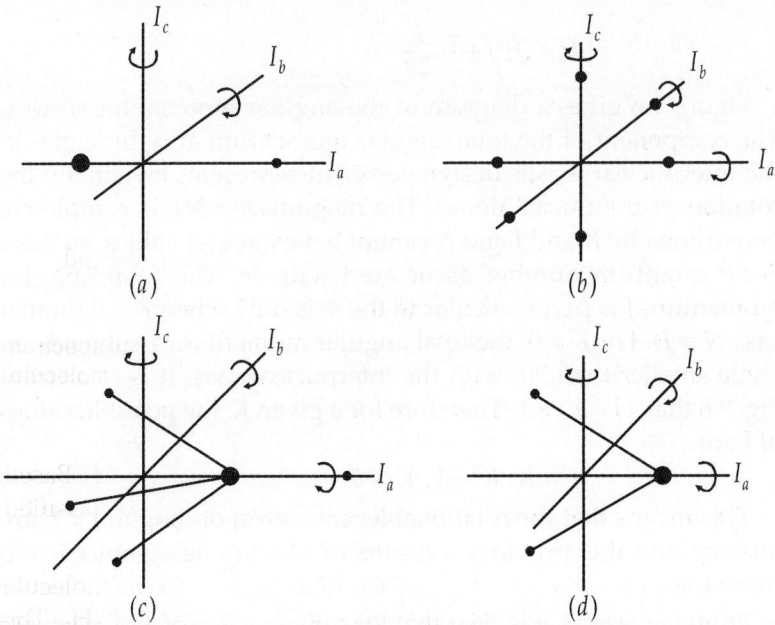

Fig. 9.5: Rotation of molecules about the principal axes. (a) Linear, (b) Spherical top, (c) Symmetric top and (d) Asymmetric top

(*iv*) **Asymmetric Top Molecules:** In these molecules, $I_a \neq I_b \neq I_c$. A few examples are H_2O, CH_3OH. The majority of the molecules belong to this category.

9.6 THE DIATOMIC MOLECULE AS A SYMMETRIC TOP

A diatomic molecule is considered equivalent to a simple rigid rotator with the tacit assumption that the moment of inertia about the line joining the nuclei is zero. Actually, due to a number of electrons revolving about the two nuclei, the moment of inertia about the line joining the nuclei is not exactly zero, rather it has a small magnitude characterised by an angular momentum K.

and $\qquad |K| = K\dfrac{h}{2\pi}$

where K is the quantum number of the angular momentum of the electrons about the internuclear axis. (The system is like a dumbbell carrying a flywheel on its axis—a more general case of a symmetric top). The total angular momentum designated by J can take only the values

$$|J| = \sqrt{J(J+1)}\,\frac{h}{2\pi}$$

Figure 9.6 gives a diagram of the angular momentum vectors. The component of the total angular momentum at right angles to the internuclear axis is designated N (it represents essentially the rotation of the nuclei alone). The magnitude $|N|$ is completely determined by K and J and N cannot have integral values, so there is no quantum number associated with N. The total angular momentum J is perpendicular to the axis only when $K = 0$ (in that case $N \equiv J$). For $K \neq 0$, the total angular momentum J subtends an angle smaller than $90°$ with the internuclear axis. It is clear from Fig. 9.6 that $|J| \geq |K|$. Therefore for a given K, the possible values of J are

$$J = K, K+1, K+2, \ldots \qquad\qquad ..(55)$$

This means that the rotational levels corresponding to $J < K$ are missing and this provides a means of identifying symmetric-top molecules.

From Fig. 9.6 (*b*), it is clear that for a given value of $|J|$, there are two modes of motion of the system, corresponding to two directions K and $-K$.

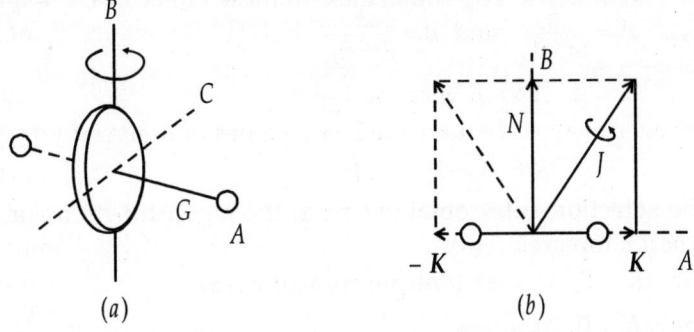

Fig. 9.6: (a) Diatomic molecule as symmetric top (b) vector diagram for the symmetric top. The curved arrow (Fig. b) indicates the rotation of whole diagram about **J**. The dotted part of the figure gives the vector diagram when the sense of the direction of **K** is reversed

Energy Levels

The total energy of rotation is given by

$$E = \frac{1}{2} I_A \omega_A^2 + \frac{1}{2} I_B \omega_B^2 + \frac{1}{2} I_C \omega_C^2$$

$$= \frac{1}{2} I_A \omega_A^2 + I_B \omega_B^2 \quad (\because I_B = I_C, \ \omega_B = \omega_C)$$

$$= \frac{L_A^2}{2 I_A} + \frac{L_B^2}{I_B}$$

Now $L_A = K\hbar$ and $L_B^2 + L_C^2 = 2L_B^2 = N^2\hbar^2 = \left[J(J+1) - K^2 \right]\hbar^2$

$$\therefore \quad E = \frac{1}{2 I_A} K^2 \hbar^2 + \frac{1}{I_B} \times \frac{1}{2} \left[J(J+1) - K^2 \right]\hbar^2$$

$$= \left[\frac{K^2}{2 I_A} + \frac{J(J+1) - K^2}{2 I_B} \right]\hbar^2 \ \text{ergs} \qquad ..(56)$$

In wave number units, we have

$$F(J) = \frac{E}{hc} = \left[\frac{K^2}{I_A} + \frac{J(J+1) - K^2}{I_B} \right]\frac{h}{8\pi^2 c} \ \text{cm}^{-1}$$

$$= \left[\frac{J(J+1)}{I_B} + \left(\frac{1}{I_A} - \frac{1}{I_B} \right)K^2 \right]\frac{h}{8\pi^2 c} \ \text{cm}^{-1} \qquad ..(57)$$

or $\qquad F(J) = BJ(J + 1) + (A - B)K^2 \qquad ..(58)$

where $\quad A = \dfrac{h}{8\pi^2 c I_A}$ and $B = \dfrac{h}{8\pi^2 c I_B}$

and $\qquad A \gg B$ because $I_A \ll I_B$.

For every value of J there are $2J + 1$ values of K given by

$$K = 0, \pm 2, .., \pm J \qquad ..(59)$$

The selection rules obtained from the eigen functions of the symmetric top are:

For $\quad K = 0, \Delta J = \pm 1$ (simple rotator case)

For $\quad K \neq 0; \Delta J = 0, \pm 1$

i.e. with $K \neq 0$, in addition to the rotation vibration transitions with $\Delta J = \pm 1$, the transition $\Delta J = 0$ also appear. The $\Delta J = \pm 1$ transitions produce an $R-$ and $P-$branch while $\Delta J = 0$ produces an additional branch called $Q-$branch.

9.7 ASYMMETRIC TOP

For asymmetric top molecules, the $\pm K$ degeneracy which exists for the symmetric top molecule is broken in general. It is, in general, not possible to derive an expression for the rotational energy of an asymmetric rotator. Hence, their spectra are very complex. The prolate and oblate symmetric rotators represent the two extreme limits of an asymmetric rotator. The moment of inertia relations are

$$\begin{aligned} I_a &\neq I_b \neq I_c \quad \text{(asymmetric rotator)} \\ I_a &< I_b \simeq I_c \quad \text{(near prolate asymmetric rotator)} \\ I_a &\simeq I_b < I_c \quad \text{(near oblate asymmetric rotator).} \end{aligned} \qquad ..(60)$$

9.8 STARK EFFECT IN MOLECULES

In presence of an external electric field, each energy level (in rotational spectra) splits up and is shifted. This is known as Stark effect. The field exerts a torque on the molecular dipole moment thereby changing its rotational motion which leads to Stark splitting.

If the dipole moment has a component along the angular momentum J, a first order Stark effect is observed. If the dipole moment is $\perp$ to the angular momentum J, a second order Stark effect is obtained.

In symmetric top molecules, the permanent dipole moment μ is always parallel to the symmetry axis. The component of dipole moment μ_c along the angular momentum vector will be

$$\mu_c = \frac{\mu K}{\sqrt{J(J+1)}} \qquad ..(61)$$

(except $K = 0$). The energy E of such a molecule in an electric field ε is given by

$$E = -\mu_c \varepsilon \cos\theta$$
$$= -\frac{\mu K \varepsilon \cos\theta}{\sqrt{J(J+1)}} \qquad ..(62)$$

where θ is the angle between J and ε. Further,

$$M_J \hbar = \sqrt{J(J+1)} \cos\theta \, \hbar$$

$$\Rightarrow \qquad \cos\theta = \frac{M_J}{\sqrt{J(J+1)}} \qquad ..(63)$$

So, the energies are given by

$$E = -\frac{\mu \varepsilon K M_J}{J(J+1)} \qquad ..(64)$$

Transitions occur for selection rules

$$\Delta J = \pm 1, \; \Delta K = 0 \qquad ..(65)$$

If the radiation field and electric field are parallel, then $\Delta M_J = 0$; if they are perpendicular then $\Delta M_J = \pm 1$.

For a $J \rightarrow J + 1$ transition, absorption frequency will be

(i) For $\Delta M_J = 0$,

$$\bar{\nu} = 2B_e(J+1) + \frac{2M_J K \mu \varepsilon}{J(J+1)(J+2)h} \qquad ..(66)$$

(ii) For $\Delta M_J = \pm 1$

$$\bar{\nu} = 2B_e(J+1) + \frac{(2M_J \mp J)K \mu \varepsilon}{J(J+1)(J+2)h} \qquad ..(67)$$

This is first order Stark effect. The above relations hold under the assumption that dipole moment is independent of the electric field.

9.9 MICROWAVE SPECTROMETER

A microwave spectrometer requires following parts: (i) source (ii) measurement of frequency (iii) guidance of radiation to absorbing substance (*iv*) sample cell and (*v*) detector. Schematic representation of a microwave spectrometer is shown in Fig. 9.7.

Source

A Klystron emitting monochromatic microwave radiation may be used. In place of Klystron, backward wave oscillator which can be tuned electronically over a wide range of frequencies can also be used.

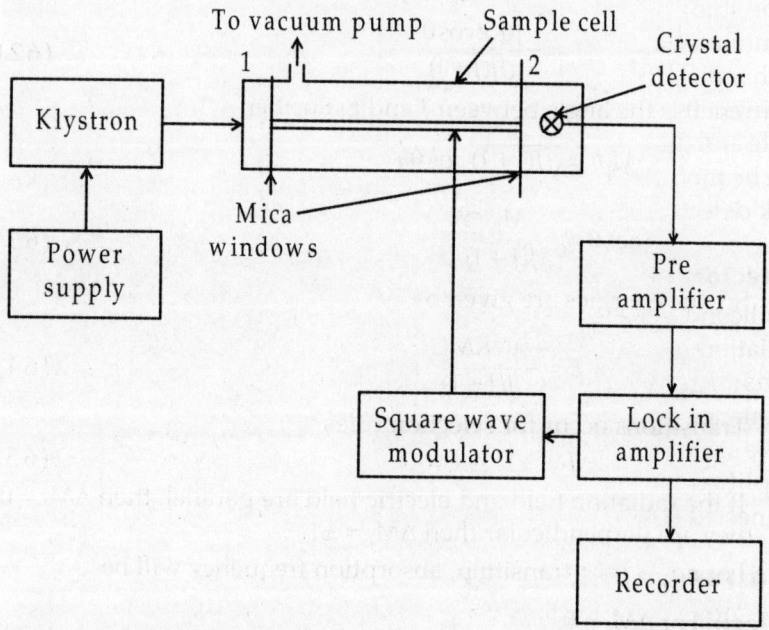

Fig. 9.7: Schematic representation of a microwave spectrometer

Frequency Measuring Device

Cavity wavemeters (accuracy ~±1 to ±5 MHz) are used to get a rough estimate of the frequency of microwave radiation.

Guidance of the Radiation to the Cell

Microwave radiation from the source can be transmitted to the sample cell through rectangular wave guides. Waveguides of different dimensions are needed depending on the range of microwave radiation used, e.g. *X* band (8–12.4 GHz), *K* band (12.4–18 GHz). A small fraction of power is to be coupled out from the main transmission line for frequency measurement or monitoring power. For this purpose, directional coupler or magic-tre may be used.

Sample Cell

One of the commonly used type of cell is the Stark cell. It consists of long (3 m to 4 m) rectangular waveguide with ends sealed off by means of mica windows (1 and 2). It can be evacuated and sample to be studied can be placed inside it. In Stark cell, a flat metal strip is mounted half way between the broad faces of the waveguide with a dielectric insulating the plate from the metallic waveguide. By injecting a 0–2000 V square wave potential between the metallic plate and the waveguide, the resonant frequencies of the molecules can be modulated via the Stark effect. This is followed by a phase-lock detector which responds only to molecular resonances.

Detector

A silicon crystal detector is most commonly used. The incoming radiation gives rise to a DC current. In Stark spectrometers, the transmitted radiation is modulated only when a molecular resonance (i.e. absorption) occurs. On absorption, a small square wave is imprinted on the top of the DC signal. This AC signal is amplified and detected by a phase sensitive detector which is connected to a recorder (or oscilloscope).

9.10 INFORMATION FROM ROTATIONAL SPECTRA

Molecular Structure

We can derive details of molecular structure. In a diatomic molecule, there is only one observable principal moment of inertia which gives the interatomic distance in terms of the atomic masses:

$$I = \mu r^2, \quad r = \left[I \left(\frac{1}{m_1} + \frac{1}{m_2} \right) \right]^{1/2}$$

Here m_1 and m_2 are the masses of the two atoms.

The force constant of a bond can also be calculated from the values of rotational constant B and rotation distortion constant D.

$$\overline{v}^2 = \frac{4B^3}{D}, \quad \overline{v} = \frac{1}{2\pi c} \sqrt{\frac{k}{\mu}}$$

The force constant is given by

$$K = \frac{16\pi^2 \mu B^3 c^2}{D} \qquad ..(68)$$

Dipole Moment

Precise determination of electric dipole moment is possible from microwave spectroscopy by measurement of the Stark effect. Stark effect data gives the value of dipole moment in the appropriate vibrational level. Dipole moments can also give information regarding nature of molecular bonds.

Atomic Mass

The very high resolution possible in microwave spectroscopy enables us to determine the atomic masses. The rotational spectra of molecules and isotopes give B and B':

$$B = \frac{h}{8\pi^2 Ic}; \quad B' = \frac{h}{8\pi^2 I' c}$$

$$\frac{B}{B'} = \frac{I'}{I} = \frac{\mu'}{\mu}$$

From the known value of μ, we can get μ' (the mass of the isotope).

Nuclear Quadrupole Moment

Measurement of the quadrupole hyperfine structure in molecules gives the quadrupole coupling constant (using microwave spectrometers).

Questions

9.1 What is microwave spectroscopy? What is the difference between infrared and microwave spectroscopy?

9.2 Discuss the theory of microwave spectroscopy.

9.3 Distinguish between symmetric top, spherical top and asymmetric top molecules

9.4 What is centrifugal distortion? Explain the effect of centrifugal distortion on the moment of inertia and energy of a diatomic molecules.

9.5 Explain the effect of isotopic substitution on the rotational spectra of molecules.

9.6 The bond length of HF molecule is 0.0927 nm. What is the moment of inertia of the HF molecule? What is the value of rotational constant in joules and in cm^{-1}?

$$(\textbf{Ans:} \ 1.356 \times 10^{-47} \ \text{kg m}^2, \ 40.97 \times 10^{-23} \ \text{J}, \ 20.61 \ \text{cm}^{-1})$$

9.7 Discuss clearly the principal features of the rotation band spectrum of a diatomic molecule.

9.8 The moment of inertia of HCl is 2.66×10^{-40} gm-cm^2. Estimate the energy difference between the rotational levels $J = 0$ and $J = 1$ ($h = 6.62 \times 10^{-27}$ erg-s). $\hspace{1cm}$ (**Ans:** 2.6×10^{-3} eV)

Vibration of Molecules

10.1 THE VIBRATING DIATOMIC MOLECULE AS A SIMPLE HARMONIC OSCILLATOR

We are familiar with the vertical oscillations of a mass m connected to a stretched spring of force constant k whose other end is fixed at the top. The mass executes simple harmonic motion with a characteristic frequency

$$\nu_0 = \frac{1}{2\pi}\sqrt{\frac{k}{m}} \qquad \qquad ...(1)$$

We now extend this to the vibration of a system of two masses m_1 and m_2 connected by a spring which is the analogue of a diatomic molecule. If the spring is compressed and released, the system executes simple harmonic motion with fundamental frequency which depends on the reduced mass μ of the system and the force constant but is independent of amount of distortion

i.e., $$\nu_0 = \frac{1}{2\pi}\sqrt{\frac{k}{\mu}} \ \text{Hz}$$

or $$\bar{\nu}_0 = \frac{1}{2\pi c}\sqrt{\frac{k}{\mu}} \ \text{cm}^{-1} \qquad \qquad ...(2)$$

$$= 5.3\times10^{-12}\sqrt{\frac{k}{\mu}} \ \text{cm}^{-1}$$

where $\mu = \dfrac{m_1 m_2}{(m_1 + m_2)}$ ..(3)

Here k is in Newtons per meter, μ in kg and c is the velocity of light in cm/s.

Quantum mechanically, the vibrational energies of such a harmonic system are given by

$$E_v = \left(v + \frac{1}{2}\right)h\nu_0 \qquad (v = 0, 1, 2 ..)$$

..(4)

where v is the vibrational quantum number. Expressing the energy in cm^{-1} which is the commonly used unit of frequency in vibrational spectroscopy, we have

$$E_v = \frac{E_v}{hc} = \left(v + \frac{1}{2}\right)\frac{\nu_0}{c} = \left(v + \frac{1}{2}\right)\bar{\nu}_0 \ \text{cm}^{-1} \qquad ..(5)$$

as the only energies allowed to a simple harmonic vibrator. Some of these are shown in the Fig. 10.1.

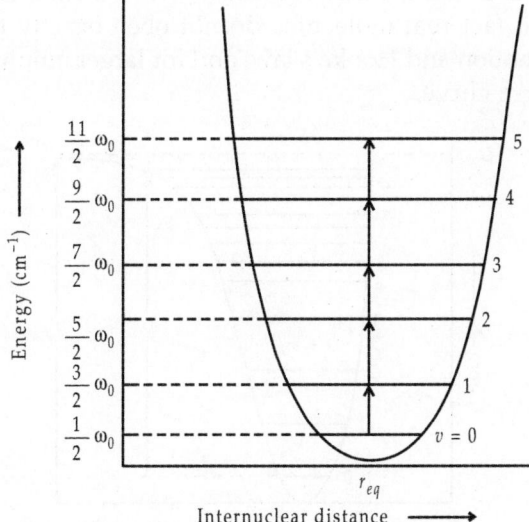

Fig. 10.1: The allowed vibrational energy levels and transitions between them for a diatomic molecule undergoing simple harmonic motion

In particular, we note the lowest vibrational energy (obtained by putting $v = 0$ in eqn (4) or (5) is

$$E_0 = \frac{1}{2} h\omega_0 \qquad \text{Joules } (\omega_0 \text{ in Hz}) \qquad ..(6)$$

or $$\varepsilon_0 = \frac{1}{2} \bar{v}_0 \text{ cm}^{-1} \left(\bar{v}_0 \text{ in cm}^{-1} \right) \qquad ..(7)$$

This implies that the diatomic molecule can never have zero vibrational energy, the atoms can never be completely at rest relative to each other. The quantity $\frac{1}{2} h \bar{v}_0$ joules or $\frac{1}{2} \bar{v}_0$ cm^{-1} is known as the *zero point energy*, it depends only on the classical vibrational frequency and hence on the strength of the chemical bond and the atomic masses. (This fact has been confirmed by experiments as well).

In diatomic molecules, the actual potential energy curve is not of the simple harmonic type but is given by

$$U(r) = D_e \{1 - \exp[-a(r - r_e)]\}^2 \qquad ..(8)$$

where D_e is the dissociation energy, a is a constant for the given molecule and r_e is the internuclear distance corresponding to $U(r) = 0$. This potential energy function was suggested by P. M. Morse and is called the *Morse function*. This function is shown in Fig. 10.2. In fact real molecules do not obey exactly the simple harmonic motion and Hooke's law, and for larger amplitudes they follow Morse curve.

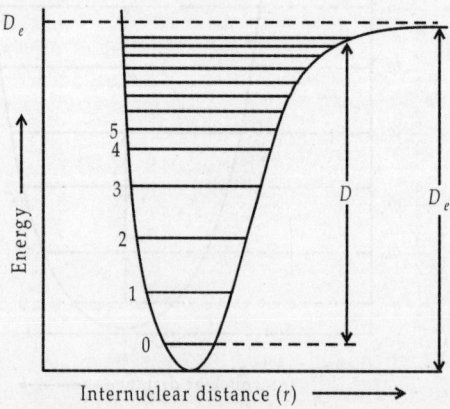

Fig. 10.2

The Fig. 10.2 corresponds to the case when one atom is considered fixed at $r = 0$ while the other oscillates between the two branches of the Morse curve. With the potential in eqn (8), the Schrodinger equation of such an anharmonic oscillator gives the allowed vibrational energy eigen values

$$\varepsilon_v = \left(v + \frac{1}{2}\right)\overline{\omega}_e - \left(v + \frac{1}{2}\right)^2 x_e\overline{\omega}_e + \left(v + \frac{1}{2}\right)^3 y_e\overline{\omega}_e + ... \quad ..(9)$$

$$(v = 0, 1, 2, ..)$$

where we have written $\overline{\omega}_e$ instead of $\overline{\nu}_e$. $\overline{\omega}_e$ is the oscillation frequency of the anharmonic system in cm^{-1}, x_e and y_e are small positive (anharmonicity) constants for bond stretching vibrations. Retaining, the first anharmonic term only, we get

$$\varepsilon_v = \left(v + \frac{1}{2}\right)\overline{\omega}_e - \left(v + \frac{1}{2}\right)^2 x_e\overline{\omega}_e \ cm^{-1} \quad ..(10)$$

$$(v = 0, 1, 2, ..)$$

As x_e is positive, the effect of anharmonicity is to crowd more closely the vibrational levels. These energy levels are also shown in Fig. 10.2. The exact zero point energy is

$$\varepsilon_0 = \frac{1}{2}\left(1 - \frac{1}{2}x_e\right)\overline{\omega}_e \quad ..(11)$$

Rewriting eqn (10) as

$$\varepsilon_v = \left[1 - \left(v + \frac{1}{2}\right)x_e\right]\left(v + \frac{1}{2}\right)\overline{\omega}_e$$

and equating it to that given in eqn (5), we get

$$\overline{\omega}_0 = \overline{\omega}_e\left[1 - \left(v + \frac{1}{2}\right)x_e\right] \quad ..(12)$$

Here $\overline{\omega}_0 (\equiv \nu_0)$ is the oscillation frequency of the harmonic oscillator having the same vibrational energy as the anharmonic one. In other words, the anharmonic oscillator behaves like the harmonic oscillator but with an oscillation frequency that decreases with increasing v. If we set $v = -1/2$, $\varepsilon_v = 0$ and $\overline{\omega}_0 = \overline{\omega}_e$. Thus, we may consider $\overline{\omega}_e$ as the hypothetical equilibrium oscillation frequency of the anharmonic system. The transition between the vibrational levels is generally observed in the region 50–4000 cm^{-1} as infrared (IR) or Raman bands when absorption or scattering of radiant energy causes a change in the energy of a molecular vibration.

Example 10.1: Calculate the approximate wave number of the fundamental absorption peak due to the stretching vibration of a carbonyl group $C = O$ given that the force constant for this bond is $\approx 1 \times 10^3$ N/m.

Solution: The mass of the carbon atom in kilogram is

$$m_1 \approx \frac{12 \times 10^{-3} \text{ kg/mole}}{6 \times 10^{23} \text{ atoms/mole}} \times 1 \text{ atom}$$

$$= 2 \times 10^{-26} \text{ kg}$$

Similarly for oxygen, $m_2 = \dfrac{16 \times 10^{-3}}{6 \times 10^{23}} = 2.7 \times 10^{-26}$ kg

and the reduced mass

$$\mu = \frac{m_1 m_2}{m_1 + m_2} = \frac{2 \times 10^{-26} \times 2.7 \times 10^{-26}}{(2 + 2.7) \times 10^{-26}} \text{ kg}$$

$$= 1.1 \times 10^{-26} \text{ kg}$$

$$k = 1 \times 10^3 \text{ N/m}$$

Thus $\bar{\nu} = 5.3 \times 10^{-12} \sqrt{\dfrac{k}{\mu}}$ cm^{-1}

$$= 5.3 \times 10^{-12} \sqrt{\frac{1 \times 10^3 \text{ N/m}}{1.1 \times 10^{-26} \text{ kg}}} \text{ cm}^{-1}$$

$$= 1.6 \times 10^3 \text{ cm}^{-1}$$

which is close to the experimental value.

Example 10.2: In the near infrared spectrum of HCl molecule, there is a single intense band at 2886 cm^{-1}. Assuming that it is due to the transition between vibrational levels, show that the force constant k is 4.8×10^5 dynes/cm, given $M_H = 1.68 \times 10^{-24}$ gm.

Solution: $\bar{\omega} = 2886$ cm$^{-1} = \dfrac{\nu_{osc}}{c}$

where ν_{osc} is the fundamental frequency. Further,

$$\nu_{osc} = \frac{1}{2\pi} \sqrt{\frac{k}{\mu}}$$

$\Rightarrow$ $k = 4\pi^2 \nu_{osc}^2 \, \mu = 4\pi^2 \mu \, \bar{\omega}^2 c^2$

Reduced mass $\qquad \mu = \dfrac{M_H M_{Cl}}{M_H + M_{Cl}} = \dfrac{1 \times 35}{1 + 35} M_H$

$$= \frac{35}{36} \times 1.68 \times 10^{-24} \text{ gm} = 1.63 \times 10^{-24} \text{ gm}$$

$\therefore \qquad\qquad k = 4 \times (3.14)^2 \times (1.63 \times 10^{-24} \text{ gm}) \times (2886 \text{ cm}^{-1})^2$
$$\times (3 \times 10^{10} \text{ cm s}^{-1})^2$$

$$= 4.8 \times 10^5 \text{ dynes/cm}.$$

10.2 VIBRATIONS OF POLYATOMIC MOLECULES

The study of vibrations of polyatomic molecules provides a background for an understanding of the absorption spectrum in the infrared region.

Number of Independent Vibrations: Imagine that a molecule contains N atoms which are held together by extremely weak bonds, the collective motion being described in terms of nearly independent motions of the constituent N atoms. If x_i, y_i, z_i are the coordinates of ith atom, there will be a total of $3N$ such coordinates, either of these will represent a degree of freedom, so that there will be $3N$ degrees of freedom.

For actual molecules, however, the $3N$ coordinates are not in fact convenient. It is convenient to describe their motions in terms of the motion of their centre of mass. The centre of mass will have 3 coordinates needed to describe the rotation of the molecule about the centre of mass. If one imagines a gradual strengthening of the bonds between the atoms as one goes from a set of independent to the rather firmly held (compact) set that constitutes the molecule, one will find that none of the $3N$ degrees of freedom will be destroyed. Then if 3 translational and 3 rotational degrees of freedom are recognised there will be remaining $3N - 6$ degrees of freedom which must be accounted for by the internal coordinates of the molecular system. Then displacements and velocities in accordance with these $(3N - 6)$ coordinates constitute vibrations of the molecule.

If the molecule is linear, molecular rotations can occur only about the two axes that can be drawn perpendicular to the molecular axis. For such molecules, three overall translations and two overall rotations are to be subtracted from the total $3N$ degrees of freedom and so there remain $(3N - 5)$ internal degrees of freedom for linear molecules.

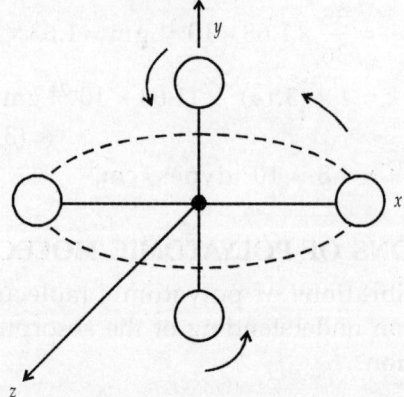

Fig. 10.3

For example, H_2O is a non-linear molecule and has 3 modes of vibration and 3 modes of rotation; CO_2 is a three-atom linear molecule and has four modes of vibration and two modes of rotation.

10.3 THE NORMAL VIBRATIONS, NORMAL COORDINATES (AND NORMAL MODES)

A molecular system may be considered to be a collection of many mass units interconnected by springs which undergo infinite vibrations depending on initial displacements given to the molecules of the system. The general motion of such a system can be analysed in terms of above mentioned $3N - 6$ (or $3N - 5$) independent internal coordinates. Apart from this, there exist furthermore a set of coordinates called *normal coordinates* which is specially convenient for the description of vibrations of a system. In order to understand the nature of normal coordinates, we now consider the following example:

Let a mass-particle (mass = m) be held by springs and for convenience we consider two dimensional motion.

The potential energy

$$U = \frac{1}{2}k_x x^2 + \frac{1}{2}k_y y^2 \qquad ..(13)$$

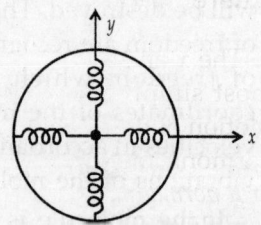

Fig. 10.4: A one particle (two dimensional) system

and
$$T = \frac{1}{2}m\dot{x}^2 + \frac{1}{2}m\dot{y}^2 \qquad \qquad ..(14)$$

The springs are assumed to obey Hooke's law (small displacements).

Forces
$$f_x = -\left(\frac{\partial U}{\partial x}\right) = -k_x x$$

$$f_y = -\left(\frac{\partial U}{\partial y}\right) = -k_y y \qquad \qquad ..(15)$$

The motion of the particle can be deduced by applying Lagrange's equation. For the x-coordinate

$$\frac{d}{dt}\left(\frac{\partial T}{\partial \dot{x}}\right) + \frac{\partial U}{\partial x} = 0 \qquad \qquad ..(16)$$

This gives
$$m\ddot{x} + k_x x = 0 \qquad \qquad ..(17)$$

This characterises a vibrational motion with displacement (given by)

$$x = A_x[A_n(2\pi\nu_x t + \phi_x)] \qquad \qquad ..(18)$$

where $\nu_x = \dfrac{1}{2\pi}\sqrt{\dfrac{k_x}{m}}$

Similarly for y-coordinate
$$m\ddot{y} + k_y y = 0 \qquad \qquad ..(19)$$

$\Rightarrow \qquad \qquad y = A_x \sin[2\pi\nu_y t + \phi_y] \qquad \qquad ..(20)$

where $\nu_y = \dfrac{1}{2\pi}\sqrt{\dfrac{k_y}{m}}$

The x and y coordinates allow the motion of the particle to be most simply described and are termed the *normal coordinates*. Motion of the system along a normal coordinate consists of simple harmonic motion and such motion is said to be a normal vibration or a *normal mode*. Similar normal coordinates will exist for a more complicated system. In a general (molecular) system, the normal coordinate will be a complicated combination of atomic displacements. In terms of a normal coordinate, the vibrational motion of a molecular system will be most simply described and a displacement of the system (according to the normal coordinate) will lead to a simple motion in which all the particles move in phase with the same frequency (if Hooke's law is obeyed) and will execute simple harmonic motion.

10.4 TYPES OF NORMAL MODES

Normal modes of vibrations are generally categorized in two types:

1. Stretching vibrations
2. Bending or deformation vibrations.

1. Stretching Vibrations: In this type of vibration, the atoms of the molecule move essentially along the bond-axis, so that the bond length increases or decreases at regular intervals but the atoms remain along the bond-axis. Such a mode does not cause any change in dipole moment in symmetrical molecules and so they are not IR active. Stretching vibrations are of two types:

   (i) Symmetric Stretching: In this type of stretching with respect to a particular atom, other atoms in a molecule move in the same way, e.g. in case of methylene group (H — C — H), the two H atoms move away from the central carbon atom without change in the bond angle (Fig. 10.5 (a)).

   (ii) Asymmetric Stretching: In this type of stretching, one atom moves away from the central atom while the other moves towards the central atom (Fig. 10.5 (b)).

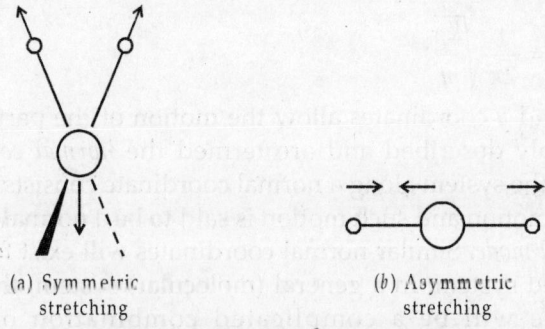

(a) Symmetric
stretching

(b) Asymmetric
stretching

Fig. 10.5: Stretching vibrations

2. **Bending or Deformation Vibrations:** Such vibrations may consist of a change in bond angle between bonds with a common atom or the movement of a group of atoms with respect to the remainder of the molecule without movement of the atoms in the group with respect to one another. These are of four types:

(*i*) Scissoring (*ii*) Wagging (*iii*) Twisting and (*iv*) Rocking

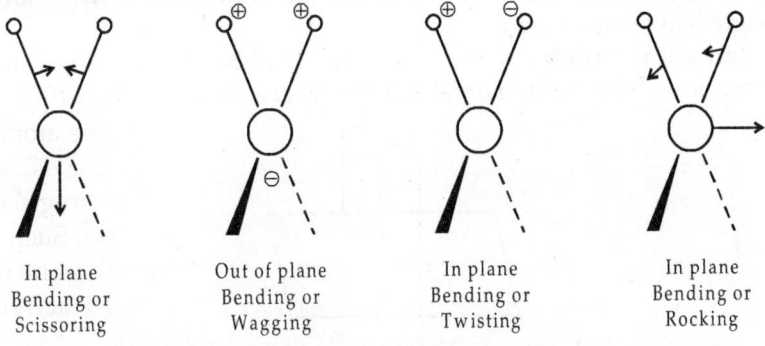

Fig. 10.6: Bending or deformation vibrations. ⊕ indicates motion from plane of paper towards reader. ⊖ indicates motion from plane of paper away from reader

(*i*) **Scissoring:** In scissoring, the two atoms (concerned to another atom) move towards or away from each other with deformation of the valency angle.

(*ii*) **Wagging:** In wagging, the structural unit swings back and forth out of the plane of the molecule.

(*iii*) **Twisting:** In twisting, the structural unit rotates about the bond which joins it to the remainder of the molecule.

(*iv*) **Rocking:** In rocking, the structural unit swings back and forth in the plane of the molecule.

As the symmetric stretching vibration produces no change in the dipole moment of the molecule, it is not seen (∴ inactive) in the infrared spectra. On the other hand, in the asymmetric stretching, e.g. in linear CO_2 molecule, one oxygen atom approaches the carbon atom as the other moves away. As a consequence, the asymmetric stretching vibration appears in the infrared spectrum region.

10.5 NORMAL VIBRATIONS

10.5.1 Normal Vibrations of CO_2 Molecule

CO_2 is a linear triatomic molecule. For this molecule, there are two sets of symmetryaxes. There is an infinite number of two fold axes C_2 passing through the carbon atom at right angles to the bond direction and there is one more an infinite fold axis C_∞ passing through the bond axis. This is referred to as infinite fold since rotation of the molecule about the bond axis through any angle leaves no change.

Since it is a triatomic molecule, there will be $3N - 5 = 4$ normal vibrations. The fundamental vibrations are shown in Fig. 10.7.

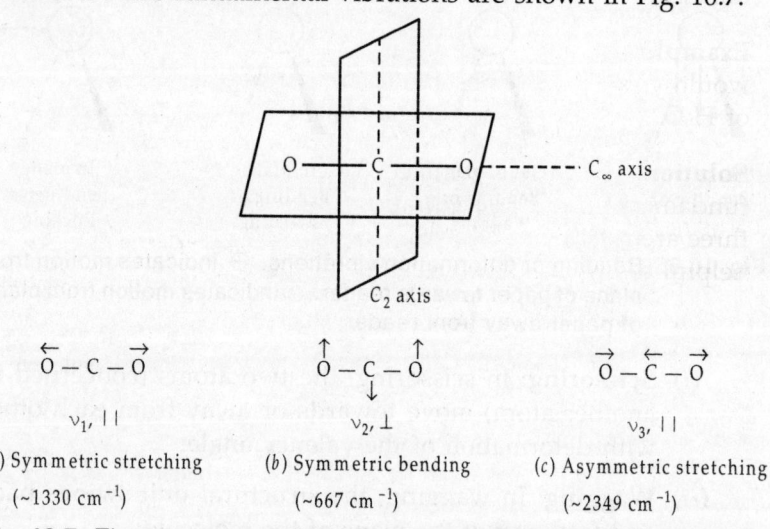

Fig. 10.7: The symmetry and fundamental vibrations of CO_2 molecule. The arrows indicate the direction of motion of atoms at a particular instant

By convention, the vibrations are labelled in decreasing frequency with their symmetry type. Thus, the symmetric vibrations are labelled ν_1 (highest frequency), ν_2 (next highest symmetric) and the antisymmetric (or asymmetric) as ν_3.

Though we have shown three vibrational modes, however ν_2 in fact consists of two vibrations—one in the plane of the paper as drawn, and the other in which the oxygen atoms move simultaneously into and out of the plane. These two vibration-types are identical in all respects except direction and are, therefore, doubly degenerate. The observed normal vibrational frequencies are given in brackets.

In general, any normal mode is an independent synchronous motion of atoms or groups of atoms that may be excited without leading to the excitation of any other normal mode.

10.5.2 Normal Vibrations of Water Molecule

Let us consider the H_2O molecule. It is a non-linear triatomic molecule. There are $3N - 6 = 3$ allowed vibrational modes and are shown in Fig. 10.8. The arrows attached to each atom show the direction of its motion during half of the vibration. Motion is described as stretching or bending depending on the nature of the change in molecular shape.

Example 10.3: How many fundamental vibrational frequencies would you expect to observe in the infrared absorption spectrum of H_2O?

Solution: The bent H_2O molecule will have $3N - 6 = 3 \times 3 - 6 = 3$ fundamental modes of vibration. In order to decide whether all three are infrared active and whether any degeneracy exists, it is helpful to sketch the vibrational modes as follows:

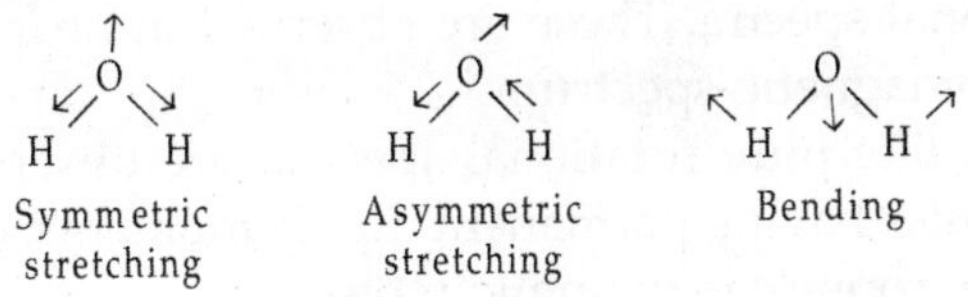

| Symmetric | Asymmetric | Bending |
| stretching | stretching | |

Fig. 10.8

No other fundamental mode can be diagrammed. All three modes of vibrations are infrared active (undergo a change in dipole moment) and are non-degenerate. The observed absorption occur at 3652 cm^{-1}, 3756 cm^{-1} and 1595 cm^{-1}.

Example 10.4: How many fundamental vibrational frequencies would you expect to observe in the infrared absorption spectrum of CO_2?

Solution: Since CO_2 is linear, it has $3N - 5 = 4$ fundamental vibrational modes which may be represented as follows:

(*i*) $\overset{\leftarrow}{O} = C = \vec{O}$: Symmetric stretching

(*ii*) $\vec{O} = \overset{\leftarrow}{C} = \vec{O}$: Asymmetric stretching

(iii) $\downarrow O = \underset{\downarrow}{C} = O \uparrow$: Bending (in the plane of paper)

(iv) $\underset{+}{O} = \underset{-}{C} = \underset{+}{O}$: Bending ($\perp$ to the plane of the paper)

Of the above four vibrational modes, the symmetric stretch will be infrared inactive since there is no change in dipole moment during vibration.

The two bending modes (iii and iv) are degenerate and will absorb at the same frequency. The actual vibration will be a combination of both.

Thus, CO_2 would exhibit two fundamental absorption bands: one due to asymmetric stretch (ii) and another due to doubly degenerate bending modes (iii and iv).

In practice, these two are observed at 2349 cm^{-1} and 667 cm^{-1} respectively.

10.6 VIBRATIONAL–ROTATIONAL SPECTRA

10.6.1 Salient Features

1. When there are transitions between the vibrational states of the same electronic state of a molecule, we get vibrational rotational spectra. These are observed in near IR region of electromagnetic spectrum.

2. These, like pure rotational spectra, are observed only for molecules having permanent dipole moments, e.g. HCl, HF, i.e. heteronuclear diatomic molecules.

3. Due to relative vibration of the nuclei of such molecules, these molecules possess an oscillating dipole moment causing them to radiate at frequencies lying in the near infrared region. These molecules may exhibit both, emission and absorption spectra. However, in practice, the vibration rotation spectra are observed in absorption.

10.6.2 Fine Structure of Vibration–Rotation Bands

Observed fine structure of rotation bands give ample evidence that a simultaneous rotation and vibration do occur. As a first approximation, we may consider that a diatomic molecule can execute rotations and vibrations quite independently, i.e. no interaction between rotational and vibrational energies. Then

$$E_{\text{total}} = E_{\text{rot}} + E_{\text{vib}} \qquad ..(21)$$

If we assume the diatomic molecule to behave as a linear harmonic oscillator as well as a rigid rotator, the combined energy will be

$$E_{vr} = \left(v + \frac{1}{2}\right)hc\omega + \frac{h^2}{8\pi^2 I}J(J+1) \qquad ..(22)$$

Suppose a simultaneous transition from the vibrational level v' to the level v'' and from the rotational level J' to J'' occurs. Then change in energy, using eqn (21) is

$$E'_{vr} - E''_{vr} = (v' - v'')hc\omega + \frac{h^2}{8\pi^2 I}\{J'(J'+1) - J''(J''+1)\} \qquad ..(23)$$

and the corresponding frequency of radiation arising due to this transition will be

$$\nu = \frac{E'_{vr} - E''_{vr}}{hc}$$

$$= (v' - v'')\omega + B\{J'(J'+1) - J''(J''+1)\} \qquad ..(24)$$

where $B = \dfrac{h}{8\pi^2 Ic}$

10.7 DIATOMIC MOLECULE AS A NON-RIGID ROTATOR (ANHARMONIC OSCILLATOR)

If we consider the diatomic molecule as the non-rigid rotator and anharmonic oscillator, the combined energy will be

$$E_{rv} = hc\{BJ(J+1) - DJ^2(J+1)^2 + HJ^3(J+1)^3$$

$$+... + \left(v + \frac{1}{2}\right)\omega_e - x\left(v + \frac{1}{2}\right)^2 \omega_r\}$$

Initially, we ignore the small centrifugal distortion constants D, H, etc. Then

$$E_{rv} = hc\left\{BJ(J+1) + \left(v + \frac{1}{2}\right)\omega_e - x\left(v + \frac{1}{2}\right)^2 \omega_r\right\} \qquad ..(25)$$

By neglecting D, we are treating here the molecule as rigid yet vibrating. We know that $B \leq 10$ cm^{-1} and D is about 0.01% of B. Now, since a good infrared spectrometer has a resolving power of about 0.5 cm^{-1}, it is obvious that we may neglect D to a very high degree of accuracy.

Selection rules are

$$\Delta v = \pm 1, \pm 2, \ldots$$

$$\Delta J = \pm 1 \qquad\qquad ..(26)$$

The Fig. 10.9 shows relevant energy levels and transitions, designating rotational quantum numbers in $v = 0$ state as J'' and $v = 1$ state as J' (single prime is used for upper state and double for lower state).

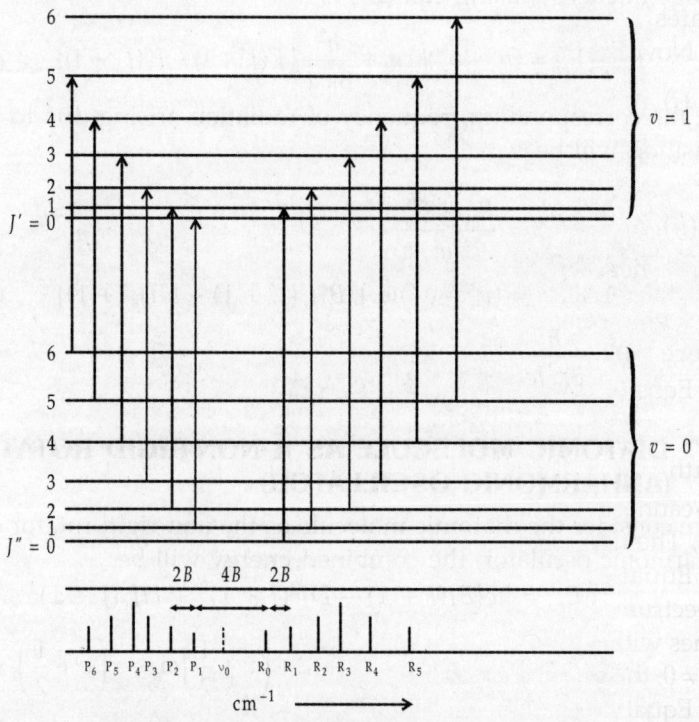

Fig. 10.9: Some transitions between the rotational vibrational energy levels of a diatomic molecule and the spectrum arising from them (absorption spectrum)

An analytical expression for the spectrum can be obtained by applying above selection rules to the energy levels of eqn (25) considering only $v = 0$ to $v = 1$ transition, we have, in general, the frequency of transition as

$$\nu = \frac{E'_{r, v=1} - E''_{r, v=0}}{hc}$$

$$= BJ'(J'+1) + \frac{3}{2}\omega_e - \frac{9}{4}x\omega_r - \left\{ BJ''(J''+1) + \frac{1}{2}\omega_e - \frac{1}{4}x\omega_r \right\}$$

$$= \nu_0 + B(J'-J'')(J'+J''+1) \text{ cm}^{-1}.$$

$$..(27)$$

where we have written ν_0 for $\omega_e(1-2x)$. It is to be noted that because we have taken rotation to be independent of vibrational changes, B will remain identical in the lower and upper vibrational states.

Now, we can have

(i) $\Delta J = +1$ i.e., $J' = J'' + 1$ or $J' - J'' = +1$

hence $\nu(R) = \nu_0 + 2B(J'' + 1) \text{ cm}^{-1}$

$J'' = 0, 1, 2, ...$ ..(28)

(ii) $\Delta J = -1$ i.e., $J'' = J' + 1$ or $J' - J'' = -1$

hence $\nu(P) = \nu_0 - 2B(J' + 1) \text{ cm}^{-1}$

$J' = 0, 1, 2, ...$ ..(29)

$= \nu_0 - 2BJ'', J'' = 1, 2, 3, ...$

Eqns (28) and (29) can be combined in the form

$$\nu = \nu_0 + 2Bm \text{ cm}^{-1} \qquad\qquad ..(30)$$

with $m = \pm 1, \pm 2, \pm 3 ...$

m cannot be zero since this would imply values of J' or J'' to be -1. The frequency ν_0 is usually called the band origin.

Equation (30) represents the combined vibration rotation spectrum. Such a spectrum will obviously consist of equally spaced lines with a *spacing* 2B on each side of the band origin ν_0 but since $m \neq 0$, the line at ν_0 itself will not appear (i.e., absence of Q branch).

Equation (30) yields a series of lines with a constant separation of 2B cm^{-1} *lying on the lower frequency (longer wavelength) side of the centre of the band (of frequency ν_0) when m has negative values. These lines constitute the fine structure of what is known as the P-branch of the vibration-rotation band.*

The frequencies of this branch are represented by eqn (29). For positive values of m, a similar series of lines with a constant frequency separation of 2B cm^{-1} appears on the higher frequency (shorter wavelength) side of the centre of the band. This series is termed as the R-branch of vibration-rotation band. The frequencies of this branch are represented by eqn (28). It is to be noticed that the value of m in eqn (30) cannot be zero, so, the line of frequency corresponding to the centre of the band should be absent. Thus lines arising from

$$\Delta J = -2 \qquad -1 \qquad 0 \qquad +1 \qquad +2$$
are called O P Q R S
branches.

These theoretical predictions are in agreement with experimental results. The Fig. 10.10 illustrates some rotational-vibrational transitions leading to formation of P, Q and R branches.

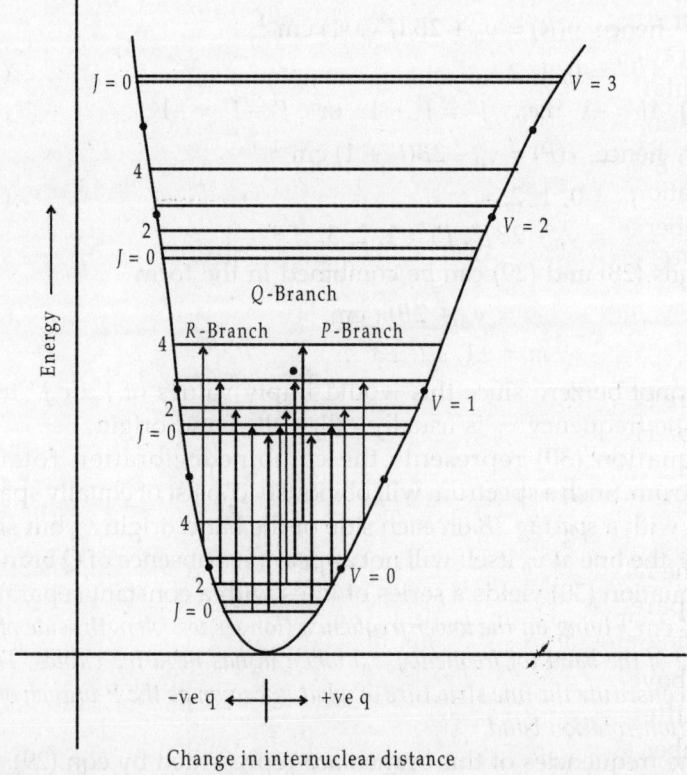

Fig. 10.10: Some rotational energy transitions for vibrational states $v = 0, 1, 2$ giving rise to P, Q and R branches. The observed fine structure suggests that simultaneous rotation and vibration do take place. The periods of vibration and rotation are $\sim 10^{-14}$ s and 10^{-12} s respectively

10.8 THERMAL DISTRIBUTION OF QUANTUM STATES IN VIBRATIONAL BANDS

The intensities of the vibrational bands depend also on the number of molecules in the particular vibrational state from where the quantum jump takes place. For the sake of simplicity, we consider here only the transitions under the conditions of thermal equilibrium. For this, we have to consider the thermal distribution of molecules in different vibration states.

The number of molecules in each vibrational state is proportional to the Boltzmann factor $e^{-E/kT} = e^{-\varepsilon_0(v)hc/kT} = e^{-\varepsilon_0(v)/0.6952T}$. The zero point energy is neglected here because it will give a constant factor for all vibrational levels and generally we consider about the ratio of population in two vibrational levels. Thus, the population would decrease exponentially as we go to higher vibrational states ($v = 0$ to $v = 1, 2, 3, ..$).

To get an idea of decrease in population as one goes to higher vibration states, consider the example of I_2 molecule. The ratio of number of molecules in the $v = 1$ state to the number in $v = 0$ state at 27°C is

$$\frac{N_{(1)}}{N_{(0)}} = e^{213.2/0.6952T} = e^{213.2/(0.6952 \times 300)}$$

where $\varepsilon_0 = 213.2$ cm^{-1} for iodine.

For HCl

$$\frac{N_{(1)}}{N_{(0)}} = e^{2885.9/(0.6952 \times 300)} = 0.00000098$$

The falling off is thus more pronounced in lighter molecules as compared to the heavier ones because the vibrational energy spacing are larger in the former case.

Above, we have considered the relative number of molecules in different vibrational levels. Now, if we consider the total number of molecules in the system, it would be proportional to the total number of molecules in each vibrational state according to

$$Q_v = 1 + e^{-\varepsilon_0(1)hc/kT} + e^{-\varepsilon_0(2)hc/kT} + ...$$

where Q_v is called the partition function or the state sum. The number of molecules in the v state is then

$$N_v = \frac{N}{Q_v}\, e^{-\varepsilon_0(v)hc/kT}$$

For all practical purposes, we can consider $Q_v = 1$ as the second and the other higher terms in expression for Q_v are small compared to the first term 1. Hence

$$N_v = Ne^{-\varepsilon_0(v)hc/kT}$$

10.9 INTERPRETATION OF VIBRATIONAL SPECTRA

The vibrational spectrum of a molecule consists of two major regions-the group frequency region and finger-print region. Group frequencies are vibrations that are associated with certain structural units such as —CH_3, —NH, —C $\equiv$ N etc. and appear in fairly constant regions in the spectrum. The approximate constancy of the position of group frequencies form the basis for the structural analysis of the compounds. The region $900-1460$ cm^{-1} is very rich in absorption bands and contains mainly bending and certain stretching vibrations. A molecular or structural moiety may often be identified by the assignments of the bands in this region. Though molecules having similar groups show very similar spectra outside this region, they show bands typical of the molecule in this region (hence the name *fingerprint region*). Some of the important group frequencies are listed in Table 10.1. More extensive data are available in the literature. It is to be noted here that confirmation of the findings based on the above concept is needed for the complete identification of compounds. Only practice can give the necessary intuition for making correct assignment of group frequencies.

10.10 APPLICATIONS OF INFRA RED SPECTROSCOPY

Infra Red Spectroscopy, and in particular, Fourier transform Infra-red spectroscopy, is one of the most widely used analytical tools available today. It has a rapidly increasing demand for a routine analysis of a wide range of compounds and the present data handling capabilities have generated this interest. Applications of infrared spectroscopy are of varied types. Some major applications are as follows.

TABLE 10.1: Characteristics frequencies of certain molecular groups

Group	Approximate Frequency (cm^{-1})	Assignment
—OH	3550-3650	ν OH (free)
—CH_3	2920-2990	ν_{as} CH_3
	2855-2900	ν_s CH_3
	1445-1475	δ_{as} CH_3
	1365-1385	δ_s CH_3
—CH_2—	2910-2945	ν_{as} CH_2
	2835-2865	ν_s CH_2
	1455-1485	δ CH_2
$\equiv$CH	3300-3325	ν CH
	620-670	δ CH
$-C\diagup^{O}_{\diagdown OH}$	2800-3000	ν OH
	1600-1750	νC$=$O
	1220-1300	ν C—O
—C$\equiv$C—	2190-2250	ν C$\equiv$C
$>$C$=$C$<$	1600-1650	ν C$=$C
$\geq$C—C$\leq$	1000-1200	ν C—C
$>$N—N$<$	3300-3450	ν N—N

10.10.1 Identification of Molecular Constituents

Identification of molecular constituents is done by assigning the experimentally observed bands to possible functional groups and stretching and bending vibrations of bonds. We know that the vibrational spectrum of a molecule consists of two major regions-the group frequency region and the finger print region. Then, a simple way of identifying a compound is by matching its *IR* spectrum to that of a known sample.

10.10.2 Elucidation of Molecular Structure

Elucidation of molecular structure is possible with the help of *IR* spectroscopy only if Raman results are also taken into account, since both are complementary to each other. However, one can restrict himself to certain cases where one can get information from IR alone. From the observed *P*, *Q*, *R* branches of the *IR* spectra of polyatomic molecules, an estimate of the rotational constants are possible from which the moments of inertia and interatomic distances can be determined. This procedure is possible even for molecules with zero dipole moment, if they exhibit rotational fine structure of the vibration band. Study of the influence of nuclear spins on the intensities of fine structure components of vibrational spectrum permits the understanding of grouping of atoms in molecules.

10.10.3 Biological Applications

There are two basic aspects which make *IR* spectroscopy potentially attractive to the biomedical field. First, the technique is truly molecular level in nature. One is able to probe directly the structure of the molecule or system under study and information is obtained about functional groups, bonding forms, conformations and environmental influences that affect the molecular frequencies. Second, *IR* spectroscopy needs relatively small quantity of the sample to be studied (micro to nanogram). Substances can be studied in the solid or liquid or gaseous state and, even water samples are being studied successfully.

10.10.4 Other Applications

IR spectroscopy is extensively used in different areas for varying purposes. Some of them are the study of inter and intramolecular interactions, distinguishing free and hindered internal rotation, chemical reaction kinetics, pollution studies, hydrogen bonding, surface chemistry, catalytic processes, establishment of molecular symmetry, etc.

Questions

10.1 Explain the effect of anharmonicity on the vibrational spectra of diatomic molecules.

10.2 Homonuclear diatomic molecules do not show vibrational spectra, why?

10.3 One of the fundamental vibration modes of H_2O occurs at 3652 cm^{-1}. What would be the frequency of the corresponding mode in D_2O?

[**Ans**. 2657 cm^{-1}]

10.4 Discuss the conditions under which the infrared spectrum of a diatomic molecule can be observed.

10.5 Discuss how the study of vibrational spectrum of a diatomic molecule enables us to determine anharmonicity constant and equilibrium frequency of vibration.

10.6 Give the theory of a vibrational-rotational spectrum of diatomic molecules.

10.7 What are P, Q and R branches in the vibration-rotation spectra? Explain their origin.

Molecular Electronic Spectra

11.1 SALIENT FEATURES OF MOLECULAR ELECTRONIC SPECTRA

1. They appear in the visible and ultraviolet regions.
2. It involves a change in all the three, viz. electronic, rotational and vibrational energies of the molecule.
3. They are studied in both emission and absorption. The bands show a fine structure usually more complicated than in vibration-rotation bands. The tendency of head formation is much stronger and the bands may be degraded either towards the red or towards the violet.
4. All molecules exhibit electronic spectra. Homonuclear molecules, e.g. H_2, O_2, N_2 which neither give rotational nor rotation-vibration spectra (as not possessing permanent dipole moment) exhibit electronic spectra (because of change in instantaneous dipole moment due to redistribution of electronic charge accompanying the (electronic) transition).
5. Molecules possessing permanent electronic dipole moments give pure rotational spectra. Vibrational spectra require a change of dipole moment. However, electronic spectra are given by all the molecules. Now, since changes in the electron distribution in a molecule are always accompanied by a dipole moment change, therefore, homonuclear diatomic molecules, though possess no permanent dipole moment, show an electronic spectra with rotational and vibrational structures.

11.2 ELECTRONIC TRANSITIONS

To a close approximation, the total energy E of the molecule in a given quantum state may be supposed as made up of the electronic energy E_e, the vibrational energy E_v and the rotational energy E_r, i.e.

$$E = E_e + E_v + E_r \qquad ..(1)$$

or in the form of energy changes

$$\frac{\Delta E}{hc} = \frac{\Delta E_e}{hc} + \frac{\Delta E_v}{hc} + \frac{\Delta E_r}{hc} \text{ cm}^{-1} \qquad ..(2)$$

The approximate orders of magnitudes of these changes are related by

$$\frac{\Delta E_e}{hc} \approx \frac{\Delta E_v}{hc} \times 10^3 \text{ cm}^{-1} \qquad ..(3)$$

$$\approx \frac{\Delta E_r}{hc} \times 10^6 \text{ cm}^{-1} \qquad ..(4)$$

This shows that vibrational energy changes will produce a "coarse structure" and rotational changes, a fine structure in the spectra of electronic transition.

When an electronic transition occurs in a molecule, a large change of energy is involved because electronic spectra involve transitions between electronic states. (These are observed in the visible and ultraviolet region of the spectrum).

It is to be noted here that the pure rotation spectra are given by molecules having a permanent electric dipole moment and vibration spectra require a change of dipole moment, consequently, each band in the electronic spectra consists of a number of finer lines due to simultaneous changes in vibrational and rotational energies.

The selection rules for electronic spectra are quite complicated and in general the following rules are used. However, these are not always strictly followed.

1. The electronic transitions in which the spin of the two states are involved must be the same for the allowed transitions.

2. Transitions between symmetric and antisymmetric states are allowed. Transitions in the same symmetry are not allowed. That is, $s \to s$, $p \to p$ and $d \to d$ are not allowed but $s \to p$ is allowed.
3. Transitions in which the change in the component of orbital angular momentum along the internuclear axis is either 0, +1 or −1, are allowed.

11.2.1 The Vibrational Structure of Electronic Bands and Franck-Condon Principle

The way in which the potential energy of a diatomic molecule might vary, as a function of the internuclear distance for two different electronic states of the molecule is shown in Fig. 11.1.

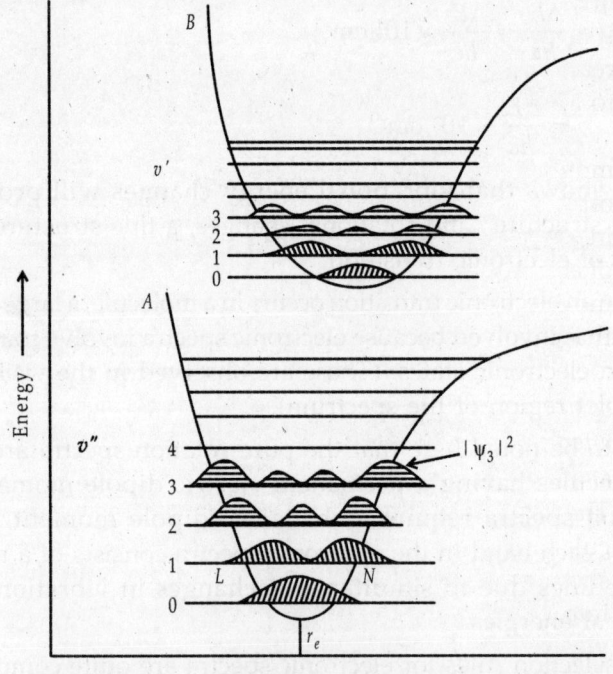

Fig. 11.1: Variation in potential energy of a diatomic molecule as a function of internuclear distance for two different states

The curves shown in the figure are typical of a ground and an excited electronic state. The potential energy curves of this figure imply that for each of the two electronic arrangements, the molecule will vibrate and horizontal lines represent the energies of the allowed vibrational states. Further, the probabilities of the molecule being found at various internuclear distances (i.e. $|\psi_v|^2$) are also shown on the energy levels to which they refer.

Now one may be interested in seeking an answer to the question which transitions are expected between the vibrational levels labelled $v'' = 0, 1, 2, ..$, the lower electronic state and the levels labelled $v' = 0, 1, 2, ...,$ of the upper electronic states. In this respect, following three factors are to be borne in mind:

1. As observation of the vibrational structure of electronic bands will confirm, there are no general restrictions on the changes in v for a transition going from one electronic state to another. This is in contrast to the rule $\Delta v = \pm 1$, operative in vibrational transitions within a given electronic state.

2. It must be remembered that electrons can move and rearrange themselves much faster than can the nuclei move to alter the internuclear distance.

   A hint to the relative times for electronic and nuclear motions is provided by the fact that an electron in a Bohr orbit of an atom completes a revolution round the nucleus in $\sim 10^{-15}$ s whereas a typical molecule vibrates with a period of about 10^{-13} s. This characteristic of electronic and nuclear motions leads to the Franck-Condon principle which states that:

   an electronic transition in a molecule takes place so rapidly compared to the vibrational motion of the nuclei that immediately afterwards, the nuclei still have very much the same relative position and velocity as before the transition.

3. The fact that the probability of the molecule being at a particular internuclear distance is a function of the distance (as shown by the probability curves of Fig. 11.1) must be taken into account. An electronic transition must be expected, in view of the transition moment integral defined by

$$\left|\mu_{xlm}\right| = \int_{-\infty}^{+\infty} \psi_m^* \mu_x \psi_l \, dx \qquad ..(5)$$

to be most favoured if it occurs while the molecule has an internuclear distance such that the transition connects probable states of the molecule. This rule gets simplified in view of the shapes of the probability functions of the

Fig. 11.1 by considering transitions to be relatively more probable if they begin or end at the middle of the $v = 0$ level or either end of any of the higher vibrational levels.

Every electronic state of a molecule has its own potential energy curve as indicated by curves X' and X'' of Fig. 11.2 for two states of a molecule. These are the Morse curves along with the probability densities $|\psi_v|^2$. For $v = 0$ the atom is most likely to be found at the centre of its motion (i.e. r_e). For $v = 1, 2, 3, ..$, the most probable position steadily approaches the extremities as in harmonic oscillator. This is because the nuclei spend at these positions the longest time on account of vibrational kinetic energy being zero there. Also quantum mechanically probability $|\psi_v|^2$ is

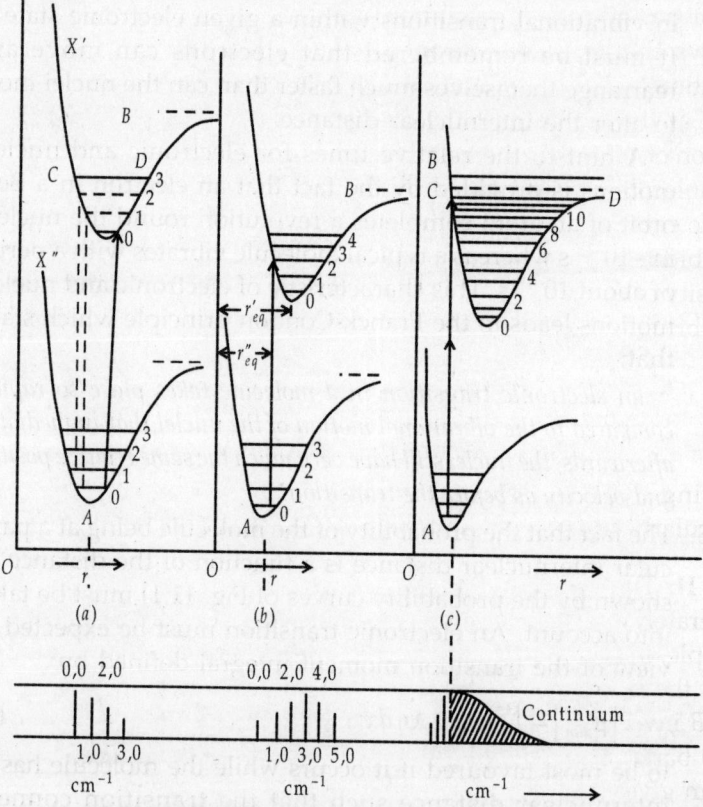

Fig. 11.2: Potential energy curves explaining the intensity distribution in the absorption spectra according to Franck-Condon principle

greatest at the extreme positions. It means that (*i*) the observed transitions between two states should start from extreme positions of vibrational levels and (*ii*) they should be represented by vertical lines.

The consequences of the Franck-Condon principle can be illustrated if we consider the following three situations:

(*i*) $r'_{eq} = r''_{eq}$ (*ii*) $r'_{eq} > r''_{eq}$ (*iii*) $r'_{eq} >> r''_{eq}$

As a consequence we have following three cases:

Case I: The potential energy minima lies very nearly one above the other as, for example, in the near ultraviolet spectrum of cyanide radical (Fig. 11.2 (*a*)). If zero point energy is disregarded, and the molecule is initially at the minimum of the lower curve, then a transition to the minimum of the upper curve (0-0 band) follows the Franck-Condon principle. On the other hand, a transition to a higher vibrational state *CD* would be possible only when at the moment of the electronic jump, either the nuclear position or velocity or both alter to an appreciable extent. At point *E*, the molecule has kinetic energy *EB*. Only at the extremities is the kinetic energy the same at the initial state. Thus such a transition is forbidden. Therefore 0-0 and 1-0 transitions show a relative intensity of ~10 : 1 for *CN* radical with the following values of the equilibrium bond lengths and vibrational frequencies for the 0-0 band

$$r'_{eq} = 0.1151 \text{ nm}, \quad \overline{\nu}'_e \approx 2164 \text{ cm}^{-1}$$

$$r''_{eq} = 0.1172 \text{ nm}, \quad \overline{\nu}''_e \approx 2069 \text{ cm}^{-1}$$

showing that the potential energy curves are actually similar particularly in the neighbourhood of minima.

Case II: The minimum of the upper potential energy lies at a moderately greater r_{eq} value (Fig. 11.2 (*b*)) than the lower as, for example, in the case of CO spectra. Therefore, the 0-0 band in CO is not the most probable, but the vertical upward transition from *A* to *B* involving no change in internuclear distance and velocity is 4-0 band. The 2-0 and 3-0 bands are the strongest. This band system gives the values

$$r'_{eq} = 0.1235 \text{ nm}, \quad \overline{\nu}'_e \approx 1516 \text{ cm}^{-1}$$

$$r''_{eq} = 0.1128 \text{ nm}, \quad \overline{\nu}''_e \approx 2170 \text{ cm}^{-1}$$

Case III: The potential energy minima in the two curves (Fig. 11.2 (c)) lie at a considerably different internuclear distance. ($r'_{eq} \gg r''_{eq}$). The transition point in the upper curve at B lies asymptotic to this curve, thus giving rise to a continuous spectrum. (This leads to dissociation of the molecule without any vibration and since, the atoms formed may taken up any value of kinetic energy, therefore, transitions are not quantized, and result into a continuum).

Thus Franck-Condon principle is able to account for the intensities of lines in the vibrational electronic spectra.

11.3 VIBRATIONAL COARSE STRUCTURE

An electronic transition involves a change in all the three—electronic, vibrational and rotational energies of the molecule. When rotational energy changes are negligible, the energy changes accompanying an electronic transition will be

$$E' - E'' = (E'_e - E''_e) + (E'_v - E''_v) \qquad ..(6)$$

The frequency of the spectrum arising due to the electronic transition is then

$$\nu_0 = \frac{E'_e - E''_e}{hc} + \frac{E'_v - E''_v}{hc} \qquad ..(7)$$

We also know that total energy for a diatomic molecule (neglecting rotational energy) is given by

$$\varepsilon'_v = \frac{E'_v}{hc} = \left(v' + \frac{1}{2}\right)\overline{\omega}'_e - \left(v' + \frac{1}{2}\right)^2 x'_e \overline{\omega}'_e + \left(v' + \frac{1}{2}\right)^3 y'_e \overline{\omega}'_e \qquad ..(8)$$

and $\varepsilon''_v = \dfrac{E''_v}{hc} = \left(v'' + \dfrac{1}{2}\right)\overline{\omega}''_e - \left(v'' + \dfrac{1}{2}\right)^2 x''_e \overline{\omega}''_e + \left(v'' + \dfrac{1}{2}\right)^3 y''_e \overline{\omega}''_e \qquad ..(9)$

for v' or $v'' = 0, 1, 2, 3, \ldots$

Here, single prime corresponds to the upper state and a double prime to the lower state. We get,

$$\nu_0 = \nu_e + \left[\left(v' + \frac{1}{2}\right)\overline{\omega}'_e - \left(v' + \frac{1}{2}\right)^2 \overline{\omega}'_e x'_e\right]$$

$$- \left[\left(v'' + \frac{1}{2}\right)\overline{\omega}''_e - \left(v'' + \frac{1}{2}\right)^2 \overline{\omega}''_e x''_e\right] \qquad ..(10)$$

The change in vibrational quantum number accompanying an electronic transition are not governed by the selection rule of vibrational spectroscopy and any $v'' \rightarrow v'$ transition has some definite probability. However, transitions originating from the $v'' = 0$ state are of considerable intensity and are given in Fig. 11.3. These are labelled according to their (v', v'') values. The corresponding energy levels have been indicated. Since the excited electronic state corresponds to a weaker bond, the vibrational levels in the excited state are more clearly spaced than that in the ground state.

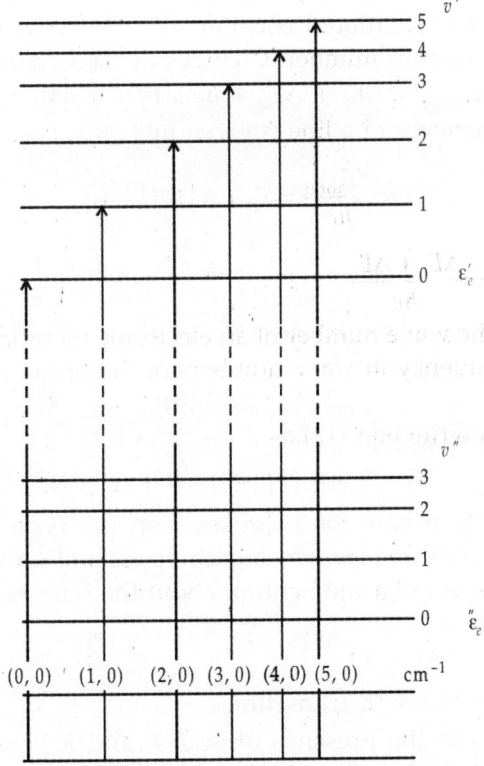

Fig. 11.3: The vibrational coarse structure of electronic absorption spectrum

Since the rotational energy change is taken zero, eqn (10) gives the frequencies v_0 of the centres (or origins) of the series of vibrational bands building the electronic spectrum. The set of lines corresponding to these transitions is called a v' *progression* since the value of v' increases by unity for each line in the set. Further, there will be a crowding of the lines on the high frequency side.

11.4 ROTATIONAL FINE STRUCTURE OF ELECTRONIC VIBRATION TRANSITIONS

While discussing the vibrational coarse structure, we disregard the rotational energy contribution. But, however, under high resolution each spectral line is observed broad consisting of a cluster of many close lines which is the rotational fine structure.

From Born-Openheimer approximation, the total energy of a diatomic molecule is

$$E_{total} = E_e + E_v + hcBJ(J + 1) \qquad ..(11)$$

where B is the rotational constant ($B = h/(8\pi^2 Ic)$) and J is the rotational quantum number. Changes in total energy

$$\Delta E_{total} = \Delta E_e + \Delta E_v + hc\Delta[BJ(J + 1)] \qquad ..(12)$$

The frequency ν of a line is given by

$$\bar{\nu} = \frac{\Delta E_{total}}{hc} = \bar{\nu}_0 + \Delta\{BJ(J + 1)\} \qquad ..(13)$$

where $\bar{\nu}_0 = \dfrac{\Delta E_e + \Delta E_v}{hc}$ $\qquad ..(14)$

represents the wave number of an electronic vibrational transition (it is the frequency in wave numbers of the origin of a particular band).

We can rewrite eqn (12) as

$$\bar{\nu} = \bar{\nu}_0 + B'J'(J' + 1) - B''J''(J'' + 1) \qquad ..(15)$$

The selection rule for J depends on the type of electronic transition. Transition for which both upper and lower states have no electronic angular momentum about the internuclear axis, the selection rule is

$$\Delta J = \pm 1.$$

i.e. only for $^1\Sigma \rightarrow {}^1\Sigma$ transitions.

This leads to the presence of both P and R branches. For all other transitions, i.e. when there is angular momentum about the bond axis, we have

$$\Delta J = 0, \pm 1$$

with a restriction

$$J = 0 \leftrightarrow\!\!\!\!/\!\!\!\!\rightarrow J = 0 \qquad ..(16)$$

For such transitions, in addition to P and R branches, Q branch will also be present (because $\Delta J = 0$).

We shall now consider the rotational contribution in eqn (15) in detail.

The Spectrum

Here B' and B'' refer to different electronic states and different vibrational states unlike vibration-rotation spectra where difference $(B' - B'')$ was small, i.e. now B' and B'' are quite different from one another.

(i) **P-branch:** $\Delta J = -1$ or $J' - J'' = -1$

$$\therefore B'J'(J' + 1) - B''J''(J'' + 1) = B''(J'' - 1)J'' - B''J''(J'' + 1)$$
$$= -(B' + B'')J'' + (B' - B'')J''^2 \text{ cm}^{-1}$$
$$(J'' = 1, 2, 3, ..)$$

$$\bar{\nu}(P) = \bar{\nu}_0 - (B' + B'')(J' + 1) + (B' - B'')(J' + 1)^2$$
$$(J' = 0, 1, 2, 3, ..) \quad ..(17)$$

As the lowest value of $J' = 1$ for P-branch, there will not be any spectral line at $\bar{\nu}_0$.

(ii) **R-branch:** $\Delta J = +1$ or $J' - J'' = +1$

$$\bar{\nu}(R) = \bar{\nu}_0 + (B' + B'')(J'' + 1) + (B' - B'')(J'' + 1)^2$$
$$(J'' = 0, 1, 2, 3, ..) \quad ..(18)$$

(iii) **Q-branch:** $\Delta J = 0$ or $J'' = J'$

$$\bar{\nu}(Q) = \bar{\nu}_0 + (B' - B'')J'' + (B' - B'')J''^2$$
$$(J'' = 1, 2, 3, ..) \quad ..(19)$$

The frequency $\bar{\nu}_0$ is referred to as *band origin*. J'' cannot be zero as it would mean a transition from $J = 0$ to $J = 0$. Further, there would be no spectral line at $\bar{\nu}_0$ in the R and Q-branches also. Equations (17) and (18) can be combined into a single equation as

$$\bar{\nu} = \bar{\nu}_0 + (B' + B'')m + (B' - B'')m^2 \text{ cm}^{-1}$$
$$(m = \pm 1, \pm 2, ..)$$

where $m = (J' + 1)$ in eqn (17) and $(J'' + 1)$ in eqn (18). m takes positive values for R-branch lines and negative values for P-branch lines.

11.5 ROTATIONAL FINE STRUCTURE OF A PARTICULAR VIBRATIONAL ELECTRONIC BAND

We now discuss the appearance of the above mentioned branches in a vibrational electronic band. There are two possibilities:

(i) $B'-B''$ may be negative or

(ii) positive. Thus there arise two cases as mentioned below:

Case I: $B' < B''$. Then

(i) P-branch lines appear on the low wave number side. Spacing between the lines increases with m.

(ii) R-branch lines appear on the high wave number side. Spacing decreases rapidly with m. The R-branch converges to a line (zero separation) called the band-head and then begins to return to the low wave number side with increasing spacing. Such a band is said to be degraded (or shaded) towards the red because the tail of the band when intensity falls off, points towards the red (i.e. low frequency) end of the spectrum. The band-head appears on the high wave number (violet) side of the spectrum.

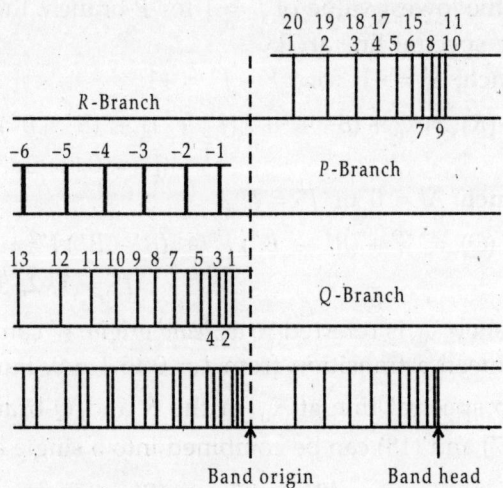

Fig. 11.4: Rotational fine structure of a vibrational electronic transition of a diatomic molecule

(iii) Q-branch lines appear close to low wave number side of the band origin and their spacing increases with the value of J''.

Case II: $B' > B''$. The above arguments are reversed completely.

(i) P-branch lines crowd together to form a band head at the red end (i.e. low frequency side). Such a band is shaded towards the violet.

(ii) R-branch lines though still appear on the high wave number side of band origin, show increase in separation with m.

(iii) The Q-branch will be on the high wave number side of band origin with spacing increasing with J''. When $B' = B''$, the spectrum resembles that of a pure vibration-rotation spectrum with no band head.

11.6 THE FORTRAT DIAGRAM

Frequencies of the lines in P, Q and R branches (eqns (17), (18) and (19)) can be written with continuous variables p and q as

$$\bar{v}_{P,R} = \bar{v}_0 + (B' + B'')p + (B' - B'')p^2 \qquad ..(20)$$

$$\bar{v}_Q = \bar{v}_0 + (B' - B'')q + (B' - B'')q^2 \qquad ..(21)$$

where p takes both positive and negative values and q takes only positive values. Equations (20) and (21) are equations of parabolae and depicted in Fig. 11.5. These are referred to as Fortrat

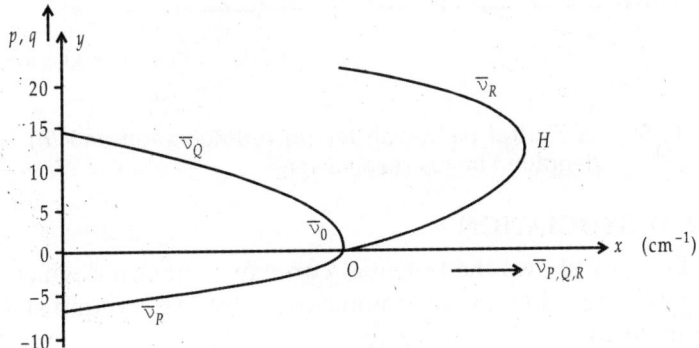

Fig. 11.5: The Fortrat parabola for $B' < B''$. (H is band head, O is band origin)

parabolae. The PR parabola intersects the x-axis at O, the band origin. Its vertex H is the band head. As vertex H is a turning point, so

$$\frac{d\bar{v}_{P,R}}{dp} = 0 = B' + B'' + 2(B' - B'')p \qquad ..(22)$$

i.e. the position of the band head corresponds to

$$p_H = \frac{-(B' + B'')}{2(B' - B'')} \qquad ..(23)$$

Substituting this in eqn (20), we get

$$\overline{\nu}_{P,R} - \overline{\nu}_0 = \frac{-(B' + B'')^2}{2(B' - B'')} + \frac{(B' - B'')(B' + B'')^2}{4(B' - B'')^2} \qquad ..(24)$$

or $\qquad \overline{\nu}_{P,R} - \overline{\nu}_0 = -\dfrac{(B' + B'')^2}{4(B' - B'')} \qquad ..(25)$

The right hand side of this equation is positive if $B' < B''$, i.e. the head of a band is at a higher frequency than the band origin. This means that the band head appears in the R-branch (with positive p values).

If $B' > B''$, the band head occurs in the P-branch with negative p values. $\overline{\nu}_H - \overline{\nu}_0$ is positive for a band degraded to the red and negative for the one degraded to the violet.

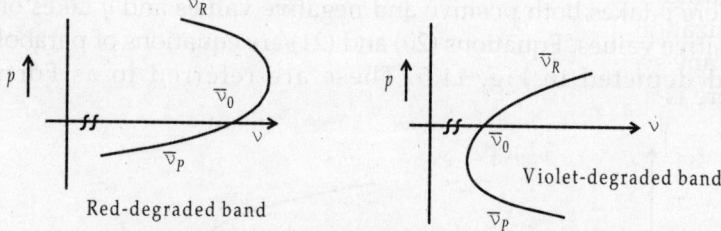

Fig. 11.6: The Fortrat parabolae for (a) red-degraded and (b) violet degraded bands respectively

11.7 DISSOCIATION

The Fig. 11.7 shows the potential curve of a molecule which can be approximated to an anharmonic oscillator, e.g. ground state of HCl molecule.

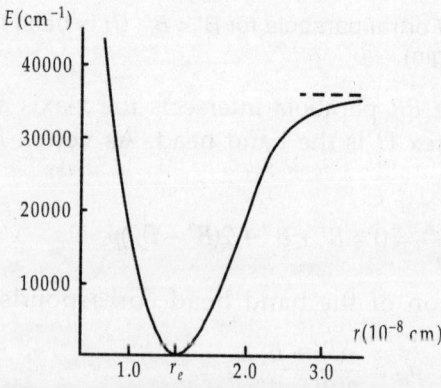

Fig. 11.7: Potential curve of a molecule

When an anharmonic oscillator with a potential curve of this type receives more energy than corresponding to the horizontal asymptote, the mass particle will be completely removed from its equilibrium position and will not return to it. This situation in the molecule corresponds to atoms flying completely away from each other ($r \rightarrow \infty$), i.e. the molecule dissociates. If the energy imparted just corresponds to the asymptote, the atoms at a great distance from each other will have zero velocity with increasing energy above that of asymptote, the atoms (at a great distance apart) will have increasing kinetic energy. This kinetic energy is not quantized. Hence, above asymptote, a continuous term spectrum, corresponding to dissociation, joins onto the discrete vibrational term series (as shown in Fig. 11.8). This is quite analogous to the continuum that adjoins the atomic term series corresponding to ionization. It is in fact a consequence of quantum mechanics that for any system for which potential at infinity has a finite value $V(\infty)$, there is a continuous range of energy values above $E = V(\infty)$.

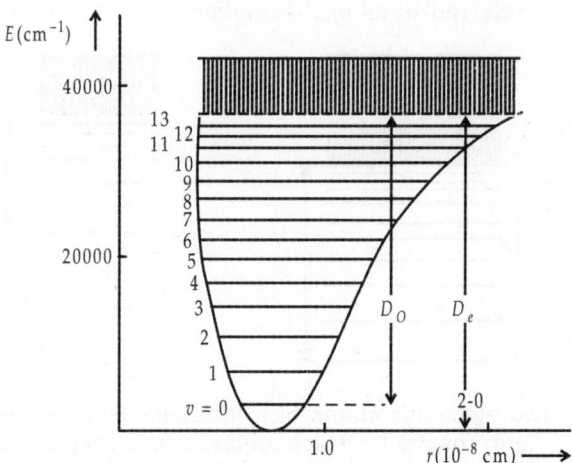

Fig. 11.8: Potential curve of the H_2 ground state with vibrational levels and continuous term spectrum

The Fig. 11.8 shows the potential curve for the special case of H_2 in its electronic ground state, with the vibrational levels and the adjoining continuum. The height of the asymptote (i.e., the beginning of the continuum) above the lowest vibrational level is equal to the work that must be done in order to dissociate the molecule (i.e., the *dissociation energy*). It is designated D_O. The energy difference D_e, between the minimum of potential energy and the asymptote is a little greater than D_O by an amount equal to the zero point energy, i.e.,

$$D_e = D_O + \varepsilon_O.$$

11.8 PREDISSOCIATION

It was realized by Bonhoeffer, Farkas and Kronig that in the case of many diffuse molecular spectra, the Auger effect is responsible for the diffuseness. In a molecule, the overlapping of discrete energy levels by a continuous range of levels necessary for the occurrence of this process is very often present and a *pre-ionization* is possible, (exactly as for atoms). In Fig. 11.9, the vibrational levels of the upper state B from $v = 4$ onward are overlapped by the continuum of the lower (i.e. A) state. The system, thus can go over (without radiating) from the discrete state into the continuous state (lying at the same height), i.e. the molecule (after the radiationless transition) dissociates. This process of radiationless transition is referred to as *predissociation*.

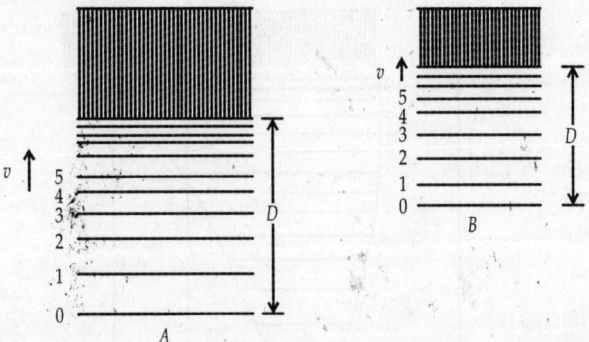

Fig. 11.9: Two electronic states of a molecule with their continuous term spectra for which predissociation is possible

11.8.1 Occurrence of Predissociation

In electronic spectra, a continuum at the high wave number side corresponds to ordinary dissociation. However, often, the rotational fine structure is diffused for intermediate changes. Occurrence of such diffuse structure (or complete continuum) below the true dissociation limit is the predissociation. It was first observed in the electronic spectrum of S_2.

11.8.2 Different Types of Predissociation

Corresponding to the three forms of energy of the molecule (electronic, vibrational or rotational), three kinds of overlapping of molecular energy levels by a dissociation continuum, i.e., three cases of predissociation are possible:

1. Overlapping of a certain electronic state (i.e. of its vibrational or rotational levels) by the dissociation continuum belonging to another electronic state (radiationless transition into this other dissociated electronic state).

2. Overlapping of the highest vibrational levels of an electronic state of a polyatomic molecule by a dissociation continuum joining onto a lower dissociation limit of the same electronic state (predissociation by vibration).

3. Overlapping of the higher rotational levels of a given vibrational level of a diatomic molecule by the dissociation continuum belonging to the same electronic state-(predissociation by rotation).

Case (1) is most important for diatomic molecules. Case (2) applies to polyatomic molecules only. Case (3) can occur for those vibrational levels of an electronic state that lie in the neighbourhood of the dissociation limit (since the higher discrete rotational levels of such vibrational levels can lie above the dissociation limit). This is most readily observed when the dissociation energy of the electronic state is small, e.g. for AlH and HgH.

Predissociation is responsible for two characteristic phenomena observed in molecular spectra viz.

(i) diffuseness of bands and

(ii) breaking-off of band structure.

(i) **Diffuseness of Bands:** In some molecular absorption spectra, such as of S_2, it has been observed in passing along a given v' progression ($v'' =$ constant) that the first few bands of the progression show a normal type of rotational fine structure consisting of sharp lines; but beyond a certain value of v', the rotational lines of the bands become diffuse.

(ii) **Breaking-off of Band Structure:** In some emission bands, it is found that the band structure breaks-off suddenly at a certain point, for example, in the (0, 0) band of the $C^2\Sigma \rightarrow$

$X^2\Sigma$ system of CaH, the R-branch lines run smoothly upto $R(9)$, and then terminate abruptly. The P-branch behaves similarly, ending abruptly at $P(11)$.

The origin of the above two phenomena is attributed to the same cause, viz. predissociation. This can be understood as follows:

Consider Fig. 11.10, which shows two excited states of a molecule which are intersecting, state B has a minimum and usual set of vibrational levels while state C is having no minimum and, therefore, is unstable.

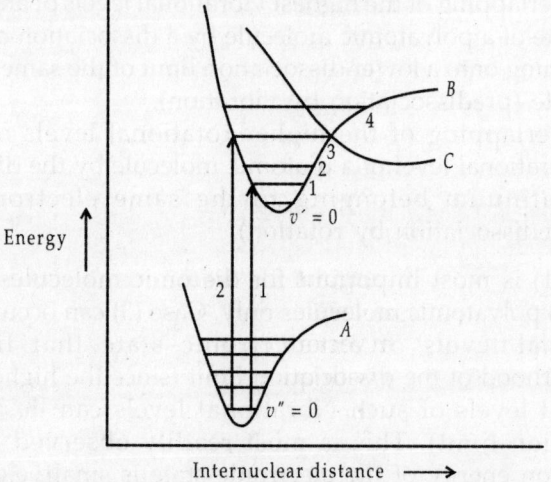

Fig. 11.10: Two intersecting excited states of molecule

The transition 1 from the ground state A to the excited state B would give a normal absorption band with sharp vibrational structure. However, transition 2 is quite different. The energy absorbed by the molecule in transition 2 is not large enough to dissociate it in the state B but if the molecule can cross over (without radiation) to state C, it can predissociate. This takes place in a time which is long compared with that required for vibration but smaller than that required for rotation. Then, the molecule finds enough time to vibrate but not to rotate, before it dissociates in state C. Thus, in transition 2, the vibrational energy remains quantized but the rotational energy is not strictly quantized. Consequently, the vibrational structure of the band system is not influenced by predissociation but the rotational structure of the bands becomes diffuse. Thus in absorption, diffuseness of bands is preceded by sharp bands in the same progression.

11.9 CLASSIFICATION OF MOLECULAR ELECTRONIC STATES

The atomic nuclei in a molecule are held together by the electrons. Therefore, as for atoms, we expect different electronic states of the molecule depending on the orbitals which the electrons occupy. These molecular electronic states have almost the same order of energy difference (1–20 eV) as in case of atoms and are likewise classified according to the value of electronic orbital and spin angular momenta and total angular momentum, etc.

11.9.1 Electronic Orbital Angular Momentum Vector, L

In a diatomic molecule, the electrons move in an axially symmetric electrostatic field produced by the two nuclei. Therefore the component of the orbital angular momentum of the electrons along the internuclear axis is a constant of the motion. The electronic orbital angular momentum vector L precesses about the field direction (i.e. the internuclear axis) with quantized components $M_L(h/2\pi)$ where M_L can take on the values:

$$M_L = L - 1, L - 2, ..., - L \qquad ..(26)$$

In most molecules, the electrostatic field due to the nuclei is so strong that L precesses very rapidly and loses its meaning as angular momentum; only its component M_L along the field remains well-defined. Further, in an electrostatic field, reversing the direction of motion of all electrons does not change energy of the system and it is simply equivalent to changing M_L to $-M_L$. Therefore, in diatomic molecules, the states corresponding to $+M_L$ and $-M_L$ are degenerate (i.e. have the same energy).

The absolute value of M_L is designated by a quantum number $\wedge$, i.e.

$$\wedge = |M_L| = 0, 1, 2, .., L \qquad ..(27)$$

(only positive values).

The corresponding angular momentum vector $\wedge$ represents the component of the electronic orbital angular momentum along the internuclear axis and its magnitude is $\wedge(h/2\pi)$, i.e. it is the axial component of orbital angular momentum. The precession of L is illustrated in Fig. 11.11.

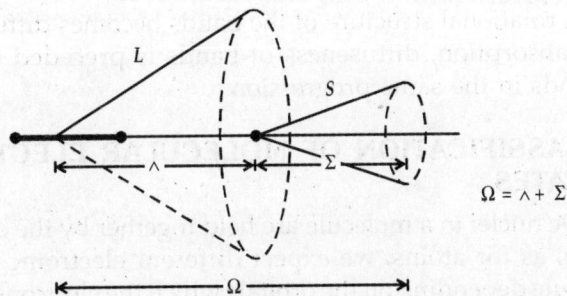

Fig. 11.11: Total electronic angular momentum

The electronic states of the molecule are classified according to the values of $\wedge$, designated as

for	$\wedge =$	0	1	2	3
states are		Σ	π	Δ	ϕ
	$M_L =$	0	± 1	± 2	± 3

Σ states are non-degenerate ($M_L = 0$) whereas π, Δ, ϕ, .., etc. states are doubly degenerate ($\pm M_L$). The various states Σ, π, Δ, etc. have widely different energies since the electric field which causes this splitting is very strong.

11.9.2 Electronic Spin Angular Momentum

Electron spin momentum is not greatly affected by the electric field of the two nuclei in a diatomic molecule. The orbital motion of the electrons produces a weak internal magnetic field along the internuclear axis (except the Σ-state in which the electron cloud is cylindrically symmetrical about the axis). This magnetic field causes the electronic spin angular momentum vector S to precess about the internuclear axis with quantized components $\Sigma(h/2\pi)$, where Σ^* can take on the values

$$\Sigma = S, S - 1, ..., -S \qquad ..(28)$$

$$(2S + 1 \text{ values})$$

In contrast to $\wedge$, Σ can be integral or half integral and positive or negative. It is not defined for states with $\wedge = 0$, i.e. Σ states which have no internal magnetic field.

---

*For molecules, the magnetic Σ denotes the magnetic spin quantum number (in analogy with M_s for atoms). This Σ must not be confused with Σ used in (l) for electronic states with $\wedge = 0$.

11.9.3 Total Electronic Angular Momentum

The total electronic angular momentum about the internuclear axis is denoted by Ω and has a magnitude $\Omega(h/2\pi)$. It is obtained as

$$\Omega = \wedge + \Sigma \qquad \qquad ..(29)$$

Both $\wedge$ and Σ lie along the same direction (i.e. internuclear axis) hence an algebraic addition is sufficient. Consequently, this corresponding quantum number is

$$\Omega = |\wedge + \Sigma| \qquad \qquad ..(30)$$

where Ω will assume integral or half integral values depending upon whether Σ is integral or half integral.

$\wedge$-Σ Interaction

For a given value of $\wedge$ ($\neq 0$), there are $(2S + 1)$ different values of Ω which correspond to slightly different energies of the resulting molecular states. This means that as a result of $\wedge$-Σ interaction, an electronic state with a given $\wedge$ ($\neq 0$) splits into $(2S + 1)$ components, i.e. $2S + 1$ is the multiplicity and assigned as a left superscript to the state symbol. Further, the value of Ω ($= \wedge + \Sigma$) is also added as a subscript.

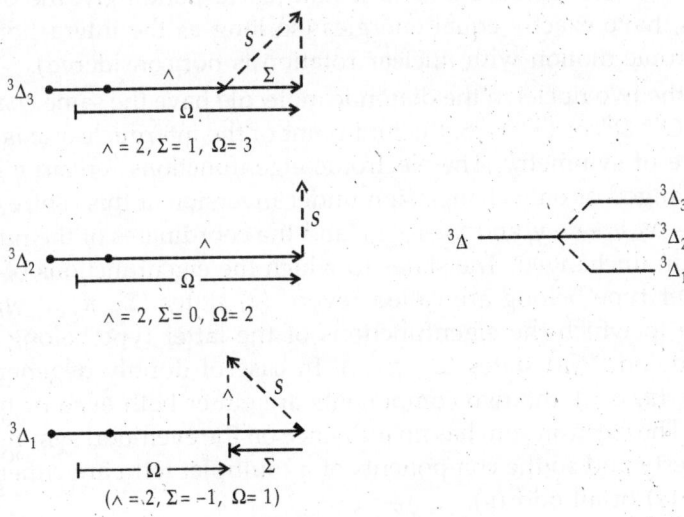

Fig. 11.12: $\wedge$-Σ interaction

As an example, we consider the electronic state $^3\Delta$. For this state, we have

$$\wedge = 2, \quad 2S + 1 = 3 \quad \text{so,} \quad S = 1$$

$$\therefore \qquad \Sigma = S,\ S-1,\ \ldots -S$$

$$= 1,\ 0,\ -1. \qquad\qquad ..(31)$$

$$\Omega = |\wedge + \Sigma| = |2+1|,\ |2+0|,\ |2-1|$$

$$= 3,\ 2,\ 1 \qquad ..(32)$$

Thus, the electronic state $^3\Delta$ has three components $^3\Delta_3$, $^3\Delta_2$, $^3\Delta_1$. These are depicted in Fig. 11.12.

11.10 SYMMETRY CLASSIFICATION OF ELECTRONIC STATES

In a diatomic molecule, the electric field due to the two nuclei is axially symmetric about the internuclear axis. Thus any plane through the internuclear axis is a plane of symmetry. The electronic eigenfunction ψ_e of a non-degenerate state (Σ state) either remains unchanged or changes sign when reflected at such a plane, i.e.

$$\sigma \psi_e = \psi_e \text{ or } -\psi_e$$

where σ is the reflection operator. In the first case, the state is called a Σ^+ state and in the second, it is called a Σ^- state. However, it is superfluous to make the +, – distinction for degenerate states (π, Δ ...states) because π^+ and π^- and correspondingly, the other pairs, have exactly equal energies (so long as the interaction of electronic motion with nuclear rotation is not considered).

If the two nuclei in the diatomic molecule have the same charge (e.g. $O^{16} O^{16}$ or $O^{16} O^{18}$), the mid point of the internuclear axis is a centre of symmetry. The electronic eigenfunctions remain either unchanged or only change sign under inversion at this centre (i.e. $x_i \rightarrow -x_i$, $y_i \rightarrow -y_i$ and $z_i \rightarrow -z_i$), and the coordinates of the nuclei remain unchanged. The states to which the eigenfunctions of the former type belong are called "even" (g) states (Σ_g, π_g ..) while those to which the eigenfunctions of the latter type belong are called "odd" (u) states (Σ_u, π_u ..). In case of doubly degenerate states (π, Δ ..), the two components are either both even or both odd. The electron spin has no influence on the even-odd symmetry property and so the components of a multiplet term are either all even (g) or all odd (u).

In practice, the ground electronic state of all diatomic molecules is $^1\Sigma^+$ or $^1\Sigma_g^+$ (if the nuclei have the same charge), except for molecules NO and O_2 whose ground electronic states are $^2\pi$ and $^3\Sigma_g^-$ respectively.

The selection rules for the electronic transitions are as follows:

(*i*) $\Delta\Lambda = 0, \pm 1$; thus, $\Sigma \leftrightarrow \Sigma, \Sigma \leftrightarrow \pi$ but $\Sigma \nleftrightarrow \Delta$

(*ii*) $\Sigma^+ \leftrightarrow \Sigma^+; \Sigma^- \leftrightarrow \Sigma^-$ but $\Sigma^+ \nleftrightarrow \Sigma^-$

(*iii*) $\Delta S = 0$; thus $^1\Sigma \leftrightarrow {}^1\Sigma$ but $^1\Sigma \nleftrightarrow {}^2\Sigma$

(*iv*) $g \leftrightarrow u$, but $g \nleftrightarrow g$ and $u \nleftrightarrow u$

11.11 SYMMETRY PROPERTIES OF ROTATIONAL LEVELS

The rotational levels of a given electronic state of a diatomic molecule are classified according to the behavior of the total coordinate eigenfunction (not of the rotational eigenfunction alone) with respect to inversion at the origin of a space-fixed system of axes (not of a molecule-fixed system). A rotational level is called positive or negative according as the total coordinate eigenfunction $\psi(= \psi_e\, \psi_v\, \psi_r)$ remains unchanged or changes sign when inverted at such an origin (i.e. when the signs of the coordinates of all the electrons and the nuclei are reversed). By considering the symmetry properties of the functions ψ_e, ψ_v and ψ_r separately, it may be seen that

in a Σ^+ state   :   even-*J* levels are positive and
                                        odd-*J* levels are negative,

in a Σ^- state   :   even-*J* levels are negative and
                                        odd-*J* levels are positive.

This is shown in Fig. 11.13.

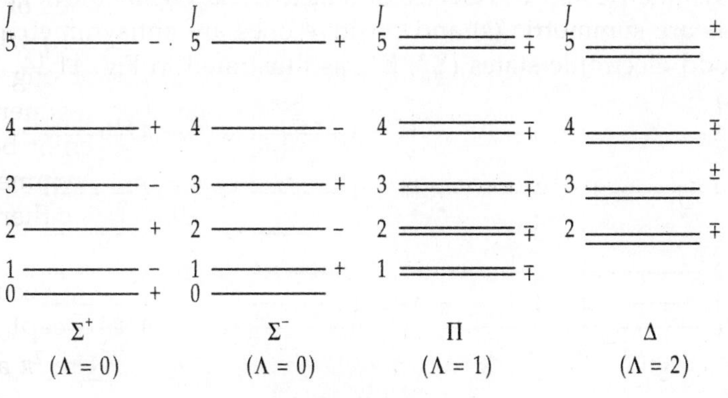

Fig. 11.13

In π, Δ ...states, ($\wedge = 1, 2, ..$), for each value of J, there is a positive and negative rotational level of equal energy. For these states, levels with $J < \wedge$ do not occur. Moreover, the degenerate levels are shown slightly splitted. Actually such a splitting appears when interaction of electronic motion with molecular rotation is taken into account.

The selection rule for transitions accompanied by dipole radiation is that *'positive levels combine only with negative levels and vice versa*. Symbolically,

$$+ \leftrightarrow -, + \not\leftrightarrow +, - \not\leftrightarrow -$$

It can be seen from the above figure that this selection rule supports the dipole-radiation selection rules for J, viz. $\Delta J = \pm 1$ for $\wedge = 0$; and $\Delta J = 0, \pm 1$ for $\wedge > 0$.

The opposite selection rule holds for Raman transitions, i.e. *'positive levels combine with positive levels and negative levels combine with negative levels'*. Symbolically,

$$+ \leftrightarrow +, - \leftrightarrow -, + \not\leftrightarrow -$$

Again, this rule supports the Raman selection rules for J viz. $\Delta J = 0, \pm 2$ for $\wedge = 0$ and $\Delta J = 0, \pm 1, \pm 2$ for $\wedge > 0$.

For homonuclear molecules having identical nuclei (such as $O^{16} O^{16}$ but not $O^{18} O^{18}$), the wave equation remains unchanged if the two nuclei are mutually exchanged. This additional symmetry operation provides a further classification of the rotational levels. The levels are called symmetric (s) or antisymmetric (a) with respect to the nuclei according as the total coordinate eigenfunction ψ remains unchanged or only changes sign when the coordinates of the two nuclei are interchanged. It can be seen that the positive rotational levels are symmetric and the negative ones are antisymmetric for even electronic states (Σ_g^+, Σ_g^-) while the negative levels are symmetric (s) and positive ones are antisymmetric (a) for odd electronic states (Σ_u^+, Σ_u^-) as illustrated in Fig. 11.14.

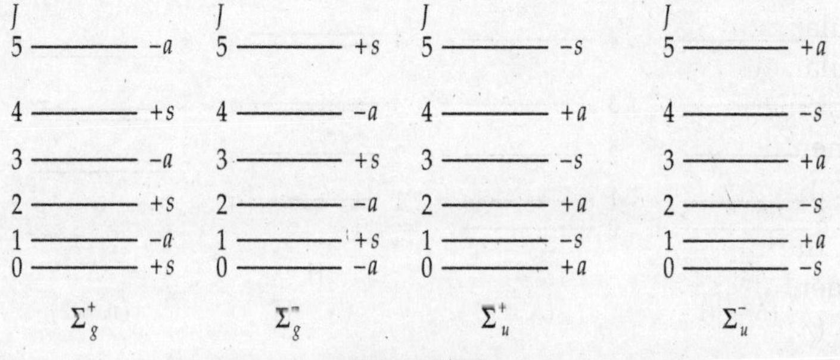

J		J		J		J	
5	$-a$	5	$+s$	5	$-s$	5	$+a$
4	$+s$	4	$-a$	4	$+a$	4	$-s$
3	$-a$	3	$+s$	3	$-s$	3	$+a$
2	$+s$	2	$-a$	2	$+a$	2	$-s$
1	$-a$	1	$+s$	1	$-s$	1	$+a$
0	$+s$	0	$-a$	0	$+a$	0	$-s$
Σ_g^+		Σ_g^-		Σ_u^+		Σ_u^-	

Fig. 11.14

It can be noted that the even-J levels are symmetric (s) and odd -J levels are antisymmetric (a) when the electronic state is even and positive (Σ_g^+) or odd and negative (Σ_u^-). On the other hand, even-J levels are antisymmetric (a) and odd-J levels are symmetric (s) when the electronic state is even and negative (Σ_g^-) or odd and positive (Σ_u^+).

If the identical nuclei have zero nuclear spin, or if the interaction of the nuclear spin with the rest of the molecule is neglected, then a transition between a symmetric and an antisymmetric level is strictly prohibited. Symbolically,

$$a \not\leftrightarrow s,$$

i.e. $$a \leftrightarrow a \text{ and } s \leftrightarrow s$$

This rule holds not only for transitions with dipole radiation but also for transitions with quadrupole radiation, Raman effect, etc.

The above rule supports the absence of infrared rotation or vibration-rotation spectra of homonuclear molecules because the transitions $\Delta J = \pm 1$ necessary for these spectra are in contradiction of the rule. However, Raman transitions $\Delta J = 0, \pm 2$ do take place in homonuclear molecules because they are in accordance of the above rule.

11.12 COUPLING OF NUCLEAR ROTATIONAL AND ELECTRONIC INTERACTIONS

In an actual molecule the nuclei vibrate relative to each other and also rotate as a whole about the centre of mass of the molecule. These nuclear motions and the electronic motion occur simultaneously and interact with each other. In general, the total angular momentum of a diatomic molecule is the resultant of three angular momenta—the electronic orbital angular momentum, the electronic spin angular momentum and nuclear rotational angular momentum (it does not include nuclear spin), and is represented by $\vec{J}$ having a magnitude $\sqrt{J(J+1)}\,\hbar$.

If the molecule is in a $^1\Sigma$ state, i.e. the electronic angular momentum as well as the electronic spin angular momentum is zero ($\wedge = 0$, $S = 0$); the nuclear rotational angular momentum is identical with the total angular momentum $\vec{J}$. This corresponds to the "vibrating-rotator" model (Fig. 11.15), whose rotational-energy is expressed by

$$F(v, J) = B_v \, J(J + 1)$$

neglecting the term for the non-rigidity of the molecule.

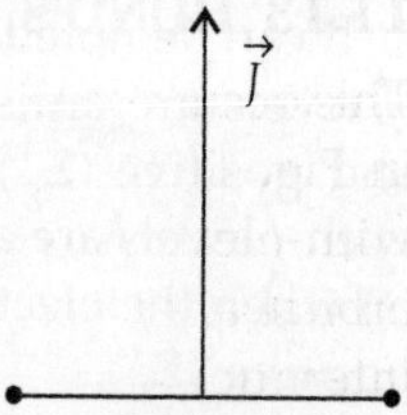

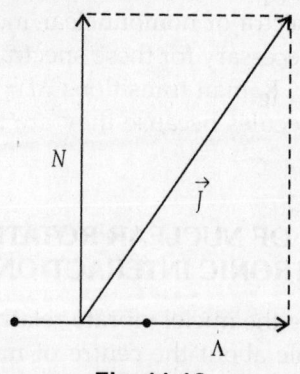

Fig. 11.15

If the molecule is in a singlet ($S = 0$) state, corresponding to $\wedge > 0$, i.e. in one of the $^1\pi$, $^1\Delta$ ... states, it then corresponds to the symmetric top model (Fig. 11.16). In this case, the nuclear rotational angular momentum $N^{(*)}$ perpendicular to the internuclear axis and the electronic orbital angular momentum component $\wedge$ along the internuclear axis form the total angular momentum $\vec{J}$. The rotational energy of the molecule is expressed by

$$F(v, J) = B_v \, J(J + 1) + (A - B_v)\wedge^2$$

neglecting the centrifuged term.

Fig. 11.16

In all other electronic states, the rotational energy depends upon the modes of the coupling between the various angular momenta. F. Hund in 1926 described four types of coupling depending upon which of the different interactions predominate. These are described subsequently. (These however, represent limiting situations and most molecular states show an intermediate type of behavior in practice.)

$^{(*)}$ N is not a quantum number, its value is determined by the value of the quantum numbers J and $\wedge$.

11.13 HUND'S COUPLING CASE (a)

This occurs only for states corresponding to $\wedge > 0$ and is illustrated in Fig. 11.17. In this case, the interaction of the nuclear rotation with electronic orbital and spin motion is weak. The electronic orbital angular momentum vector $\bar{L}$ precesses rapidly about the internuclear axis with quantized component $\wedge$. Further, in this case, the magnetic field generated by the orbiting electrons is sufficiently strong to couple the spin vector $\bar{S}$ to the internuclear

axis, i.e. S also precesses rapidly about this axis with quantized component Σ. (The magnitude is $\Sigma h/(2\pi)$, where the quantum number Σ has $2S + 1$ values ranging from $+S$ to $-S$. The total component of electronic angular momentum about the internuclear axis has a magnitude $\Omega h/(2\pi)$, where the quantum number Ω is the algebraic sum $|\wedge + \Sigma|$ and takes $2S + 1$ values ranging from $\wedge + S$ to $L - S$.

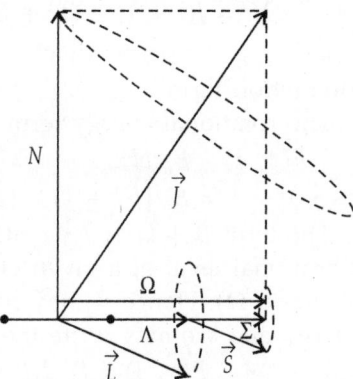

Fig. 11.17

The electronic angular momentum component Ω and the nuclear rotational angular momentum component N (for the whole molecule), form the total angular momentum $\bar{J}$. Thus, this Hund's case (*a*) is identical to the symmetric top model except that here, we have Ω in place of $\wedge$. The vector $\bar{J}$ is constant in magnitude and direction. Both Ω and N rotate about this vector (nutation). This rotation is however, much slower than the precession of $\bar{L}$ and $\bar{S}$ about the internuclear axis.

Since Ω is the component of $\bar{J}$ along the internuclear axis therefore J cannot be smaller than Ω and the quantum number J has the possible values

$$J = \Omega, \ \Omega + 1, \ \Omega + 2 \ ...$$

Ω and hence J is integral or half integral according as Σ is integral or half integral. Thus, associated with each electronic substate (with a given Ω) there is a set of rotational levels with $J = \Omega$, $(\Omega + 1)$, $(\Omega + 2)$ ...Levels with $J < \Omega$ do not occur.

In order to calculate these rotational energy levels, we must find the nuclear rotational angular momentum $N\hbar$. Here N is not a quantum number. Its value is fixed by the values of the quantum numbers J and Ω by (*see* Fig. 11.17):

$$N^2 = J(J + 1) - \Omega^2$$

This is, however, an approximate relation because, the value of N will be affected by the components of $\vec{L}$ and $\vec{S}$ perpendicular to the internuclear axis. When these effects are included, we have

$$N^2 = J(J + 1) - \Omega^2 + L_p^2 + S(S + 1) - \Sigma^2 + \phi(J)$$

where L_p is the perpendicular component of $\vec{L}$ and $\phi(J)$ is a small interaction term.

The rotational energy term is given by

$$F(v, J) = B_v N^2$$
$$= B_v [J(J + 1) - \Omega^2 + L_p^2 + S(S + 1) - \Sigma^2] + B_v \phi(J)$$

The term $B_v [-\Omega^2 + L_p^2 + S(S + 1) - \Sigma^2]$ is constant for a given vibrational level of a given electronic substate (for given values of v and Ω). Hence it may be regarded as part of the vibronic energy and we may write the purely rotational terms as

$$F(v, J) = B_v J(J + 1) + \phi_i(J)$$

where we have put $B_v \phi(J) \equiv \phi_i(J)$.

As an example, let us construct the rotational levels of a $^3\pi$ state in case (*a*). For this state, we have

$$\wedge = 1, S = 1$$

As a result of spin orbit ($\Sigma - \wedge$) interaction, the given state is split up into $(2S + 1) = 3$ substates (multiplet components) $^3\pi_0$, $^3\pi_1$ and $^3\pi_2$ distinguished by $\Omega = |\wedge + S| = 0, 1, 2$. Associated with each of these electronic substates is a pile of rotational levels, each having a characteristic J-value ($J = \Omega, \Omega + 1, \Omega + 2$ ..) and an energy $F(v, J)$ given by the above expression (Fig. 11.18).

Since the minimum value of J is Ω, there is one missing level ($J = 0$) in the pile associated with $^3\pi_1$ and two missing levels ($J = 0$, 1) in the pile associated with $^3\pi_2$. The small interaction term ϕ_i differs for the multiplet components. Since B_v is almost independent of Ω, the separation between levels with the same value of J in each pile are the same.

When $\wedge = 0$ and $S \neq 0$, the spin vector $\vec{S}$ cannot be coupled to the internuclear axis which is a necessity for case (*a*). Hence Hund's case (*a*) cannot be applied to Σ-states.

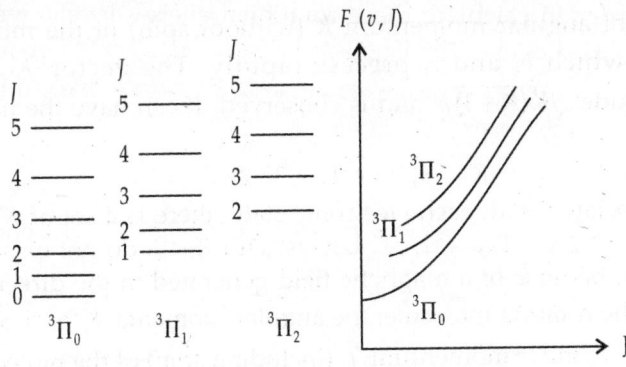

Fig. 11.18

11.14 HUND'S COUPLING CASE (b)

In this case (b), the vector $\vec{L}$ precesses rapidly round the internuclear axis with quantized component $\wedge$ but unlike case (a), the magnetic field generated by orbiting electron is so weak that the spin vector $\vec{S}$ is no longer coupled to the internuclear axis, consequently, the quantum number Σ is undefined, therefore case (b) always applies to Σ states ($\wedge = 0$) and may also apply to other electronic states when the molecule contains relatively few electrons. The case (b) is illustrated in Fig. 11.19.

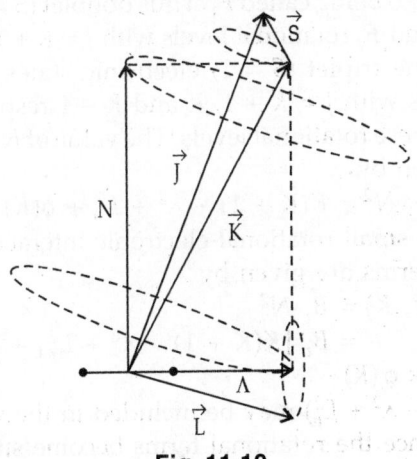

Fig. 11.19

The electronic orbital angular momentum component $\wedge$ and the nuclear rotational angular momentum component N form a resultant angular momentum K (without spin) of the molecule about which N and $\wedge$ precess rapidly. The vector $\vec{K}$ has a magnitude $\sqrt{K(K+1)}\hbar$ and is conserved. K can have the integral values

$$K = \wedge, \ \wedge + 1; \ \wedge + 2, \ \dots$$

i.e. associated with each electronic state, there is a set of K-levels with $K = \wedge, \ \wedge + 1, \ \wedge + 2, \ \dots$ Levels with $K < \wedge$ do not exist.

Now, because of a magnetic field generated in the direction of $\vec{K}$ by the rotating molecule, the angular momenta $\vec{K}$ and $\vec{S}$ form the total angular momentum $\vec{J}$ (including spin) of the molecule in a fixed direction about which both $\vec{K}$ and $\vec{S}$ precess slowly. The total quantum number J takes values given by

$$J = (K + S), \ (K + S - 1), \ \dots \ |K - S|$$

i.e. each K-level of a given electronic state consists of $2S + 1$ (when $K > S$) or $2K + 1$ (when $K < S$) component levels. These are called 'spin multiplet components' differing in J-values which may be integral or half integral depending upon S. The slight splitting of the levels with different J but same K is due to the relatively weak magnetic spin-rotation $(\vec{S} - \vec{K})$ coupling, and increases with increasing K. This splitting is called case (b) spin-splitting.

Mulliken designated the component levels as $F_1, F_2, F_3 \dots$, those for which $J = K + S$ being called F_1. Thus doublet ($S = 1/2$) electronic states have F_1 and F_2 rotational levels with $J = K + 1/2$ and $K - 1/2$ respectively. The triplet ($S = 1$) electronic states have F_1, F_2, F_3 rotational levels with $J = K + 1$, K and $K - 1$ respectively. Let us now calculate these rotational levels. The value of N (not a quantum number) is given by

$$N^2 = K(K + 1) - \wedge^2 + L_p^2 + \phi(K)$$

where $\phi(K)$ is a small rotational-electronic interaction term. Thus the rotational terms are given by

$$F(v, \ K) = B_v \ N^2$$
$$= B_v \ [K(K + 1) - \wedge^2 + L_p^2] + \phi_i(K)$$

where $B_v \ \phi(K) \equiv \phi_i(K)$

The part $B_v \ (- \wedge^2 + L_p^2)$ may be included in the vibronic part of the energy. Hence the rotational terms become simply

$$F(v, \ K) = B_v \ K(K + 1) + \phi_i(K)$$

As an example, let us consider the rotational levels of a $^2\Sigma$-state. For this, we have

$$\Lambda = 0, S = \frac{1}{2}$$

$$\therefore \quad K = 0, 1, 2, 3, \dots$$

and
$$J = \frac{1}{2}; \left(\frac{3}{2}, \frac{1}{2}\right); \left(\frac{5}{2}, \frac{3}{2}\right); \left(\frac{7}{2}, \frac{5}{2}\right) \dots$$

Thus the rotational levels of $a^2\Sigma$-state are doubled (spin-doubling), with the exception of the $K = 0$ level, as shown in Fig. 11.20.

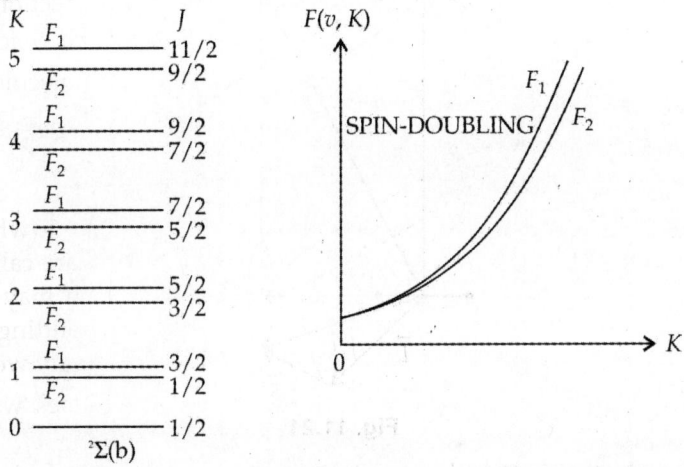

Fig. 11.20

Hund and Van Vleck have calculated the values of $\phi_i(K)$ for the $^2\Sigma$-states and have shown that the rotational term values are given by

$$F_1(v, K) = B_v K(K + 1) + \frac{1}{2}\gamma K$$

$$F_2(v, K) = B_v K(K + 1) - \frac{1}{2}\gamma(K + 1)$$

where the splitting constant $\gamma \ll B_v$. The splitting $F_1 - F_2$ is $\gamma\left(K + \frac{1}{2}\right)$ which increases linearly with K.

11.15 HUND'S COUPLING CASE (c)

This case is rare and found among the less stable excited states of heavier molecules, particularly the halogens. It occurs when the interaction between $\vec{L}$ and $\vec{S}$ is stronger than the interaction with the internuclear axis i.e., $\wedge$ and Σ are not defined. In this case (Fig. 11.21), $\vec{L}$ and $\vec{S}$ first form a resultant $\vec{J}_a$ about which they precess. The vector $\vec{J}_a$ is coupled to, and precesses round the internuclear axis with quantised component Ω. The quantum number Ω can take the possible values

$$J_a, J_a - 1 \; .., \; \frac{1}{2} \; \text{or} \; 0.$$

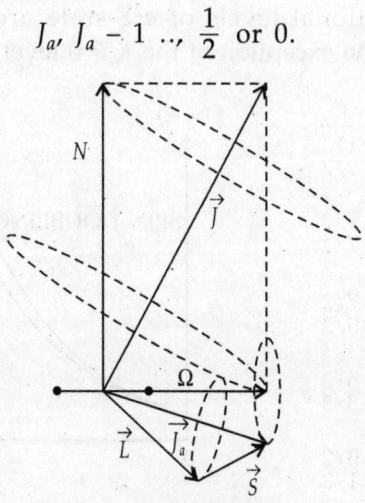

Fig. 11.21

The electronic angular momentum component Ω (along the internuclear axis) and the nuclear rotational angular momentum component N ($\perp$ to the internuclear axis) then combine vectorially to give a resultant $\vec{J}$ in a fixed direction about which both Ω and N precess.

It is clear that in case (c), an electronic state cannot be denoted by the symbols Σ, π, Δ ...etc because the quantum number $\wedge$ is not defined in this case. It can only be denoted by its Ω-value 0, $\frac{1}{2}$, 1 ...The rotational energy values are roughly represented as

$$F(v, J) = B_v \, [J(J + 1) - \Omega^2]$$

where $\qquad\qquad J = \Omega, \; \Omega + 1, \; \Omega + 2 \; ...$

11.16 HUND'S COUPLING CASE (d)

This coupling case occurs in some very highly excited states of H_2 and He_2 when the outer electron generating the orbital angular momentum $\vec{L}$ is far removed from the nuclei and inner electrons.

Then the nuclei plus the inner electrons act like a point charge so that the orbiting electron is not influenced by the electric field along the internuclear axis i.e. vector $\vec{L}$ is not coupled to the internuclear axis. In this case, the angular momentum of nuclear rotation called $\vec{R}$ is separately quantized with magnitude $\sqrt{R(R+1)}\,\hbar$ where the quantum number R takes the values

$$R = 0, 1, 2, \ldots$$

The angular momenta $\vec{R}$ and $\vec{L}$ combine vectorially to give a resultant $\vec{K}$ about which both $\vec{R}$ and $\vec{L}$ precess (Fig. 11.22). The quantum number K for a given R can have the values

$$K = (R + L), (R + L - 1) \ldots |R - L|$$

Thus there are $(2L + 1)$ different K values for each R, except when $R < L$. The rotational energy terms are given by

$$F(v, R) = B_v\, R(R + 1) + \text{small interaction terms causing}$$
$$\text{splitting into } 2L + 1 \text{ components.}$$

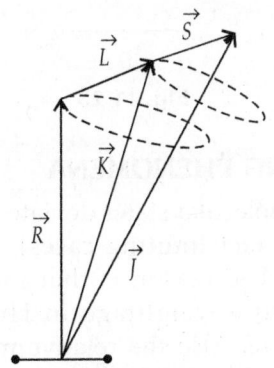

Fig. 11.22

Thus each main rotational level (fixed by the value of R) is splitted into $2L + 1$ sublevels, each characterised by a value of K.

Finally when $S > 0$, the vectors $\vec{K}$ and $\vec{S}$ couple to give $\vec{J}$ where J takes $2S + 1$ values between $K + S$ and $K - S$. Due to this coupling, each sublevel of a given K-value is further splitted into $2S + 1$ still closer components, each characterised by a J-value. This splitting is however very-very small and is usually negligible.

In this case, the molecular quantum numbers $\wedge$ and Σ are not defined. Hence each electronic state is identified as S, P, D according to the value of L, as in atoms.

As an example, let us consider the electronic state $^1D(L = 2,$ $S = 0)$. With this state are associated a number of rotational levels $R = 0, 1, 2, ...$ Each of these has a pile of $2L + 1 = 5$ sublevels each characterised by K. For $R < L$ however, the number of sublevels is $2R + 1$. This is shown in Fig. 11.23.

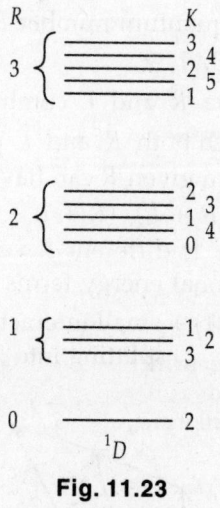

Fig. 11.23

11.17 UNCOUPLING PHENOMENA

In actual case, most molecular states deviate from Hund's coupling cases (which are in fact limiting cases), and belong to some intermediate cases. The reason is that interactions which are ignored or given little weightage in Hund's cases have an appreciable magnitude. Also the relative magnitudes of various interactions change with increasing rotation. Hence there are cases when, with increasing rotation, a transition takes place from one Hund's coupling case to another. There are in particular important uncoupling phenomena which give rise to coupling transitions:

    (*i*) **Orbital uncoupling:** When $\vec{L}$ is uncoupled from the internuclear axis, a transition from case (*a*) or (*b*) to case (*d*) takes place. This phenomenon is associated with Λ-type doubling.

   (*ii*) **Spin uncoupling:** When $\vec{S}$ is uncoupled from the internuclear axis, a transition takes place from case (*a*) to case (*b*).

Both types of uncoupling may be present together in actual molecular states.

Questions

11.1   State and explain Franck-Condon principle.

11.2   What are Fortrat parabolae?

11.3   Explain band origin and band head in relation to the rotational fine structure of electronic vibrational spectra.

11.4   In the rotational fine structure of electronic vibration spectra, in certain molecules, the band head appears on the violet side of the spectrum and in certain other molecules, the band head appears at the red end of the spectrum, why?

11.5   Explain predissociation.

11.6   Discuss the principal features of the electronic spectrum of a diatomic molecule.

11.7   What do you understand by the coarse structure of an electronic transition?

11.8   State Franck-Condon principle and give its wave-mechanical interpretation. How does it help in understanding the intensity distribution in the vibrational structure of the electronic transitions of a diatomic molecule?

11.9   Write a short note on Franck-Condon principle.

Raman Spectra

12.1 CLASSICAL THEORY OF RAMAN EFFECT

Raman spectroscopy is a branch of molecular spectroscopy which is based on the Raman effect. This is a molecular light scattering phenomenon in which a change of frequency takes place. In 1928, Raman showed that if the incident light is monochromatic, the scattered spectrum will also consist of other frequencies in addition to radiation of incident frequency. This is known as *Raman scattering*. It uses light in the visible region of the spectrum, the frequency of which is deliberately chosen so that it is not absorbed by the system under investigation. What is observed is the spectrum of the light after it has been scattered by the molecules of the sample. Frequency shifts occur which (in the vibrational Raman effect) are found to be equal to normal vibrational frequencies of the scattering species. In this, there is no absorption of light at all. Vibrational transitions that are forbidden in infrared absorption, may be permitted in the Raman effect and *vice-versa*. Though Raman shift falls in far and near infrared regions, Raman spectra are quite different from infrared spectra as shown in Table 12.2.

The classical theory of the Raman effect though not wholly adequate, is worth considering since it leads to an understanding of a basic concept important for this form of spectroscopy, viz. the polarisability of a molecule.

12.2 MOLECULAR POLARISABILITY

A molecule is made up of positively charged nuclei embedded in a cloud of negative electrons, so it is electrically polarisable. When molecule is put into a static electric field, it suffers some distortion, the positively charged nuclei being attracted towards the negative pole of the field, and the electrons, to the positive pole.

This separation of charge centres causes an induced electric dipole moment to be set up in the molecule and the molecule is said to be polarised. The amount of the induced dipole P depends on the magnitude of the applied field E, and on the ease with which the molecule can be distorted:

$$P = \alpha E \qquad \qquad ..(1i)$$

where α is the electric polarizability of the molecule. In general, the vector P will have a different direction from that of the vector E, i.e. α is not a simple scalar quantity.

Consider, for example, a diatomic molecule, say H_2 as shown in Fig. 12.1 (a). The polarizability is *anisotropic*, i.e. the electrons forming the bond are more easily displaced by an electric field applied along the bond axis than across this direction. We can represent the polarizability in various directions by drawing a *polarizability ellipsoid* as in Fig 12.1 (b). It is a three-dimensional surface whose distance from the electrical centre of the molecule is proportional to $1/\sqrt{\alpha_i}$, where α_i is the polarizability along the line joining point i on the ellipsoid with the electrical centre. Thus where the polarizability is greatest, the axis of the ellipsoid is least and *vice versa* (This is analogous to the momentum ellipsoid defined by $1/\sqrt{I_i}$, I_i being moment of inertia about an axis i). In the directions at right angle to the bond axis, the ellipsoid has a circular cross-section. All diatomic molecules and linear polyatomic molecules have ellipsoids of the same general shape, differing only in the relative sizes of their major and minor axes. In general, α is a tensor.

The magnitudes of the components of P are related to the magnitudes of the components of applied electric field E by the relation:

$$\left.\begin{array}{l} P_x = \alpha_{xx}E_x + \alpha_{xy}E_y + \alpha_{xz}E_z \\ P_y = \alpha_{yx}E_x + \alpha_{yy}E_y + \alpha_{yz}E_z \\ P_z = \alpha_{zx}E_x + \alpha_{zy}E_y + \alpha_{zz}E_z. \end{array}\right] \qquad ..(1ii)$$

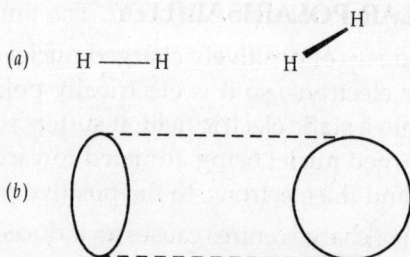

Fig. 12.1: The H_2 molecule and its polarizability ellipsoid seen from two directions at right angles

The nine coefficients α_{ij} involved are called the components of the polarizability tensor α. We can rewrite the above equations in following matrix form:

$$\begin{pmatrix} P_x \\ P_y \\ P_z \end{pmatrix} = \begin{pmatrix} \alpha_{xx} & \alpha_{xy} & \alpha_{xz} \\ \alpha_{yx} & \alpha_{yy} & \alpha_{yz} \\ \alpha_{zx} & \alpha_{zy} & \alpha_{zz} \end{pmatrix} \begin{pmatrix} E_x \\ E_y \\ E_z \end{pmatrix} \qquad ..(2)$$

12.3 THE CLASSICAL THEORY

Now, the polarizability α will, in general, be a function of all the normal coordinates of vibration Q_k's associated with some fundamental frequency ν_{mol} of the scattering molecule.

We may therefore expand α as a Taylor series with respect to these coordinates and neglect powers higher than the first. Thus,

$$\alpha = \alpha_0 + \sum_k \left\{ \left(\frac{\partial \alpha}{\partial Q_k} \right)_0 Q_k \right\} \qquad ..(3)$$

α_0 is the polarizability tensor in the equilibrium configuration of the molecule and $(\partial \alpha / \partial Q_k)_0$ is the so called derived polarizability (also at the equilibrium configuration) for the kth normal mode. The subscript 0 indicates that these are to be taken at equilibrium configuration.

In Raman spectroscopy, the electric field is applied by irradiating the molecule with monochromatic light of a frequency ν_0, which usually lies in the visible region and is chosen so as to ensure that no absorption by the molecule can occur. The time dependence of the electric field E of the incident radiation is given by

$$E = E_0 \cos \nu_0 t \qquad ..(4)$$

where E_0 is the amplitude. The time dependence of the normal coordinate of vibration Q_k is given in simple harmonic approximation by

$$Q_k = Q_0 \cos \nu_{mol} t \qquad ..(5)$$

where ν_{mol} is the rotational or vibrational or rotational-vibrational frequency of the molecule.

Using equations (1), (3), (4), and (5), we get the following expression for time dependence of polarizability P

$$P = \left[\alpha_0 + \left(\frac{\partial \alpha}{\partial Q_k} \right)_0 Q_0 \cos \nu_{mol} \, t \right] E_0 \cos \nu_0 t$$

$$..(6)$$

Using the relation

$$\cos A \cos B = \frac{1}{2} \{ \cos(A + B) - \cos(A - B) \}$$

we get

$$P = \alpha_0 E_0 \cos \nu t = \frac{1}{2} Q_0 E_0 \left(\frac{\partial \alpha}{\partial Q_k} \right)_0 \left[\cos(\nu_0 + \nu_{mol}) t - \right.$$

$$\left. \cos(\nu_0 - \nu_{mol}) t \right] \quad ..(7)$$

Consider the first term on the right hand side. Since every component of α_0 is simply a molecular constant and every component of E oscillates with the incident light frequency ν_0, it follows that the corresponding part of every component of P must oscillate with this frequency. Thus light of the incident frequency ν_0 will be emitted and will be observed in directions which differ from that of the incident light. This phenomenon is known as *Rayleigh scattering*.

Consider now the second term, the contribution to P by this term is characterized by two new frequencies $(\nu_0 + \nu_{mol})$ and $(\nu_0 - \nu_{mol})$. This is sometimes, referred to as the effect of *optical beating* (similar to sound beats)—light of the two 'beat frequencies' will be emitted by the molecule and will be observed in directions different from that of the incident light. (They constitute the contribution to the kth normal mode to the Raman spectrum of the molecule. Thus superimposed on the oscillating dipole are two new frequency components $\nu_0 \pm \nu_{mol}$ and the frequency shifts (relative to the incident frequency) are known as *Raman frequencies*. The scattered lines which are formed on the long wavelength side are called the *Stokes lines*. On the other hand, those which are observed on the short wavelength side of the incident beam are called the *anti-Stokes lines*.

Experimentally it was observed that the Raman frequencies fall in the range 100–4000 cm^{-1} and hence belong to the infrared region of the spectrum. It is considered that the energy difference between the incident and the scattered beam is used up in changing the vibrational energy levels of the molecule. Thus *Raman frequencies are clearly the normal vibrational frequencies of the molecules* of the sample in question as might be observed directly in the infrared (if active in absorption).

Sometimes, Raman lines with very small differences in frequencies are observed. These energy changes are considered to correspond to *pure rotational transitions*.

It must be noted here that if the vibration or rotation does not alter the polarizability of the molecule then the Raman effect will not be observed since then $(\partial\alpha/\partial Q_k)_0 = 0$. Thus we have the following *selection rule*:

$$\left(\frac{\partial\alpha_{ij}}{\partial Q_k}\right)_0 \neq 0 \qquad\qquad ..(8)$$

for at least one of the components of the polarizability tensor α. This means that in order to be Raman active, a molecular rotation or vibration must cause some change in a component of the molecular polarizability. A change in polarizability is obviously reflected by a change in either the magnitude or the direction of the polarizability ellipsoid.

In general, the Raman effect is of the order of a thousand times less intense than the Rayleigh scattering which itself is feeble. Also, long exposure times (commonly ~ 20–40 hrs) are needed to detect Raman spectra with ordinary light. With the advent of lasers, easily operable Raman spectrometers have come to stay.

12.4 QUANTUM THEORY OF RAMAN EFFECT

The quantum mechanical explanation of the Raman effect is as follows: when the incident light quantum $h\nu'$ collides with a molecule, it either scatters elastically, in that case its energy (and therefore its frequency) remains unaltered (This is Rayleigh scattering) or, it scatters inelastically; then it either gives up part of its energy to the scattering system or takes energy from it. It is quite natural that the light quantum can give to or take from the system only amounts of energy that are equal to the energy difference between the stationary states of the system. Let $\Delta E = E' - E''$ denote such an energy difference. Then if the system is initially in the lower state E'', it may be brought to the upper state E' by the scattering of a light quantum (the energy ΔE being taken from the light quantum). Consequently, after the scattering, the energy of the light quantum is $h\nu' - \Delta E$. On the other hand, if the system was initially in the state E' and then transfers to E'' by scattering, the energy of the light quantum after scattering is $h\nu' + \Delta E$. Hence as a result, the frequencies $\nu' - \Delta E/h$ and $\nu' + \Delta E/h$ appear in the scattered light as well as the undisplaced frequency ν'. This is the *Raman effect*. If frequencies and energies are measured in wave number units, the *Raman shifts* give directly the energy differences of the system in cm^{-1}. The Raman lines displaced toward longer wavelengths are called *Stokes lines* and those displaced toward shorter wavelengths are called *anti-Stokes lines*. The Fig. 12.2 illustrates quantum mechanical explanation of Raman effect.

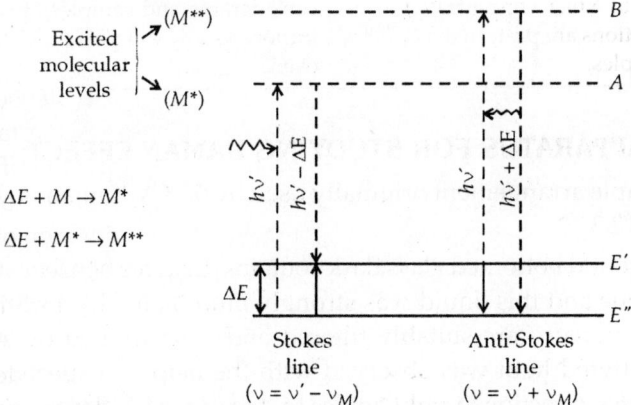

Fig. 12.2: Quantum theory of Raman effect

In the figure, the levels indicated by broken lines (A and B) do not correspond to any possible energy states of the system but only give the energy of the molecule above the initial state. The heavy arrows give the transitions actually taking place in the system.

In stokes lines, the molecule accepts the light quantum and frequency of scattered radiation is lowered. In anti-Stokes lines, the energy is given to the photon by the molecule and scattered frequency is higher.

The Raman effect has to be distinguished from fluorescence which is the re-emission of light at various frequencies after absorption of exciting light. A comparison is given in Table 12.1.

TABLE 12.1: Comparison between Raman spectra and fluorescence

Raman Spectra	Fluorescence
1. Emitted spectral lines have frequencies greater and lesser than the incident frequency.	Line frequency is always less than incident frequency.
2. Frequencies of Raman lines are not determined by the scatterer but by incident frequencies.	Frequencies are determined by the nature of the scatterer.
3. Raman lines are strongly polarized.	Lines are not polarized.
4. Raman lines are weak in intensity due to which concentrated solutions are preferred as samples.	The intensity of fluorescence lines is considerable and samples at concentrations as low as 1 part in 10^9 may be used.

12.5 APPARATUS FOR STUDYING RAMAN EFFECT

The simple arrangement originally used by Sir CV Raman is shown in Fig. 12.3.

The round bottomed glass flask contains dust free benzene (C_6H_6) or toluene and this liquid was strongly illuminated by 4358 Å line from a mercury arc suitably filtered and concentrated by a lens. The scattered light was observed with the help of a spectroscope placed in a direction at right angles to that of the incident radiation. The scattered radiation consists of Raman lines on either side of the incident radiation, i.e. stokes and anti-stokes lines. The antistokes lines are feeble than stokes lines because the energy is given from the liquid to the scattered radiation.

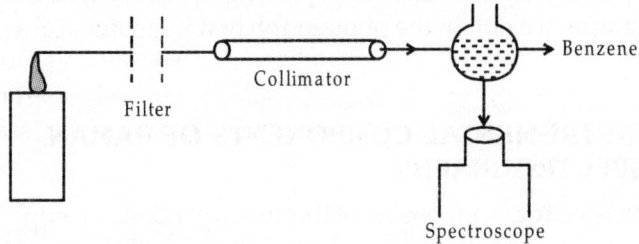

Fig. 12.3: Raman apparatus

The original simple arrangement used by CV Raman was not quite efficient. The modified apparatus used nowadays is due to Wood. It is shown in Fig. 12.4.

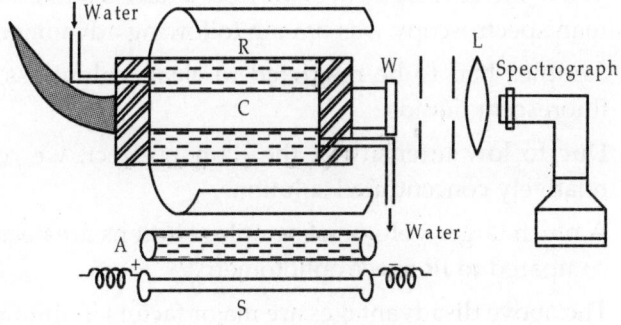

Fig. 12.4

It consists of a tube C of pyrex glass (called the Raman tube) which is filled with the liquid to be investigated. One end of this tube W is plane which transmits the scattered light and the other end is projected outwards in the shape of a horn and is painted black. The tube C is surrounded by a water jacket in which cold water is circulated to prevent overheating of liquid due to proximity of the heated arc. S is mercury arc tube which acts as a source of light. From this mercury arc, single wavelength can be obtained by the use of suitable filters. For example, to obtain 4358 Å line, slightly acidulated quinine sulphate solution contained in a novial glass vessel F is used as a filter which cuts off all the other lines except 4358 Å. R is a reflector around the tube C which increases the intensity of light focused on the tube. The scattered light is focused at the slit of the spectrograph by a lens L. In the spectrograph, we obtain the photograph of Raman lines on a photo plate.

12.6 INSTRUMENTAL COMPONENTS OF RAMAN SPECTROGRAPHY

A Raman spectrograph consists of the undermentioned components:

1. **Source of Light:** The Raman effect is relatively weak, therefore, it is essential to have a source of high intensity. The mercury arc is generally used. It is employed for obtaining a single wavelength (4358 Å) using suitable filters.

    Before the development of the laser as an excitation source, Raman spectroscopy was having following advantages:

    (*i*) Samples had to be restricted to clear, colourless, non-fluorescent liquids.

    (*ii*) Due to low intensity of the Raman effect, we require relatively concentrated solutions.

    (*iii*) A much larger volume of sample solutions are needed as compared to IR spectrophotometry.

    The above disadvantages are major factors in limiting the use of Raman spectroscopy, but the use of He-Ne laser has solved the above shortcomings.

    The laser ($\lambda \sim 6328$ Å) employs a high radio frequency voltage excitation (initial voltage ~ 5–10 kV DC). The advantages of using laser are:

    (*i*) A single intense frequency source replaces the polychromatic mercury lamp. No filtering is thus required.

    (*ii*) The linewidth of a laser line is smaller than the mercury excited line. So, the resolution will be better.

    (*iii*) It offers greater ease in focusing and collimating the radiation since laser light is highly coherent.

(*iv*) A large number of exciting frequencies are available and so, it is possible to study coloured solutions without employing any electronic transitions. This is particularly useful in the study of solutions of inorganic salts.

2. **Filters:** In case of polychromatic incident light, there will be overlapping of Raman shifts which will make the interpretation of Raman spectrum difficult. Therefore it is a must to utilise monochromatic radiation. This necessitates the use of filters. They may be made of nickel oxide glass or quartz glass.

3. **Spectrograph:** The spectrograph necessary for the study of Raman spectrum should possess the following characteristics:

(*i*) It should have large gathering power.

(*ii*) Special prisms of high resolving power should be employed.

(*iii*) A short focus camera should be employed.

A lens in front of the plane window directs the scattered radiation upon the slit of the photograph and the Raman lines can be recorded. Due to weak nature of Raman lines, photographic recording is preferred instead of automatic recording. Photomultiplier tubes are employed as detectors.

A block diagram of the Perkin-Elmer Raman spectrometer is shown in Fig. 12.5.

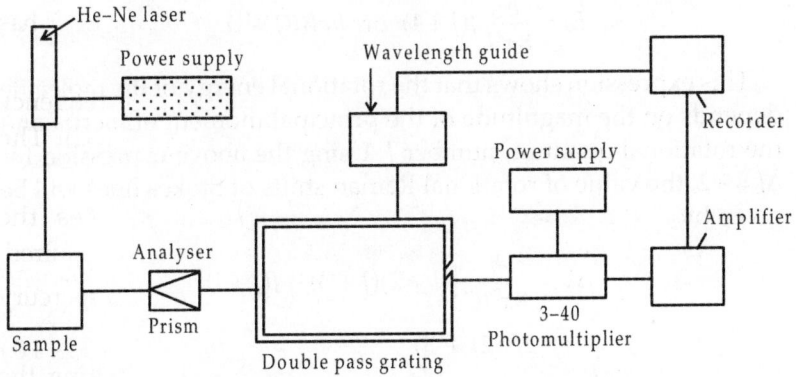

Fig. 12.5

Light from the He-Ne laser beam enters the sample-compartment horizontally. The Raman scattering from the sample-cell is then focused on the monochromator entrance slit. If we require depolarisation measurements, then Raman emission is first allowed to pass through an analyser prism before entering a monochromator (which is a double-pass Lithrow-mounted grating type).

A 13 Hz chopper is used between the first and second passes and the detector is allowed to respond only to decrease interference from stray radiation.

12.7 ROTATIONAL RAMAN SPECTRA

These spectra arise from transitions of the molecule from one rotational energy state to the other of the same vibrational state. They appear on both sides of the Rayleigh line (corresponding to $\Delta J = 0$ which defines the centre of the rotational spectrum). The selection rule is

$$\Delta J = +2$$

The transitions $\Delta J = +2$ give stokes lines (longer wavelengths) while $\Delta J = -2$ give anti-Stokes lines (shorter wavelengths). The two quantum rotational jump behaviour stems from the fact that transitions depend on polarizability, involving two dipole transitions, one for the photon coming in and one for the photon going out. The quantized rotational energy levels of a linear molecule are given as

$$E_r = \frac{h^2}{8\pi^2 I} J(J+1) \text{ or } hcBJ(J+1) \qquad ..(9)$$

This expression shows that the rotational energy of the molecule depends on the magnitude of the principal moment of inertia and the rotational quantum number J. Using the above expression for $\Delta J = +2$, the value of rotational Raman shifts of Stokes lines will be given by

$$\Delta\bar{\nu} = \frac{h}{8\pi^2 IC}\left[(I+2)(J+3) - J(J+1)\right]$$

$$= 2B(2J+3) \qquad ..(10)$$

where $B = \dfrac{h}{8\pi^2 IC}$

For anti-Stokes lines, it can be shown that

$$\Delta \overline{\nu} = -2B(2J + 3) \qquad \qquad ..(11)$$

The Raman effect can be put in the form

$$\Delta \overline{\nu} = \pm 2B(2J + 3); \quad (J = 0, 1, 2...) \qquad ..(12)$$

The wave numbers of the corresponding spectral lines are given by

$$\overline{\nu} = \overline{\nu}_{ex} - \Delta \overline{\nu} \qquad \qquad ..(13)$$

where $\overline{\nu}_{ex}$ is the wave number of exciting radiation. Figure 12.6 shows allowed transitions and Raman spectrum. It can be seen that frequency separation of successive lines is $4B$ (it is $2B$ cm^{-1} in far infrared spectra) whereas Raman study of the first Stokes or anti-Stokes line from the exciting line is $6B$ cm^{-1}.

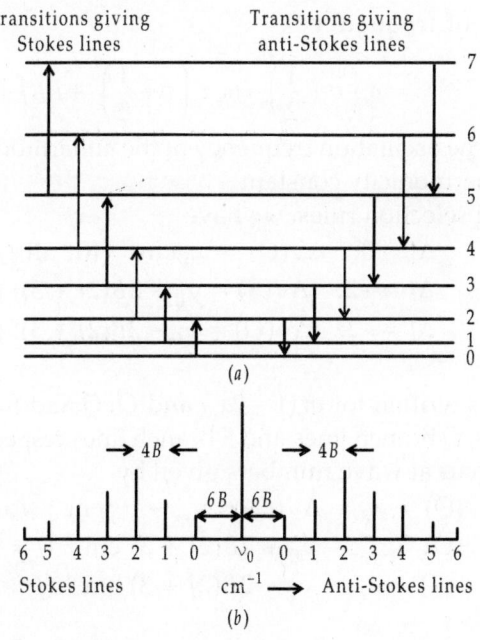

Fig. 12.6: The rotational levels of a diatomic molecule and the rotational Raman spectra

Homonuclear diatomic molecules which have no permanent dipole moment give neither vibrational nor rotational spectra but do give rotational Raman spectra. Thus, the Raman scattering permits the study of non-polar molecules such as N_2O, CO_2, C_2H_2, etc.

12.8 VIBRATION-ROTATION RAMAN SPECTRA

Vibration-rotation Raman spectra are rarely resolved except in case of diatomic molecules. The selection rules are

$$\Delta v = \pm 1 \text{ and } \Delta J = 0, \pm 2.$$

We know that a molecule having n atoms, there are $3n - 6$ ($3n - 5$ if linear) normal modes of vibrations and these different modes are anharmonic. For a diatomic molecule, vibration-rotation energy levels are given by

$$E = hc\left[\omega_e\left(v + \frac{1}{2}\right) - \omega_e x_e\left(v + \frac{1}{2}\right)^2\right] + BhcJ(J + 1) \text{ cm}^{-1}$$

$$(v = 0, 1, 2, .., J = 0, 1, 2, ..) \quad ..(14)$$

or in terms of frequency

$$\nu = \omega_e\left(v + \frac{1}{2}\right) - \omega_e x_e\left(v + \frac{1}{2}\right)^2 + BJ(J + 1) \quad ..(15)$$

where ω_e is the oscillation frequency of the anharmonic system and x_e is the anharmonicity constant.

Applying selection rules, we have

$$\Delta J = 0: \quad \Delta\nu(Q) = \nu_0 \text{ cm}^{-1} \text{ (for all } J)$$
$$\Delta J = +2: \quad \Delta\nu(S) = \nu_0 + 2B(2J + 3); \ (J = 0, 1, 2, ..)$$
$$\Delta J = -2: \quad \Delta\nu(O) = \nu_0 - 2B(2J + 3); \ (J = 2, 3, 4, ..)$$
$$..(16)$$

Here ν_0 is written for $\omega_e(1 - 2x_e)$ and O, Q and S refer to the O branch lines, Q branch lines and S branch lines respectively. Stokes lines will occur at wave numbers given by

$$\nu(Q) = \nu_{ex} - \Delta\nu(Q) = \nu_{ex} - \nu_0 \text{ cm}^{-1} \text{ (for all } J)$$
$$\nu(O) = \nu_{ex} - \nu_0 + 2B(2J + 3) \text{ cm}^{-1} \ (J = 2, 3, 4, ..)$$
$$\nu(S) = \nu_{ex} - \nu_0 - 2B(2J + 3) \text{ cm}^{-1} \ (J = 0, 1, 2, ..)$$
$$..(17)$$

The spectrum is given in Fig. 12.7.

It is to be noted here that unlike near infrared spectra, here (in Raman spectrum) Q branch is present.

Weaker anti-Stokes lines will occur at the same distance from but to high frequency side of the exciting line.

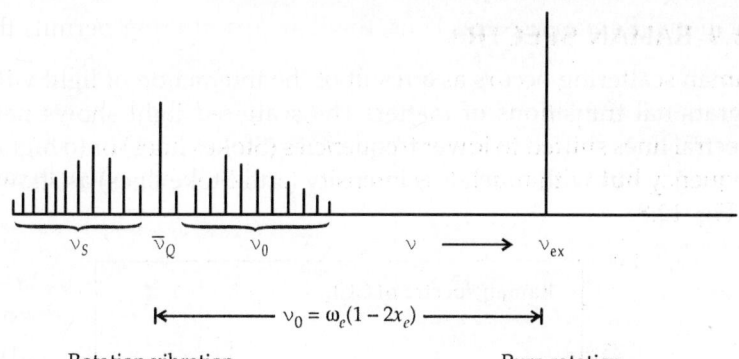

Rotation vibration   Pure rotation

Fig. 12.7: Rotational structure of a Stokes Raman lines of a diatomic molecule having a vibrational frequency $\omega_e(1 - 2x_e)$. Rotational structure near the exciting line ν_{ex} is not shown in the figure

TABLE 12.2: Comparison between Raman and infrared spectra

Raman Spectra	Infrared Spectra
1. It is due to scattering of light by the vibrating molecules	1. It is due to absorption of light by the vibrating molecules
2. Polarisability of the molecule will decide whether the Raman spectra will be observed or not.	2. The presence of a permanent dipole moment may be regarded as a criterion.
3. It can be recorded in one exposure only.	3. It requires at least two separate runs. (with different prisms) to cover the whole region of infrared.
4. Water can be used as a solvent.	4. Water cannot be used as a solvent because it is opaque to infrared radiation.
5. As Raman lines are weaker in intensity, concentrated solutions must be utilised to increase the intensity of Raman lines.	5. Generally dilute solutions are preferred.
6. Homonuclear diatomic molecules are often found to be active.	6. Homonuclear diatomic molecules are not found to be active.
7. Substances under investigation must be pure and colourless.	7. This condition is not rigid.
8. Vibrational frequencies of large molecules can be measured.	8. Vibrational frequencies of very large molecules cannot be measured.

12.9 RAMAN SPECTRA

Raman scattering occurs as a result of the interaction of light with vibrational transitions of matter. The scattered light shows new spectral lines shifted to lower frequencies (Stokes lines) or to higher frequency but with much less intensity (anti-Stoke lines) as shown in Fig. 12.8.

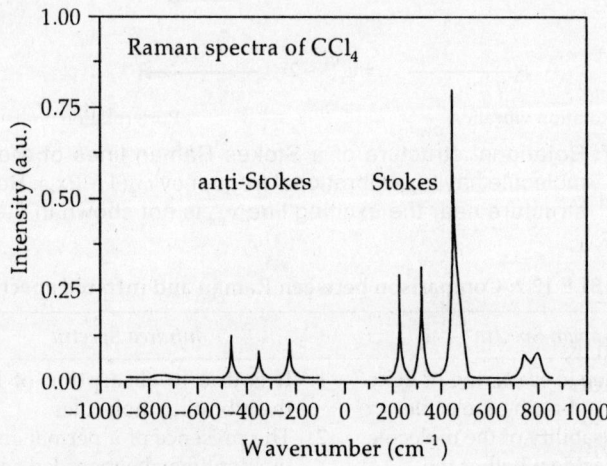

Fig. 12.8: Frequency spectrum of Raman scattering in CCl_4 measured with a conventional spectrometer based on argon laser excitation

Thus, Raman scattering is used in Raman spectroscopy for characterization of the structure of matter, for analysis and is non-linear optics, e.g. for coherent Raman spectroscopy and for frequency conversion of laser radiation as stimulated Raman scattering.

The frequency of the Raman scattered light $\nu_{R\,scatt.}$ follows from the frequency of the incident light ν_{inc} and the frequencies $\nu_{vib,\,m}$ of the (molecular) vibrations by

$$\text{Raman frequency } \nu_{R\,scatt,\,p} = \nu_{inc} \mp (p_{vib}\,\nu_{vib,\,m})$$

with $m = 1, 2$ ...limited by the number of normal vibrations of the system and the number $p_{vib} = 1, 2$ ...limited by the number of vibrations upto the ionization limit of the system.

The Raman effect requires a sufficient interaction of the electric field which oscillates 1000 times faster with the vibration of the material via the non resonant interaction of the electrons in the matter. Thus, only a few vibration of the matter are usually Raman active and only a few of these frequencies show detectable intensities. In Table 12.3 some typical Raman active vibrations of some materials are given. Traditionally, the vibrational frequencies have been measured as wave numbers in cm^{-1}, which is $1/$(wavelength) and these values are also given in the Table.

TABLE 12.3: Raman active vibrations of some gases

Substance	ν_{vib} (THz)	ν_{vib} (cm^{-1})	$d\sigma/d\nu$ $(cm^2/ster.)$
N_2	69.90	2330	2.3×10^{-30}
CH_4	87.42	2914	1.9×10^{-30}
HF	118.86	3962	3.0×10^{-30}
H_2	124.65	4155	5.1×10^{-30}

With Raman scattering, an excitation (Stokes scattering) or depletion of an excited (anti-Stokes scattering) vibration of the matter takes place (Fig. 12.9). The interaction of the electric field of light with the vibration of the matter is effected via the induced polarization in the substance. Thus the oscillating dipole moment μ_R (ν_{vib}, t) can be additionally modulated by ν_{inc}:

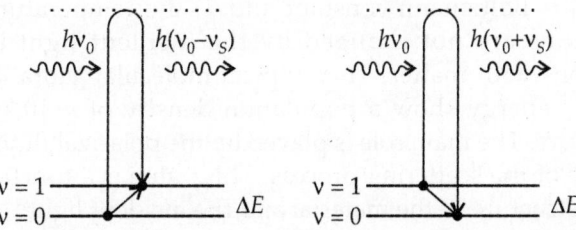

Fig. 12.9: Interaction of incident light with matter vibration generating stokes (left) and antistokes (right) lines in the Raman-process. The thick horizontal lines represent the vibrational energy levels of the material

$$\mu_R = \in_0 \alpha_0 E_0 \sin (2\pi \, \nu_{inc} \, t) + \in_0 \alpha_1 E_0 \sin (2\pi \, \nu_{inc} \, t)$$
$$\times \sin (2\pi \, \nu_{vib} \, t)$$

$$= \in_0 \alpha_0 E_0 \sin (2\pi \, \nu_{inc} \, t) + \frac{1}{2} \in_0 \alpha_1 E_0 \, [\cos \{2\pi \, (\nu_{inc} - \nu_{vib})t\}$$
$$- \cos \{2\pi \, (\nu_{inc} + \nu_{vib})t\}].$$

With the coefficient α_1 representing the polarizability of the material via the vibration and the light field. The resulting difference or sum of the frequencies are the Stokes and anti-Stokes lines of the process.

Because of the non-harmonic vibrational potentials the energy level of the vibrations are not equidistant and therefore the Raman spectra are usually much more complicated although the basic processes are as simple as described above.

The intensity of the scattered light I_{Ram} is proportional to the incident intensity I_{inc} and the population $N_{vib,\,i}$ of the vibrational energy levels:

$$I_{Ram} \,(\nu_{inc} \mp \nu_{vib,\,i}) = \sigma_{Ram} \,(\nu_{inc} \mp \nu_{vib,\,i})\, I_{inc} \,(\nu_{inc})\, N_{vib,\,i}$$

Measurements of the Raman cross sections are also given in the Table 12.3 for the given materials.

It is obvious that anti-Stokes signals are smaller the higher the vibrational frequency. The thermally induced occupation of the excited vibrational states can be calculated using Boltzmann equation for the population density $N_{vib,\,i}$:

$$N_{vib,\,i} = N_0 \exp\left[-\frac{h\nu_{vib,\,i}}{k_B T}\right]$$

with $\qquad \sum_i N_{vib,i} = N_{total}$

where k_B is Boltzmann constant and T, the temperature. These populations are not changed by the incident light in linear interactions with matter. Thus typical molecular vibrations with 1000 cm^{-1} energy show a population density of $< 10^{-6}$ at room temperature. The main role is played by the polarizability α for the efficiency of the scattering process. This value is a function of the detailed structure of the material and the incident light.

12.10 RAMAN SPECTRA AND MOLECULAR STRUCTURE

Raman effect finds an important application in determining the structure (symmetry) of diatomic and polyatomic molecules. The vibrational and rotational Raman spectra enable us to determine the force constant and bond-length of those diatomic molecules which have no permanent dipole-moment.

In polyatomic molecules, the Raman spectra and the infrared data give information regarding shape (linear or bent) and symmetry of the molecules. There is "mutual exclusion rule" according to which *for molecules with a centre of symmetry, the frequencies observed in the infrared spectra are not observed in the Raman spectra and vice-versa.* This enables us to determine the symmetry of molecules. For instance, the molecules CO_2 and CS_2 exhibit two strong infrared bands, neither of which coincides with a Raman displacement. Hence each of these molecules must have a centre of symmetry. For a triatomic molecule this implies that the molecule is linear and symmetric. Hence CO_2 and CS_2 have the structures

$$O — C — O \qquad\qquad S — C — S$$

On the other hand, the molecules N_2O and H_2O show three strong infrared bands coincident (in many cases) with strong Raman bands. Hence these molecules do not have a centre of symmetry. (They may be either bent or linear). It is possible to distinguish between these two possibilities by observation of the rotational structure of the vibrational bands, e.g. N_2O has a non-symmetrical linear structure while H_2O has a bent structure.

$$N — N — O \qquad\qquad \overset{\displaystyle O}{\underset{H \qquad H}{\diagup \diagdown}}$$

12.11 STRUCTURE DETERMINATION USING IR AND RAMAN SPECTROSCOPY

Determination of molecular structure using infrared and Raman spectroscopies is mainly based on the application of symmetry, the vibrational selection rules, state of polarization of the lines and the observed frequencies. For diatomic molecules, there will be one totally symmetric stretching mode which is active in *IR* if the molecule is heteronuclear and inactive if it is homonuclear. On the contrary, it is allowed as a polarized line in Raman effect. The observation of the rotational fine structure of these helps one to determine the moment of inertia and subsequently the bond length of the molecule.

12.11.1 Molecules of the Type XY_2

In case of triatomic molecules, the possibilities are linear symmetric, bent symmetric and linear asymmetric. The number of distinct fundamentals predicted along with their activities are listed in Table 12.4. If it is the linear symmetric model, it has to obey the rule of mutual exclusion as centre of symmetry is present in the molecule.

If it is the bent symmetric on linear asymmetric type, all the three distinct modes are active in both *IR* and Raman. In addition, if the molecule is non-linear, it cannot give rise to infrared bands with *P*, *R* contours.

TABLE 12.4: Predictions for fundamentals of XY₂ type molecules

Model	No. of fundamental allowed in IR	No. permitted in Raman effect	No. of IR and Raman coincidence	No. of polarized Raman lines
Linear Y—X—Y	2	1	0	1
Bent $\begin{smallmatrix} X \\ / \ \backslash \\ Y \quad Y \end{smallmatrix}$	3	3	3	2
Linear Y—Y—X	3	3	3	2

12.11.2 Molecules of the Type XY₃

XY_3 type molecules will have six fundamentals, some of them may become degenerate depending on the symmetry. The simplest of them are the planar D_{3h} point group and the pyramidal C_{3V} point group. The number of distinct fundamentals is four for both of them. For the planar D_{3h} point group, the totally symmetric stretching vibration which is forbidden in *IR* is expected as an intense polarised band in Raman whereas for pyramidal C_{3v}, the symmetric stretching and bending modes are expected to be polarized in Raman. The predictions for fundamentals of symmetrical XY_3 molecule are listed in Table 12.5.

TABLE 12.5: Predictions of fundamentals of symmetric XY₃ molecules

Model	Number of distinct fundamentals	Number of fundamentals permitted in IR	Number permitted in Raman	Number of coincidences	Number of polarized Raman lines
Planar (D_{3h}) $\begin{smallmatrix} Y \\ | \\ X \\ / \ \backslash \\ Y \quad Y \end{smallmatrix}$	4	3	3	2	1
Pyramidal (C_{3v}) $\begin{smallmatrix} X \\ / | \backslash \\ Y \ Y \ Y \end{smallmatrix}$	4	4	4	4	2

12.12 STIMULATED RAMAN EFFECT

The diffusion of light in the original Raman experimental corresponded to a spontaneous Raman effect. (Light was spontaneously emitted in 4π steradian and to avoid the dazzling effect of the pumping (mercury) light, the diffused light was observed in an orthogonal direction). But in 1923, Einstein introduced the possibility of a stimulated effect and an experimental demonstration was made in 1963 (at the early stage of Laser Q-switching). The two physicists, Woodbury and Ng, used a Kerr Cell filled with $C_6H_5NO_2$ (mono-nitrobenzene) to obtain giant pulses from a ruby laser (Fig. 12.10) and found that the laser light contained two spectral components: beside the usual component at 694.3 nm of a ruby laser, another component was observed at 765.8 nm which, having nothing to do with any frequency of the ruby spectrum was identified a Raman-stokes frequency of $C_6H_5NO_2$.

Like many aromatic compounds, mono-nitrobenzene also shows Raman-activity. The Raman-shift is equal to 1340 cm^{-1} and corresponds to a vibration of the radical NO_2. If we take

$$\frac{1}{\lambda_1} - \frac{1}{\lambda_2} = 1340 \text{ nm}$$

and $\lambda_1 = 694.3$ nm,

we obtain $\lambda_2 = 765.8$ nm.

The monochromatic and well-collimated characteristic of the beam emitted at this frequency clearly demonstrate the stimulated origin of the new beam.

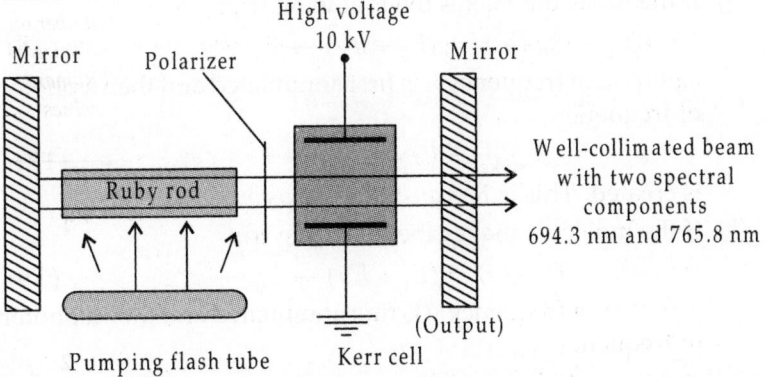

Fig. 12.10: *Q* switching of a ruby laser with a Kerr cell filled with nitrobenzene

12.13 EINSTEIN'S THEORY OF THE STIMULATED RAMAN EFFECT

Raman interaction can be either spontaneous or stimulated. Just like the case of usual interaction of radiation with a collection of two-level atoms, the stimulated effect can be given a classical or a quantum interpretation. We have already described the classical model in earlier section. This model gave a nice interpretation of the Raman diffusion (in terms of electric dipolar momentum of the molecules). The Stokes and anti-Stokes rays should have the same intensity but it is contradiction to the experimental observation. (Experimentally, it has been observed that the anti-Stokes rays are less intense than the Stokes rays). Einstein's theory anticipates different amplitudes for the two rays, the ratio of which is in good agreement with the experimental results.

Real and Virtual Energy Levels

Refer to Fig. 12.2

By absorption of a photon of frequency ν, a molecule reaches the level $E' = (E_0 + h\nu)$, (corresponding to M^*) if initially on level E_0, and level $E'' = (E_1 + h\nu)$ (corresponding to M^{**}) if initially on level E_1.

The levels E' and E'' re called *virtual levels* (as these are not allowed levels of the molecule). The molecule cannot stay for long on these levels and makes (almost immediately) a transition toward one of the two allowed states E_0 or E_1:

(a) If the molecule makes the following trip:

$$E_0 \rightarrow E' = (E_0 + h\nu) \rightarrow E_1 \qquad ..(18i)$$

a photon of frequency ν is first annihilated and then a photon of frequency

$$\nu_{stokes} = (\nu - \nu_{molec.}) \qquad ..(18ii)$$

is created. This is *Raman-Stokes diffusion.*

(b) If the molecule makes the following trip:

$$E_1 \rightarrow E' = (E_1 + h\nu) \rightarrow E_0 \qquad ..(19i)$$

a photon of frequency ν is first annihilated and then a photon of frequency

$$\nu_{anti-stokes} = (\nu + \nu_{molec.}) \qquad ..(19ii)$$

is created. This is *Raman-anti-Stokes diffusion.*

(c) If the molecule makes one of the following trip:
$$E_0 \to E' = (E_0 + h\nu) \to E_0$$
or $\qquad E_1 \to E'' = (E_1 + h\nu) \to E_1 \qquad$ ..(20)
a photon ν is first annihilated and then a new photon of same frequency ν is created. This is *Rayleigh diffusion*.

The Raman-active system may be considered in terms of an interaction of a collection of molecules at thermal equilibrium between the two states E_0 and E_1 with a pumping beam. Then, the numbers of stokes and anti-stokes diffusions are respectively, proportional to the population of the two levels. The lower level being more populated, the number of stokes photon exceeds the number of antistokes photon. The ratio of the two intensities is given by the Maxwell Boltzmann population ratio

$$\frac{I_{\text{anti-stokes}}}{I_{\text{stokes}}} = e^{-(E_1 - E_0)/kT} = e^{-h\nu \text{ molec.}/kT} \qquad ..(21)$$

Equilibrium in a Black Body

After having invented this mechanism for interaction of an electromagnetic radiation with a collection of molecules, Einstein considered as if, a sample of the Raman active material is inside a black body. The problem was then to describe the coupling mechanism between the following objects:

(i) Mode (1) of the blackbody with n_1 photons of frequency ν_1.

(ii) Mode (2) of the blackbody with n_2 photons of frequency ν_2
$= \nu_1 - \nu_{\text{molec.}}$

(iii) Mode (3) of the blackbody with n_3 photons of frequency ν_3
$= \nu_1 + \nu_{\text{molec.}}$

(iv) Collection of molecules with two energy levels E_1 and E_0 of respective populations N_1 and N_0.

We consider independently the equilibrium of the collection of molecules with mode (1) and (2) on one hand and with modes (1) and (3) on the other hand.

The interaction mechanisms are phenomenological and find their justification in giving the right statistical distributions for the two kinds of photons and for the molecules. Let u_1 and u_2 denote the electromagnetic energy densities at frequencies ν_1 and ν_2 respectively and we introduce two phenomenological coefficient A and B. We first consider the case of Antistokes interaction

$$\nu_3 = (\nu_1 + \nu_{\text{molec.}}) > \nu_1$$

and introduce the spontaneous and stimulated processes:

(*i*)   spontaneous effect: $-\left(\dfrac{dN_1}{dt}\right)_{spon} = \left(\dfrac{dN_0}{dt}\right)_{spon}$

$$= A\, N_1 u_1 - AN_0 u_3 \qquad ..(22i)$$

(*ii*)   stimulated effect: $-\left(\dfrac{dN_1}{dt}\right)_{stimul.} = \left(\dfrac{dN_0}{dt}\right)_{stimul.}$

$$= B\, N_0 u_1\, u_3 - BN_1 u_1 u_3 \qquad ..(22ii)$$

Cummulative action of the two effects gives

$$= -\left(\frac{dN_1}{dt}\right) = \frac{dN_0}{dt} = A(N_1 u_1 - N_0 u_3) + Bu_1 u_3 (N_0 - N_1)$$

$$..(23)$$

At equilibrium, the population remains constant:

$$\frac{dN_1}{dt} = 0 \Rightarrow \frac{N_1}{N_0} = \frac{Au_3 - Bu_1 u_3}{Au_1 - Bu_1 u_3} = \frac{\dfrac{1}{u_1} - \dfrac{B}{A}}{\dfrac{1}{u_1} - \dfrac{B}{A}} \qquad ..(24)$$

Now, inside a blackbody, at thermal equilibrium, u_1 and u_3 are given by

$$u_1 = \frac{8\pi h \nu_1^3}{c^3}\left(\frac{1}{e^{h\nu_1/kT} - 1}\right) \qquad ..(25)$$

$$u_3 = \frac{8\pi h \nu_3^3}{c^3}\left(\frac{1}{e^{h\nu_3/kT} - 1}\right) \qquad ..(26)$$

Here A and B are Einstein coefficients and since the two frequencies ν_1 and ν_3 are almost equal, we have

$$\therefore \quad \frac{A}{B} \equiv \frac{8\pi h \nu_1^3}{c^3} \cong \frac{8\pi h \nu_3^3}{c^3} \qquad ..(27)$$

$$\therefore \Rightarrow \frac{B}{A}\left(\frac{8\pi h \nu_1^3}{c^3}\right) \cong \frac{B}{A}\left(\frac{8\pi h \nu_3^3}{c^3}\right) = 1 \qquad ..(28)$$

$$\frac{N_1}{N_0} = \frac{\dfrac{c^3}{8\pi h \nu_1^3}\left(e^{h\nu_1/kT} - 1\right) - \dfrac{B}{A}}{\dfrac{c^3}{8\pi h \nu_3^3}\left(e^{h\nu_3/kT} - 1\right) - \dfrac{B}{A}} = \frac{\left[e^{h\nu_1/kT} - 2\right]}{\left[e^{h\nu_3/kT} - 2\right]} \cong \frac{e^{h\nu_1/kT}}{e^{h\nu_3/kT}}$$

$$= e^{-h\nu_{molec.}/kT}$$

A similar relation can be derived for stokes interaction.

Thus $\dfrac{N_1}{N_0} = e^{-h\nu_{molec.}/kT}$ ..(29)

The formula (eqn. 21) gives exactly the expected value for the ratio of the populations, which can be considered as a proof of the validity of the Einstein model and justification for the existence of spontaneous and stimulated Raman processes. The Raman light intensity is proportional to the intensity of the pumping beam n_1. The stokes light is proportional to the population of the ground level N_0 and the anti-stokes light is proportional to the population of the excited level N_1.

Table 12.6 depicts spontaneous Raman interaction. Here, one photon (ν_1) interacts with one molecule.

Table 12.7 illustrates the stimulated Raman interaction. In this case two photons (ν_1) interact simultaneously with one molecule.

The stimulated Raman effect has been used in design of Raman Laser.

TABLE 12.6: Spontaneous Raman interaction

Stokes	*Anti-Stokes*
Photon1 + molecule ↔ photon2 + molecule*	photon1 + molecule ↔ photon3 + molecule*
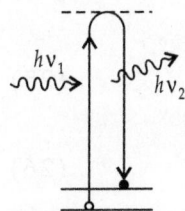	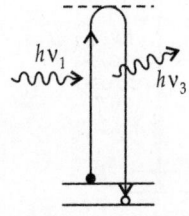
One photon, ν_1, is annihiliated. One molecule is excited. One photon, ν_2, is emitted in any direction. $$\nu_2 = (\nu_1 - \nu_{molec})$$	One photon, ν_1, is annihiliated. One molecule falls to ground level. One photon, ν_3, is emitted in any direction. $$\nu_3 = (\nu_1 + \nu_{molec})$$
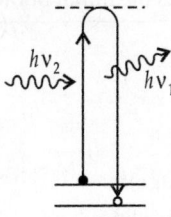	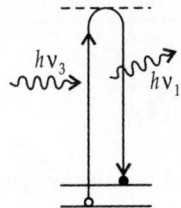
One photon, ν_2, is annihilated. One molecule falls to ground level. One photon, ν_1 is emitted in any direction.	One photon, ν_3, is annihilated. One molecule is excited. One photon, ν_1, is emitted in any direction.

TABLE 12.7: Stimulated Raman interaction

Stokes	Anti-Stokes

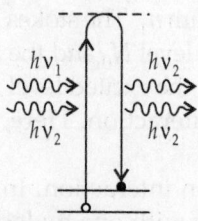 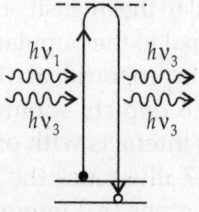

Two photons, ν_1 and ν_2, interact with a molecule at the excited state. The photon of lowest frequency, ν_1, is annihilated.
A second photon, ν_2, is created in the same mode as the initial photon, ν_2.

$$\nu_2 = (\nu_1 - \nu_{molec})$$

Two photons, ν_1 and ν_3, interact with a molecule at the excited state. The photon of lowest frequency, ν_1, is annihilated.
A second photon, ν_3, is created in the same mode as the initial photon, ν_3.

$$\nu_3 = (\nu_1 + \nu_{molec})$$

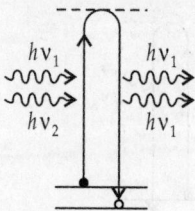 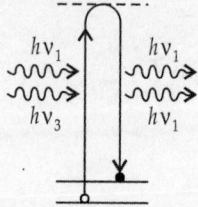

Two photons, ν_1 and ν_2, interact with a molecule at the excited state. The photon of lowest frequency, ν_2, is annihilated.
A second photon, ν_1, is created in the same mode as the initial photon, ν_1.

Two photons, ν_1 and ν_3, interact with a molecule at the excited state. The photon of lowest frequency, ν_3, is annihilated.
A second photon, ν_1, is created in the same mode as the initial photon, ν_1.

Questions

12.1   Explain Raman effect giving an energy level diagram.

12.2   Why are anti-Stokes lines less intense than Stokes lines?

12.3   Outline the advantages of using laser as a Raman source.

12.4   Discuss briefly a method for the determination of bond length of a homonuclear diatomic molecule.

12.5   How are Raman spectra studied in the laboratory?

12.6 Describe the salient features of the Raman spectrum of a heteronuclear diatomic molecule and how would you explain it. In what way does it differ from the fluorescence spectrum?

12.7 Discuss Raman spectra of diatomic molecules and point out the similarity and difference in infrared and Raman spectra.

12.8 What is Raman effect? Explain theoretically the observed characteristics of the Raman spectrum of a diatomic molecule. How is it used to explain the structure of a molecule?

13

Symmetry Elements

The symmetry present in various molecules can be described in terms of what are so called *symmetry elements* which the molecules possess in their equilibrium configurations. These symmetry elements are explained in terms of *symmetry operations* (i.e. each symmetry element has a symmetry operation associated with it).

13.1 DEFINITION OF A SYMMETRY OPERATION

A symmetry operation is a movement of a body such that after the movement has been carried out, every point of the body is coincident with the same point of the body in its original orientation (i.e. the initial and final positions are indistinguishable). Hence the effect of a symmetry operation is to take the body into an equivalent configuration which is indistinguishable from the original one.

A *symmetry element* is a geometrical entity such as a centre, a line or a plane with respect to which one or more symmetry operations may be carried out. Each symmetry element has a/ some symmetry operation(s) associated with it.

In treating molecular symmetry, five types of symmetry elements and operations need to be considered. These are given in Table 13.1.

It is to be noted that the symmetry element E is trivial, since it would be shown by all the molecules.

TABLE 13.1: Five kinds of symmetry elements and operations

Symmetry Symbol	Element Meaning	Symmetry Operation
E (or I)	Identity	(No change)
σ	Plane of symmetry (mirror plane)	Reflection through the plane
i	Centre of symmetry	Inversion through the centre
C_n	Axis of symmetry (proper axis)	Rotation about the axis by $360/n$ degrees
S_n	Rotoreflection axis (improper axis)	Rotation about the axis by $360/n$ degrees followed by reflection in a plane $\perp$ to the rotation axis

13.2 EXAMPLES OF SYMMETRY ELEMENTS

1. **Centre of Symmetry or Inversion Centre (i):** This is an imaginary point in the centre of the molecule such that reflection of each atom when carried out results in its coincidence with an equivalent atom, e.g. C_2H_4 (see Fig. 13.1).

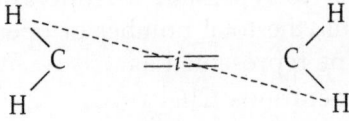

Fig. 13.1

2. **Axis of Symmetry (C_n):** If $2\pi/n$ rotation brings the molecule back to an indistinguishable position then it is said to possess an n-fold axis, e.g. in H_2O there is a C_2 axis.

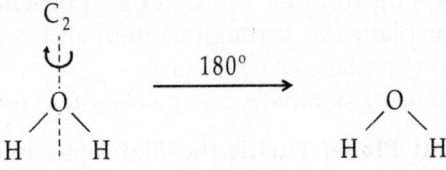

Fig. 13.2

In case of molecules with more than one axis of symmetry, the axis with the highest fold symmetry is called the *principal axis* and the axes of lower fold symmetry are called the *subsidiary axes*, e.g. in BF_3, there is C_3 principal axis and there are also subsidiary axes of two fold symmetry.

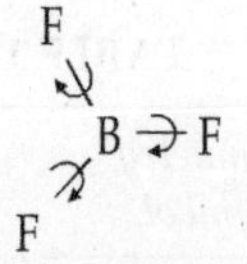

In case of Benzene (C_6H_6), there is a six fold axis (principal axis) and also two fold axes (perpendicular to the six fold axis), the subsidiary axes.

The axis of symmetry can be C_∞, i.e. an axis of infinite fold symmetry, for example, molecular axis in HCl or H_2.

Fig. 13.3

$$\text{--------} H - Cl \dashrightarrow C_\infty$$

The possible number of symmetry operations around a C_n axis is $n - 1$ because nth operation brings the molecule back to original position (as if there is no operation performed), e.g.

C_2 – one operation C_2^1 (rotation by 180°)

C_3 – two operations C_3^1 (120°), C_3^2 (240°)

C_4 – three operations C_4^1 (90°), C_4^2 (180°), C_4^3 (270°)

C_6 – five operations C_6^1 (60°), C_6^2 (120°), C_6^3 (180°),

$$C_6^4 \text{ (240°)}, C_6^5 \text{ (300°)}$$

Thus, we see that for C_2^1, C_4^2 or C_6^3 there is a rotation of 180°. Similarly for C_3^1 or C_6^2 there is a rotation of 120°. The convention is to represent the operation with smallest number. Thus, the total number of operations on C_4 and C_6 axes can be represented as

C_4 three operations $C_4^1 C_2^1 (= C_4^2)$, C_4^3

C_6 five operations C_6^1 C_3^1 $(= C_6^2)$, C_2^1 $(= C_6^3)$, C_3^2 $(= C_6^4)$, C_6^5

3. **Plane of Symmetry:** It is defined as an imaginary plane within the molecule which divides it into two parts, which are mirror images of each other, e.g. in H_2O molecule, the plane passing through the C_2 axis*, perpendicular to the molecular plane (i.e. through O atom and between the two H atoms) is a plane of symmetry.

The plane of symmetry can be classified into three types:

(*i*) **Vertical Plane:** This is the plane passing through the principal axis and one of the subsidiary axes (if present) and is represented as σ_v. A plane perpendicular to σ_v is represented by σ_v'.

---

*It is customary to set up the principal axis of symmetry vertically and, if a coordinate system is used, to have the Z-axis in this vertical direction.

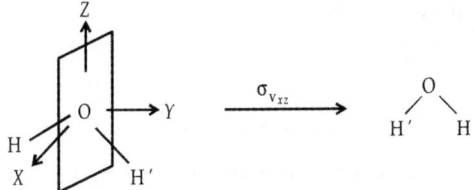

Fig. 13.4: Plane of symmetry $\sigma_{v_{xz}}$ in H_2O

(*ii*) **Horizontal Plane:** This is a plane perpendicular to the principal axis and is represented by σ_h.

(*iii*) **Dihedral Plane:** This is a plane passing through the principal axis but passes in between two subsidiary axes. It is represented as σ_d.

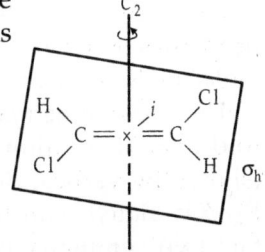

Fig. 13.5: Symmetry elements for trans-dichloroethane

Figure 13.4 shows σ_v axis in case of H_2O molecule.

Figure 13.5 shows σ_h for $C_2H_2Cl_2$ (trans-dichloroethane) and Fig. 13.6 depicts σ_d in case of allene.

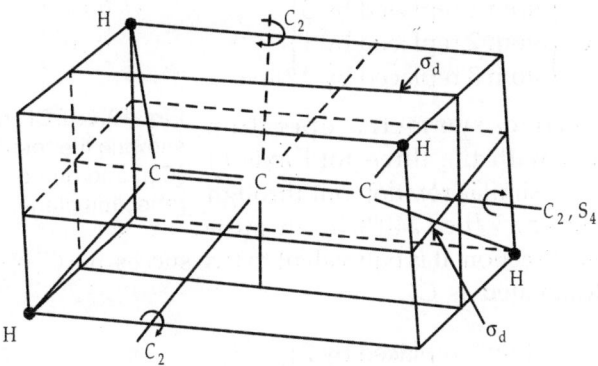

Fig. 13.6: Symmetry elements for allene

4. **Rotation Reflection Axis (Axis of Improper Rotation):** It is an imaginary axis, passing through the molecule on which the molecule has to be rotated by $2\pi/n$ angle and then reflected on a plane perpendicular to the rotation axis to attain an equivalent orientation. It is represented by S_n, e.g. in case of trans H_2O_2, we have S_2 (Fig. 13.7).

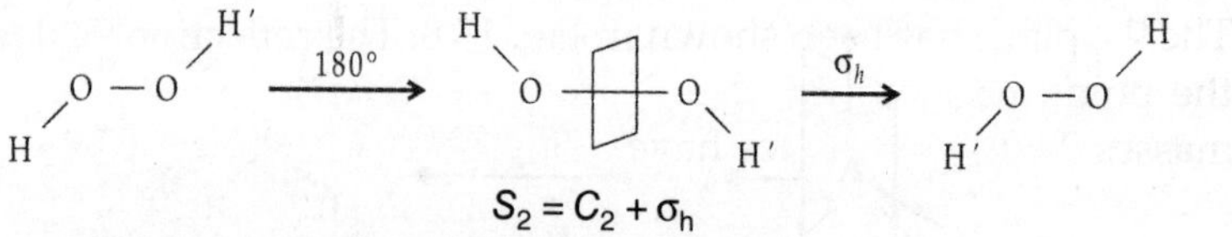

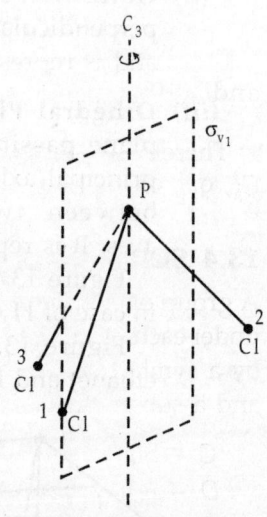

$$S_2 = C_2 + \sigma_h$$

Fig. 13.7: S_2 axis for trans H_2O_2

13.3 SYMMETRY ELEMENTS IN PCl₃

Let us now consider the symmetry properties of the PCl₃. Model of the PCl₃ molecule is shown in Fig. 13.8.

The first symmetry element is the three fold axis of rotation designated as C_3 (shown by vertical dotted line through P). Associated with this element are in fact two symmetry operations, viz. C_3 and C_3^2. C_3 is a rotation of the model around the C_3 axis through angle $2\pi/3$ (i.e. 120°) in a clockwise direction. This may be written as

$$C_3 \begin{cases} \text{atom 1 replaced by 3} \\ \text{atom 2 replaced by 1} \\ \text{atom 3 replaced by 2} \end{cases}$$

Fig. 13.8: PCl₃ molecule, showing the rotational axis (C_3), and one of the three reflection planes (σ_{v_1})

The second symmetry operation associated with the three fold axis of rotation is a similar rotation but through the angle $4\pi/3$ (i.e. 240°) in an anti-clockwise direction. It is equivalent to two successive C_3 operations hence designated as C_3^2:

$$C_3^2 \begin{cases} \text{atom 1 replaced by 2} \\ \text{atom 2 replaced by 3} \\ \text{atom 3 replaced by 1} \end{cases}$$

In both the cases point mass P remains unmoved.

We have also identity operation I.

In addition, the PCl₃ molecule (model) possesses three reflection planes or planes of symmetry (called $\sigma_{v_1}, \sigma_{v_2}, \sigma_{v_3}$) which contain the C_3 axis and the point masses labelled 1, 2 and 3 respectively.

The σ_{v_1} plane has been shown in Fig. 13.8. The reflection σ_{v_1} leaves the point mass 1 (and also P) unmoved, but interchanges point masses 2 and 3. Thus we have

$$\sigma_{v_1} \begin{cases} 2 \text{ replaced by } 3 \\ 3 \text{ replaced by } 2 \end{cases}$$

$$\sigma_{v_2} \begin{cases} 1 \text{ replaced by } 3 \\ 3 \text{ replaced by } 1 \end{cases}$$

and $\quad \sigma_{v_3} \begin{cases} 1 \text{ replaced by } 2 \\ 2 \text{ replaced by } 1 \end{cases}$

Therefore, for PCl_3, we have six symmetry elements viz. I, C_3, C_3^2, σ_{v_1}, σ_{v_2}, σ_{v_3}.

13.4 SCHÖNFLIES NOTATION FOR POINT GROUPS

A group of symmetry operations in which a point is left invariant under each operation is a point group. Each point group is identified by a symbol (Schönflies notation) that consists of a capital letter and a subscript. These are:

C – Simple rotation axis
D – n-fold rotation axis perpendicular to principal axis
S – Rotation-reflection axis
T – Symmetry based on tetrahedron
O – Symmetry based on octahedron
I – Symmetry based on icosahedron

The subscript indicates the order n of the principal axis and whether plane of symmetry occurs.

s : only plane of symmetry
i : only centre of symmetry
n : only n-fold rotation axis
nv : vertical symmetry plane σ_v that contains principal rotation axis
nh : horizontal symmetry plane σ_h perpendicular to principal rotation axis.
nd : dihedral symmetry plane σ_d perpendicular to principal rotation axis.

Order of a group is the number of elements in a group. For example, in multiplication table of H_2O, the order of group is four (E, C_2, σ_{xz}, σ_{yz}).

Conjugate Elements: If A and C are two elements of a group and

$$C^{-1}AC = B$$

where B is also in the same group. Then, we say B is the similarity transformation of A and C and that A and B are conjugate to each other.

Class: A set of elements in a group which are conjugate to each other is said to form a class of the group.

13.5 CLASSIFICATION OF GROUPS

Based on the symmetry, molecules are classified in different point groups. These are given as hereunder:

(*i*) **Group C_s, C_i:** If the molecule possesses a plane of symmetry as the only element (apart from identity) it belongs to C_s group, e.g. hypochlorus acid $\left(\begin{smallmatrix} & O \\ & \diagup \diagdown \\ H & & Cl \end{smallmatrix} \right)$

If the molecule possesses identity and centre of symmetry as the only element, it belongs to C_i group, e.g. mesotartaric acid:

$$
\begin{array}{c}
H \qquad\qquad OH \\
\diagdown \quad \diagup \\
C \text{------} C \text{-------} COOH \\
\diagup \quad \diagdown \quad \diagdown \\
HOOC \quad OH \quad H
\end{array}
$$

Mesotartaric acid

If a molecule possesses no symmetry element, it belongs to C_1 group.

(*ii*) **Group C_n:** Molecules which contain n-fold axis of symmetry belong to this group. These molecules have symmetry element E and n-fold rotation axis.

$$
\begin{array}{c}
H \qquad\qquad Cl \\
\diagdown \qquad\quad \diagup \\
H \text{--}C \text{----} C \text{--} Cl \\
\diagup \qquad\quad \diagdown \\
H \qquad\qquad Cl
\end{array}
$$

1, 1, 1, trichloroethane (C_3 group)

(*iii*) **Group C_{nv}:** Molecules which possess C_n axis and n vertical planes of symmetry are placed in this group, e.g. NH_3 belongs to C_{3v} group.

(*iv*) **Group C_{nh}:** Molecules possessing C_n axis and horizontal plane of symmetry come in this group.

$$\underset{Cl}{\overset{H}{\diagdown}} C = C \underset{H}{\overset{Cl}{\diagup}}$$

Transdichloroethylene (C_{2h} group)
(C_2 axis and σ_h plane)

(*v*) **Group D_n:** Molecules possessing C_n axis and n, 2-fold axis perpendicular to C_n axis, e.g. BF_3, PF_5 (trigonal bipyramid) C_3 is in principal axis, three C_2 are subsidiary axes, three vertical planes and molecular plane: D_{3h} group. Benzene: C_6 principal axis, six C_2 axes and σ_h plane (D_{6h} group).

(*vi*) **Tetrahedron (T_d) Group:** CH_4 and CCl_4 belong to this group. Molecules with T_d symmetry must have four axes of 3 fold symmetry passing through C and H atoms and the centre of the opposite trigonal face. Total number of operations possible are 24 ($1E$, $8C_3$, $3C_2$, $6S_4$, 6σ). Molecules that have tetrahedral geometry, e.g. $CHCl_3$ and do not possess 24 symmetry elements, do not belong to this symmetry group.

(*vii*) **Octahedral Symmetry Group (O_h):** An octahedron can be cut from a cube by joining the centre of the cube with the centres of the six faces of the cube. An octahedron has eight trigonal faces, six corners and twelve edges of equal length. Molecules with this geometry belong to this group, e.g. $CO(NH_3)_6^{3+}$ octahedral complex, SF_6. Total number of symmetry operations possible are 48($1E$, $6C_4$, $3C_2$, $6S_4$, $8C_3$, $8S_6$, $6C_2^1$, $6\sigma_v$, $3\sigma_h$, i).

(*viii*) **Icosahedron Symmetry Group (I_h):** The structure has 20 equilateral faces and 30 edges. Total number of symmetry operations possible are 120, e.g. $B_{12}H_{12}$.

13.6 MATRIX REPRESENTATION OF AN OPERATION

A rotation in X-Y plane is represented by R along the Z-direction. Consider the case when R is a rotation through an angle θ about the x_3 axis. Under certain circumstances (with appropriate values of θ), this may correspond to a molecular symmetry operation. We now choose an orthonormal basis and seek the matrix R which represents R. Obviously, rotation of a general vector around the

x_3 axis leaves its x_3 coordinate unaffected, but changes its x_1 and x_2 to different values x'_1 and x'_2. Figure 13.9 shows a projection of the vector upon the x_1-x_2 plane. The x_3 axis is normal to the plane of the paper and outwards towards the reader.

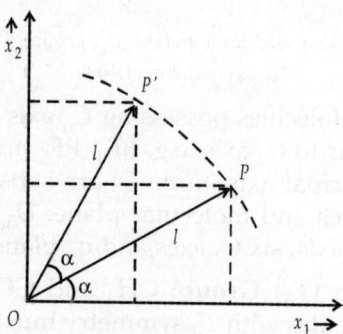

Fig. 13.9: Rotation of a vector

The lines OP and OP' (each of length l) represent the projections of the vector before and after the operation respectively. The angle θ is the angle of rotation. Let α be the angle between OP and the x_1 axis. Then

$$\left. \begin{array}{l} x_1 = l\cos\alpha \\ x_2 = l\sin\alpha \end{array} \right\} \qquad ..(1)$$

Under the operation R,

$$x_1 \xrightarrow{\ R\ } x'_1 = l\cos(\alpha + \theta)$$

$$= l\cos\alpha\cos\theta - l\sin\alpha\sin\theta \qquad ..(2)$$

and $x_2 \xrightarrow{\ R\ } x'_2 = l\sin(\alpha + \theta)$

$$= l\sin\alpha\cos\theta + l\cos\alpha\sin\theta \qquad ..(3)$$

Substituting from eqn (1) into eqns (2) and (3), we obtain

$$\left. \begin{array}{l} x'_1 = \cos\theta\, x_1 - \sin\theta\, x_2 \\ x'_2 = \sin\theta\, x_1 + \cos\theta\, x_2 \end{array} \right\} \qquad ..(4)$$

Remembering that $x'_3 = x_3$, we may now write the effect of the operation in the matrix form as

$$\begin{pmatrix} x'_1 \\ x'_2 \\ x'_3 \end{pmatrix} = \begin{bmatrix} \cos\theta & -\sin\theta & 0 \\ \sin\theta & \cos\theta & 0 \\ 0 & 0 & 1 \end{bmatrix} \begin{pmatrix} x_1 \\ x_2 \\ x_3 \end{pmatrix} \qquad ..(5)$$

The square matrix on the right hand side of this equation is the desired matrix R which represents the operation of anticlockwise rotation through θ about the x_3 axis.

For the sake of completeness, we would now also derive this matrix by considering the effect of the rotation operation upon the unit base vectors $\hat{e}_1$, $\hat{e}_2$ and $\hat{e}_3$.

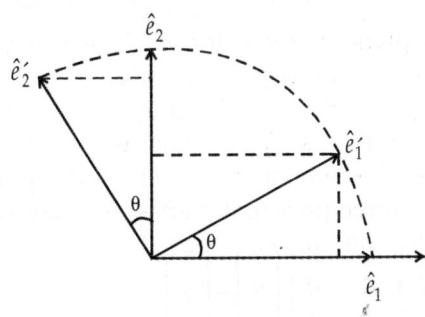

Fig. 13.10: Rotation of unit base vectors $\hat{e}_1$ and $\hat{e}_2$

From Fig. 13.10, which shows the x_1–x_2 plane, it is seen that R changes $\hat{e}_1$ into a vector whose resolved parts in the x_1 and x_2 directions are respectively $\cos\theta$ and $\sin\theta$. Thus,

$$\hat{e}_1 \xrightarrow{R} \hat{e}_1' = \cos\theta\hat{e}_1 + \sin\theta\hat{e}_2 \qquad ..(6)$$

$$\hat{e}_2 \xrightarrow{R} \hat{e}_2' = -\sin\theta\hat{e}_1 + \cos\theta\hat{e}_2 \qquad ..(7)$$

The vector $\hat{e}_3$ is unaffected:

$$\hat{e}_3 \xrightarrow{R} \hat{e}_3' = \hat{e}_3 \qquad ..(8)$$

In the preceding section, we have seen that the matrix R (which represents the operation) is the matrix which transforms the set of basis vectors when these are written as a single row matrix. The results of equations (6), (7) and (8) can now be expressed as

$$\left(\hat{e}_1', \hat{e}_2', \hat{e}_3'\right) = \left(e_1 \ e_2 \ e_3\right)\begin{pmatrix} \cos\theta & -\sin\theta & 0 \\ \sin\theta & \cos\theta & 0 \\ 0 & 0 & 1 \end{pmatrix} \qquad ..(9)$$

13.7 MATRIX NOTATION FOR SYMMETRY TRANSFORMATIONS

There are five types of operations in describing the symmetry of a molecule: E, σ, i, C_n and S_n. Each of these may be described by a matrix.

The Identity: When a point with coordinates x, y, z is subjected to the identity operation, its new coordinates are the same as the original ones viz. x, y, z. This may be expressed as

$$\begin{pmatrix} 1 & 0 & 0 \\ 0 & 1 & 0 \\ 0 & 0 & 1 \end{pmatrix} \begin{pmatrix} x \\ y \\ z \end{pmatrix} = \begin{pmatrix} x \\ y \\ z \end{pmatrix} \qquad ..(10)$$

Reflection: If a plane of reflection is chosen to coincide with a principal cartesian plane (i.e. xy, xz or yz plane), reflection of a general point has the effect of changing the sign of the coordinate (measured) perpendicular to the plane while leaving unchanged the two coordinates whose axes define the plane. Thus, for reflections in the three principal planes, we may write

$$\sigma(x, y): \begin{pmatrix} 1 & 0 & 0 \\ 0 & 1 & 0 \\ 0 & 0 & -1 \end{pmatrix} \begin{pmatrix} x \\ y \\ z \end{pmatrix} = \begin{pmatrix} x \\ y \\ \bar{z} \end{pmatrix}$$

$$\sigma(x, z): \begin{pmatrix} 1 & 0 & 0 \\ 0 & -1 & 0 \\ 0 & 0 & 1 \end{pmatrix} \begin{pmatrix} x \\ y \\ z \end{pmatrix} = \begin{pmatrix} x \\ \bar{y} \\ z \end{pmatrix}$$

$$\sigma(y, z): \begin{pmatrix} -1 & 0 & 0 \\ 0 & 1 & 0 \\ 0 & 0 & 1 \end{pmatrix} \begin{pmatrix} x \\ y \\ z \end{pmatrix} = \begin{pmatrix} \bar{x} \\ y \\ z \end{pmatrix}$$

$$..(11)$$

where a bar over x, y or z represents mirror reflection.

Inversion: To simply change the signs of all the coordinates without permitting any, we evidently need a negative unit matrix, i.e.

$$\begin{pmatrix} -1 & 0 & 0 \\ 0 & -1 & 0 \\ 0 & 0 & -1 \end{pmatrix} \begin{pmatrix} x \\ y \\ z \end{pmatrix} = \begin{pmatrix} \bar{x} \\ \bar{y} \\ \bar{z} \end{pmatrix} \qquad ..(12)$$

Proper Rotation: Defining the rotation as the z-axis, we note first that the z-coordinate will be unchanged by any rotation about the z-axis. Thus, in part, matrix is

$$\begin{pmatrix} & & 0 \\ & & 0 \\ 0 & 0 & 1 \end{pmatrix}$$

The problem of finding the four missing elements reduces to a two dimensional problem in the xy plane.

Suppose that we have a point in the xy plane with coordinates x_1 and y_1 as shown in Fig. 13.11. This point defines a vector r_1. Now suppose that this vector is rotated through an angle θ so that a new vector $\vec{r_2}$ is produced with terminus at the point x_2 and y_2. We now wish to inquire how the final coordinates x_2 and y_2 are related to the original coordinates x_1 and y_1 and the angle θ.

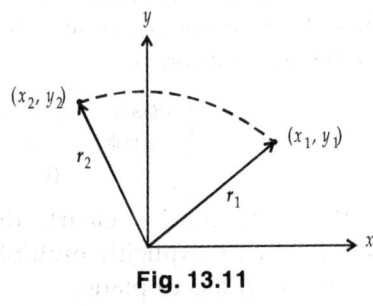

Fig. 13.11

Now, when the x-component of r_1 (i.e., x_1) is rotated by θ, it becomes a vector x' which has an x-component of $x_1\cos\theta$ and a y-component of $x_1\sin\theta$. Similarly, the y-component of r_1 (i.e., y_1) upon rotation by θ becomes a new vector y' which has an x-component of $-y_1\sin\theta$ and a y component of $y_1\cos\theta$. Now x_2 and y_2 (the components of r_2) must be equal to the sums of the x-components and y-components of x' and y' respectively, i.e.

$$\left.\begin{array}{l} x_2 = x_1\cos\theta - y_1\sin\theta \\ y_2 = x_1\sin\theta + y_1\cos\theta \end{array}\right| \qquad ..(13)$$

The transformation represented by this equation can be written in matrix notation as

$$\begin{pmatrix} \cos\theta & -\sin\theta \\ \sin\theta & \cos\theta \end{pmatrix}\begin{pmatrix} x_1 \\ y_1 \end{pmatrix} = \begin{pmatrix} x_2 \\ y_2 \end{pmatrix} \qquad ..(14)$$

This relation holds for a counterclockwise rotation. Because $\cos\phi = \cos(-\phi)$ while $\sin\phi = -\sin(-\phi)$, the matrix for a clockwise rotation through the angle ϕ must be

$$\begin{pmatrix} \cos\phi & \sin\phi \\ -\sin\phi & \cos\phi \end{pmatrix}$$

Thus, the total matrix equation for a clockwise rotation through ϕ about the z-axis is

$$\begin{pmatrix} \cos\phi & \sin\phi & 0 \\ -\sin\phi & \cos\phi & 0 \\ 0 & 0 & 1 \end{pmatrix}\begin{pmatrix} x_1 \\ y_1 \\ z_1 \end{pmatrix} = \begin{pmatrix} x_2 \\ y_2 \\ z_2 \end{pmatrix} \qquad ..(15)$$

Improper Rotation: Since an improper rotation through the angle ϕ about the z-axis produces the same transformation of the x and y coordinates as does a proper rotation through the same angles but in addition changes the sign of the z-coordinate, we infer directly from the equation just derived above that the matrix for clockwise rotation is

$$\begin{pmatrix} \cos\phi & \sin\phi & 0 \\ -\sin\phi & \cos\phi & 0 \\ 0 & 0 & -1 \end{pmatrix}$$

It is to be noticed clearly that one could have obtained this matrix also by explicitly multiplying the matrices for rotation and reflection in the xy plane.

13.8 PRODUCTS OF SYMMETRY OPERATIONS

The overall operation consisting of one symmetry operation R followed by another symmetry operation S is itself a symmetry operation, conventionally represented as the product SR which of course is one of the complete list of operations of the molecule under consideration.

We again consider the example of PCl_3 and the effect of C_3 followed by σ_{v_1}. Point mass P remains unmoved and the effects upon masses 1, 2, 3 are shown in the Fig. 13.12.

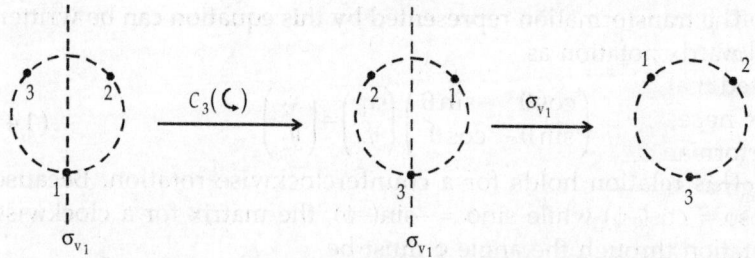

Fig. 13.12: Effects of successive symmetry operations C_3 and σ_{v_1} on the three equal point masses representing the three Cl nuclei in the PCl_3

It should be noted that the reflection plane σ_{v_i} is regarded as fixed in space in the position determined by the initial position of point mass 1 with no possibility of alteration as symmetry operations are performed upon the model. The overall effect of the two consecutive operations is to replace 1 by 3 and *vice versa* (bringing back z to its initial position). The product operation is thus identified as σ_{v_2} and so, we may write

$$\sigma_{v_1} C_3 = \sigma_{v_2}$$

Note that on the basis of this identity, the operation which is performed first is written to the right and subsequent operation on the left. Thus it may be verified that the operation SR will not be, in general, the same as operation RS. For instance, consider the operation $C_3\sigma_{v_1}$.

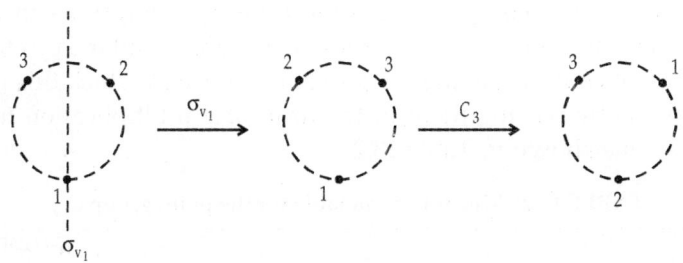

Fig. 13.13: Effect of reversing the order of operation as compared with Fig. 13.12

This is shown in Fig. 13.13 from which it is apparent that now, the effect is to replace 1 by 2 and 2 by 1. Thus, we have

$$C_3\sigma_{v_1} = \sigma_{v_3}$$

Similar considerations apply to the successive performances of more than two symmetry operations and the same convention is followed in writing the symbol for the product. It is also to be noted that two successively performed symmetry operations need not necessarily be different, e.g. $C_3C_3 = C_3^2$. Two successive performances of the reflection σ_{v_1} obviously bring us back to the original situation, accordingly

$$\sigma_{v_1}\sigma_{v_1} = I$$

A set of symmetry elements such as the symmetry operations for a molecule must, in order that they shall constitute a group, satisfy the following four requirements:

1. The set must contain the identity operation I with the property that if R be any element of the group, then

   $$RI = IR = R$$

2. The product SR of any two elements of the group must itself be an element of the group.

3. Each element R must have an inverse R^{-1} which is also an element of the group satisfying the relation

$$RR^{-1} = R^{-1}R = I$$

4. Products of elements must satisfy the associative law:

$$T(SR) = (TS)R$$

13.9 MULTIPLICATION TABLE

Now, given the complete list of symmetry operations in any particular case, it is possible to construct a table of all the products of them taken two at a time. A convenient form in which this may be done is by writing them in the form of a multiplication table such as one shown in Table 13.2.

TABLE 13.2: Multiplication table for the point group C_{3v}

C_{3v}	I	C_3	C_3^2	σ_{v_1}	σ_{v_2}	σ_{v_3}
I	I	C_3	C_3^2	σ_{v_1}	σ_{v_2}	σ_{v_3}
C_3	C_3	C_3^2	I	σ_{v_3}	σ_{v_1}	σ_{v_2}
C_3^2	C_3^2	I	C_3	σ_{v_2}	σ_{v_3}	σ_{v_1}
σ_{v_1}	σ_{v_1}	σ_{v_2}	σ_{v_3}	I	C_3	C_3^2
σ_{v_2}	σ_{v_2}	σ_{v_3}	σ_{v_1}	C_3^2	I	C_3
σ_{v_3}	σ_{v_3}	σ_{v_1}	σ_{v_2}	C_3	C_3^2	I

It is to be noted that the entry at the intersection of a certain column and a certain row is the product obtained by performing first the operation shown at the head of the column and then the operation shown at the left end of the row. Since (in general), a product SR is not the same as the product RS, so, the table is not symmetrical about the diagonal from the top left-hand corner to the bottom right-hand corner.

Now the group requirement conditions (1), (2) and (3) just mentioned before the table are satisfied by the complete set of symmetry operations for PCl_3. That requirement (4) is also satisfied may be verified by the use of the binary products in the Table 13.2. For example, we find

$$C_3^2(\sigma_{v_1}C_3) = C_3^2\sigma_{v_2} = \sigma_{v_3}$$
and $(C_3^2\sigma_{v_1})C_3 = \sigma_{v_2}C_3 = \sigma_{v_3}$

thus verifies that

$$C_3^2(\sigma_{v_1}C_3) = (C_3^2\sigma_{v_1})C_3$$

Thus, we may say that the PCl_3 molecule belongs to the point group C_{3v} or that it possesses C_{3v} symmetry.

The number of independent operations called elements in a group is called the *order* of the group, e.g. for group C_{3v}, the order is 6.

The *character* of a square matrix corresponding to a symmetry operation is defined as the sum of the diagonal elements of the matrix. It is usually given the symbol χ (Greek chi). These are listed in Table 13.3 for various symmetry operations.

TABLE 13.3: Character of the matrices of various symmetry operations

Symmetry Operation	Character of Matrix
Identity	3
Reflection	1
Inversion	-3
Proper rotation	$1 + 2\cos\phi$
Improper rotation	$-1 + 2\cos\phi$

13.10 A REPRESENTATION OF THE GROUP

Since the complete set of symmetry operations constitutes a mathematical group, it follows that a complete set of matrices representing these operations also forms such a group, for they can be combined among themselves in a manner parallel to the way in which the symmetry operations combine. Just as the individual matrices represent the individual symmetry operations, so as well the complete set of matrices is said to constitute a *representation of the group*.

13.11 GROUP MULTIPLICATION TABLE FOR THE POINT GROUP C_{2V}

The group C_{2v} consists of the operations I (or E), C_2, σ_v, σ'_v (for example H_2O molecule). Suppose the C_2 axis coincides with the z-axis of a cartesian coordinate system and let σ_v be the xz plane and σ'_v, the yz plane. The matrices representing the transformations applied on a general point can be represented by the following matrices:

$$E \equiv \begin{pmatrix} 1 & 0 & 0 \\ 0 & 1 & 0 \\ 0 & 0 & 1 \end{pmatrix}; \quad C_2 \equiv \begin{pmatrix} -1 & 0 & 0 \\ 0 & -1 & 0 \\ 0 & 0 & 1 \end{pmatrix}$$

$$\sigma_v \equiv \begin{pmatrix} 1 & 0 & 0 \\ 0 & -1 & 0 \\ 0 & 0 & 1 \end{pmatrix}; \ \sigma_v' \equiv \begin{pmatrix} -1 & 0 & 0 \\ 0 & 1 & 0 \\ 0 & 0 & 1 \end{pmatrix}$$

Now, on multiplication we will have 16 symmetry products (e.g. C_2E, $C_2\sigma_v$, $\sigma_v C_2$, .., etc.). On simplification (writing E, C_2, σ_v, σ_v' as the table heads), we get the group multiplication table as indicated in Table 13.4.

TABLE 13.4: Group multiplication table for the group C_{2v}

	E	C_2	σ_v	σ_v'
E	E	C_2	σ_v	σ_v'
C_2	C_2	E	σ_v'	σ_v
σ_v	σ_v	σ_v'	E	C_2
σ_v'	σ_v'	σ_v	C_2	E

It can be verified that the matrices multiply together in the same fashion (e.g. verify $\sigma_v C_2 = \sigma_v'$ by matrix multiplication of the representative matrices).

Similarity Transformation: In a given group it may be possible to select various smaller sets of elements, each of which are in themselves groups. If A and S are two elements of a group, then $S^{-1}AS$ will be equal to some element of the group say A'. We have

$$A' = S^{-1}AS$$

This defines a similarity transformation. We say that A' is the similarity transform of A by S.

13.12 REDUCIBLE AND IRREDUCIBLE REPRESENTATION

The complete array of the simplest sets of matrices which represent a group are called the *Irreducible Representation* of the group. It will consist of a collection of sets of simplest matrices which can't have any further simple form. On the other hand, in general, it is possible to decompose matrix representations into combinations of the simplest sets. Such representations are said to be reducible. The concept of reducible and irreducible representations is of immense importance in atomic and molecular physics.

The irreducible representations of a group are named T_1, T_2, ... T_n. The numbering is arbitrary except that T_1 is always the totally symmetric representation and the irreducible representations are numbered in increasing order of their dimensions.

The concept of reduction of a representation can be illustrated by the following considerations. Suppose that there is a matrix U such that the unitary transformation of it by X yields A

$$X^{-1} U X = A$$

X and A are matrices of the same order as U and matrix X needs no specification. The matrices X and X^{-1} are defined such that

$$X^{-1}X = C$$

where C is the unit matrix of the same dimension as X. If for same X, B' has the form

$$B' = \begin{bmatrix} B_1 & \begin{matrix} 0 & 0 & 0 \\ 0 & 0 & 0 \\ 0 & 0 & 0 \end{matrix} & | & | & | & | \\ \begin{matrix} 0 & 0 & 0 \\ 0 & 0 & 0 \\ 0 & 0 & 0 \end{matrix} & B_2 & | & | & | & | \\ \begin{matrix} 0 & 0 & 0 \\ 0 & 0 & 0 \end{matrix} & \begin{matrix} 0 & 0 & 0 \\ 0 & 0 & 0 \end{matrix} & B_3 & | \end{bmatrix} \quad ..(16)$$

i.e. there is a set of smaller matrices entered on the principal diagonal and zero elsewhere. (This is known as block diagonal form). Then it is possible to say that

$$U = B_1 + B_2 + B_3, \ ...$$

Then U is said to be reduced into the components B_1, B_2, .., etc. (i.e. U is a *reducible matrix*). Then the representation Γ consisting of matrices A', B' is called a *reducible representation*.

Such an operation can be performed upon each of the matrices of a reducible representation to give linear combinations of the matrices of the irreducible representations B_1, B_2, .., etc. The linear combinations of matrices of irreducible representations is the same for each of the matrices of the reducible representation. The whole representation is reduced into a linear combination of the irreducible representations.

13.13 CLASS OF A GROUP

A *class* of a group is a set of elements in the group, which are of similar type. For example, there are three classes in C_{3v}—the E, C_3

and σ_v. This in a geometrical sense means that elements belonging to the same class (if the operations to which they correspond) are identical except for a change in the coordinate system. The formal definition of a class is as follows:

The set of elements $P...S$ form a class if, for all elements of the group X_i,

$$X_i^{-1}PX_i = \text{one of the set } P..S$$

$$....... = \text{one of the set } P..S$$
$$|$$
$$X_i^{-1}SX_i = \text{one of the set } P..S$$

In other words, the operation by any element of the group, then by a member of the class, then by the reciprocal of the first element, must yield a member of the class.

It is possible to find one and only one combination of irreducible representations of the group (Γ_i) which gives

$$\sum_i n_i \chi_{ik} = \chi_k'$$

for all classes k of the group.

Here n_i is zero or a positive integer and the summation over i includes all the irreducible representations of the group.

13.14 THE GREAT ORTHOGONALITY THEOREM

We shall now consider the great orthogonality theorem without giving any proof.

Let us denote order of a group by h. The dimension of the ith representation, which is the order of the matrices which constitute it, will be denoted by l_i. The various operations in the group is given the general symbol R. The element in the mth row and the nth column of the matrix corresponding to an operation R in the ith irreducible representation is denoted by $\Gamma_i(R)_{mn}$

The theorem may be stated as follows:

$$\sum_R \left[\Gamma_i(R)_{mn}\right]\left[\Gamma_j(R)_{m'n'}\right]^* = \frac{h}{\sqrt{l_i l_j}}\delta_{ij}\,\delta_{mm'}\,\delta_{nn'} \qquad ..(17)$$

This means that in the set of matrices constituting any one irreducible representation any set of corresponding matrix elements, one from each matrix behaves as the components of a vector in h-dimensional space such that all these vectors are mutually orthogonal and each is normalised so that the square of its length equals h/l_i. This interpretation of eqn 17 will be more obvious if we take apart this into three simpler equations each of which is contained within it. We omit the explicit designation of complex conjugates for simplicity but it is to be remembered that they must be used when complex numbers are involved. The three simpler equations are:

$$\sum_R \Gamma_i(R)_{mn} \Gamma_j(R)_{mn} = 0 \text{ if } i \neq j \qquad ..(18)$$

$$\sum_R \Gamma_i(R)_{mn} \Gamma_i(R)_{m'n'} = 0 \text{ if } m \neq m' \text{ and/or } n = n' \qquad ..(19)$$

$$\sum_R \Gamma_i(R)_{mn} \Gamma_i(R)_{mn} = \frac{R}{l_i} \qquad ..(20)$$

Thus if the vectors differ by being chosen from matrices of different representations, they are orthogonal. If they are chosen from the same representation but from different sets of elements in the matrices of this representation, they are orthogonal. Finally eqn (20) expresses the fact that the square of the length of any such vector equals h/l_i.

There are five important rules about irreducible representations and their characters.

1. The sum of the squares of the dimensions of the irreducible representations of a group is equal to the order of the group,

   $$\Sigma l_i^2 = l_1^2 + l_2^2 + l_3^2 + \ldots\ldots = h \qquad ..(21)$$

   Let $\chi_i(E)$ be the character of the representation of E in the irreducible representation. Since $\chi_i(E)$ is equal to the order of the representation

   $$\therefore \qquad \sum_i [\chi_i(E)]^2 = h \qquad ..(22)$$

2. The sum of the squares of the characters in any irreducible representation equals h, that is

   $$\sum_R [\chi_i(R)]^2 = h \qquad ..(23)$$

3. The vectors whose components are the characters of two different irreducible representations are orthogonal,

   $$\sum_R \chi_i(R)\chi_j(R) = 0 \qquad ..(24)$$

   when $i \neq j$.

4. In a given representation (reducible or irreducible) the characters of all matrices belonging to operations in the same class are identical.

5. The number of irreducible representations of a group is equal to the number of classes in the group.

Now equations (23) and (24) can be combined into a single equation.

$$\sum_R \chi_i(R)\chi_j(R) = h\delta_{ij} \qquad ..(25)$$

If the number of symmetry operations in the kth class is written as g_k and if there are n classes in all, then eqn (25) takes the form

$$\sum_{k=1} g_k \chi_i(R_k)\chi_j(R_k) = h\delta_{ij} \qquad ..(26)$$

Here R_k refers to any one of the operations in the kth class. The equation implies that the k quantities of the type $\chi_i(R_k)$ can be considered as the components of a k-dimensional vector. There will be no more than k irreducible representations in a group which has k classes.

13.15 CHARACTER TABLE AND ITS PROPERTIES (CHARACTER TABLE OF GROUP C_{3v})

The character table for a group G is given in Table 13.5.

In any table, the point group is indicated on the left hand corner below which are the irreducible representations ($IR's$). Also, in any group, the identity element I is always a class by itself and this is taken to be K_1, so that the first column of the table shows character of I in various $IR's$.

TABLE 13.5: Character table of point group G

G	K_1	K_2	K_i	K_k
Γ_1	χ_1^1	χ_2^1	χ_i^1	χ_k^1
Γ_2	χ_1^2	χ_2^2	χ_i^2	χ_k^2
$\vdots$	$\vdots$	$\vdots$	$\vdots$	$\vdots$
Γ_μ	χ_1^μ	χ_2^μ	χ_i^μ	χ_k^μ
$\vdots$	$\vdots$	$\vdots$	$\vdots$	$\vdots$
Γ_k	χ_1^k	χ_2^k	χ_i^k	χ_k^k

In the representation Γ_μ with dimensions l_μ, I is represented by a unit matrix with l_μ rows and columns. Thus $\chi_1^\mu = \chi^\mu I = l_\mu$ and the first column of the table simply gives the dimensions of the different $IR's$ of the group G.

The character table of the point group C_{3v} is shown in Table 13.6.

TABLE 13.6: Character table of point group C_{3v}

C_{3v}	I	$2C_3$	$3\sigma_v$
Γ_1	1	1	1
Γ_2	1	1	−1
Γ_3	2	−1	0

This table has the form of Table 13.5. The names of IR's are written down in the left hand column under C_{3v} and headings of the remaining columns show the various classes of the group. (These classes are K_1 containing identity operation I, K_2, the class of the two rotations C_3 and C_3^{-1} and K_3, the class of 3 reflection planes).

The properties of a character table are as follows:

1. The number of irreducible representations of a group is equal to the number of classes in the group. As in C_{3v}, there are three classes of operation and thus there will be three irreducible representations only.

2. The group G possesses k different (non-equivalent) irreducible representations whose dimensions are l_1, l_2, ..., l_k satisfying the equation

$$l_1^2 + l_2^2 + l_3^2 + ...+ l_k^2 = h \qquad ..(27)$$

where h is order of the group. Every point group has a h and a k which are such that there is only one possible set of k integers, the sum of squares of which is equal to h. As in C_{3v} group,

$h_1 = 1, h_2 = 2, h_3 = 3$

so that $h_1 + h_2 + h_3 = h$ \qquad ..(28)

i.e., $1 + 2 + 3 = 6 = h = $ order of the group.

The dimensions of IR's are

$l_1 = 1$ for Γ_1, $l_2 = 1$ for Γ_2 and

$l_3 = 2$ for Γ_3 so that

$1^2 + 1^2 + 2^2 = 6 = h$

3. The sum of squares of characters of any irreducible representation is equal to the order of the group

$$(\chi_1^k)^2 + (\chi_2^k)^2 + ...(\chi_k^k)^2 = h \qquad ..(29)$$

As for group C_{3v}

For Γ_1 : $1^2 + 2(1^2) + 3(1^2) = 6 = h$

For Γ_2 : $1^2 + 2(1^2) + 3(-1)^2 = 6 = h$

For Γ_3 : $2^2 + 2(-1)^2 + 3(0)^2 = 6 = h$

4. Characters of two different irreducible representation of a point group are orthogonal, i.e.

$$\Sigma\chi_i(R)\chi_j(R) = h\delta_{ij} \qquad ..(30)$$

When $i \neq j$, $\delta_{ij} = 0$ and it follows the statement.

When $i = j$, $\delta_{ij} = 1$ and it leads to rule (3).

13.16 CHARACTER TABLE FOR POINT GROUP C_{2V}

Now we consider the one dimensional irreducible representations of the C_{2v} point group. It has the symmetry element E, C_2, σ_v and σ_v'. Since each of these is a separate class, according to rule 5 of the irreducible representations, there are therefore 4 irreducible representations (Γ_1, Γ_2, Γ_3, Γ_4) of this group and each is of dimension unity. As per rule 1

$$l_1^2 + l_2^2 + l_3^2 + l_4^2 = 4 \qquad ..(31)$$

where l_i is the dimension of the ith representation. The only way one can select 4 positive integers satisfying the above relations is $l_1 = l_2 = l_3 = l_4 = 1$. Hence the point group C_{2v} has four one dimensional irreducible representations.

In every point group, there will be a representation in which all the operations will have the character 1. Taking it as the Γ_1 representation

$$\chi_1(E) = \chi_1(C_2) = \chi_1(\sigma_v) = \chi_1(\delta_v') = 1$$

According to eqn (22)

$$[\chi_1(E)]^2 + [\chi_2(E)]^2 + [\chi_3(E)]^2 + [\chi_4(E)]^2 = 4 \qquad ..(32)$$

The possible values of $\chi_1(E)$, $\chi_2(E)$, $\chi_3(E)$ and $\chi_4(E)$ are then 1 each. The other representations will have to satisfy eqns (23) and (24). Using eqn (23), we get

$$[\chi_1(E)]^2 + [\chi_1(C_2)]^2 + [\chi_1(\sigma_v)]^2 + [\chi_1(\sigma_v')]^2 = 4$$

$$[\chi_2(E)]^2 + [\chi_2(C_2)]^2 + [\chi_2(\sigma_v)]^2 + [\chi_2(\sigma_v')]^2 = 4$$

$$[\chi_3(E)]^2 + [\chi_3(C_2)]^2 + [\chi_3(\sigma_v)]^2 + [\chi_3(\sigma_v')]^2 = 4$$

$$[\chi_4(E)]^2 + [\chi_4(C_2)]^2 + [\chi_4(\sigma_v)]^2 + [\chi_4(\sigma_v')]^2 = 4 \qquad ..(33)$$

These equations are possible only if $\chi_i(R) = \pm1$. Simultaneous validity of eqn (24) is possible only if two characters in the representation Γ_2, Γ_3 and Γ_4 have +1 and the other two −1, which makes the sum of two terms in eqn (33) as +2 and the other two terms as −2. Consequently, the complete set of characters for the irreducible representations satisfying various rules is as given in Table 13.7.

TABLE 13.7: **Characters for the various irreducible representations of the** C_{2v} **point group**

	E	C_2	σ_v	σ_v'
Γ_1	1	1	1	1
Γ_2	1	−1	−1	1
Γ_3	1	−1	1	−1
Γ_4	1	1	−1	−1

Questions

13.1 Distinguish between proper rotations and improper rotations.

13.2 Explain the symmetry operations reflection and inversion.

13.3 What are σ_v, σ_h and σ_d symmetry operations?

13.4 What are reducible and irreducible representations?

13.5 State and explain the great orthogonality theorem.

13.6 Describe various Schönflies notations used for point groups.

13.7 Describe classification of point groups used for molecules.

14

Nuclear Magnetic Resonance Spectroscopy

The phenomenon of nuclear magnetic resonance (*NMR*) was first discovered in 1946, independently by Purcell, Pound and Torrey (in paraffin) and by Bloch, Hansen and Packard (in water). It is based on transitions between (nuclear) energy levels that arise because of the different orientations of magnetic moment of nuclei (having a non-zero spin magnetic moment) when placed in a magnetic field. The transitions are studied by means of a resonance method, hence the name.

14.1 THE NUCLEAR SPIN

For a nucleus to be magnetic, it must possess a non-zero spin angular momentum. The spin of a nucleus is given the symbol I called the (nuclear) *spin quantum number* and the angular momentum of a nucleus is given by

$$| I | = \sqrt{I(I+1)}\, \frac{h}{2\pi} \equiv \sqrt{I(I+1)}\hbar \text{ or } \sqrt{I(I+1)} \text{ units}$$

It has been observed that

1. Nuclei with both p (total number of protons) and n (total number of neutrons) even (hence even charge and even mass) have zero spin (e.g. ^{4}He, ^{12}C, ^{16}O, etc.) $I = 0$. These are of no concern for NMR spectroscopy (both A and Z even).
2. Nuclei with both p and n odd (hence charge odd but mass $= p + n$, even) have integral spin (for example $^2_1$H, $^{10}_5$B, $^{14}_7$N) (A even and Z odd).

3. Nuclei with odd mass number A (Z may be even or odd), have half integral spin, e.g. ^1_1H, $^{13}_6C$, $^{15}_7N$, etc.

The angular momentum vector I cannot have any arbitrary direction. Rather, its projection along a particular (say Z) direction can have the values

$$m_I = I, I - 1, ...0, ..(I - 1), -I \text{ (for } I \text{ integral)}$$

$$m_I = I, I - 1, ...\frac{1}{2}, -\frac{1}{2}..., -I \quad \text{(for } I \text{ half integral)}$$

These m_I states are $2I + 1$ in number and all are degenerate in the absence of an external magnetic field. When the magnetic field is applied, degeneracy is lifted and the state (of angular momentum I) splits up into $(2I + 1)$ states.

14.2 NUCLEAR MAGNETIC MOMENT

The nucleus consists of protons and neutrons. Like electrons, protons and neutrons are Fermions and have a spin. Their spin motion is quantised and the spin quantum number has a value $s = 1/2$. The spin can orient itself in two directions under the influence of an external magnetic field specified by magnetic quantum number $m_s = \pm 1/2$. As a result, a nucleus has a characteristic resultant spin angular momentum, which is characterised by *nuclear spin quantum number I*. This I depends on the number of protons and neutrons which constitute the nucleus.

Depending on the number of proton and neutrons in a nucleus, all nuclei can be divided into four groups namely:

(i) Nuclei consisting of odd number of protons and even number of neutrons (i.e. odd-even nuclei), such $_3Li^7$, $_5B^{11}$, $_{11}Na^{23}$. The resultant nuclear spin of these nuclei will be an odd integral multiple of $1/2$.

(ii) Nuclei consisting of odd number of protons and odd number of neutrons (odd-odd nuclei). Examples are $_1H^2$, $_3Li^6$, $_5B^{10}$, $_7N^{14}$. The resultant nuclear spin of such nuclei will be vector sum of the spin of odd proton and odd neutron in the nucleus and is an integral number.

(iii) Nuclei consisting of even number of protons and odd number of neutrons (even-odd nuclei) e.g. Be^9, C^{13}, O^{17}. The resultant nuclear spin of these will be an odd integral multiple of $1/2$.

(*iv*) Nuclei consisting of even number of protons and even number of neutrons (even-even nuclei) e.g. $_6C^{12}$, $_8O^{16}$, $_{10}Ne^{20}$. The resultant nuclear spin I in this case is zero because of pairing of even protons and even neutrons.

Thus, when the number of protons and neutrons in a nucleus is even number, the nucleus will have $I = 0$, but when the mass number is even, I can have a value either 0 or an integral number. For example, He^4 (2 neutrons, 2 protons) $I = 0$, H^2 (1 neutron, 1 proton) $I = 1$, N^{14} (7 neutrons, 7 protons) $I = 1$. On the other hand, the nuclei in which the mass number is odd (sum of the number of neutrons and protons is odd), I is either $1/2$ or an integral odd multiple of $1/2$. For example H^1(1 proton) $I = 1/2$, Na^{23}(11 protons and 12 neutrons), $I = 3/2$. Cl^{35} (17 protons and 18 neutrons) $I = 5/2$.

We have seen above that nucleus has a positive charge and also spin angular momentum. Thus the nucleus must have a nuclear magnetic moment (analogous to electron spin magnetic moment) which is always parallel to the spin angular momentum of the nucleus and has a magnitude

$$\mu_N = \gamma_N \, \hbar \sqrt{I \, (I + 1)}$$

where γ_N is nuclear gyromagnetic ratio

$$\gamma_N = \frac{e \, g_N}{2 m_p \, C} \text{ where } m_p = \text{mass of a proton}$$

We may define

$$\beta_N = \gamma_N \, \hbar = \frac{e \hbar}{2 m_p} = 5.05 \times 10^{-27} \text{ JT}^{-1}$$

β_N is nuclear magneton analogous to Bohr magneton.

$$\therefore \mu_N = g_N \beta_N \sqrt{I \, (I + 1)}$$

g_N is nuclear g-factor = $2 \times 2.7245 = 5.586$ for proton. Table 14.1 summarizes various magnetic moments.

TABLE 14.1: Permanent magnetic moments in molecules

Source	Angular momentum = p Magnetic moment ($\mu = \gamma p$)	Typical examples
Orbital electron	μ_B	O, NO,
Electron spin	μ_B	H, O_2, Fe^{3+}
Nuclear spin	μ_N	^3He, H_2, HF
Molecular rotation	μ_N	$H_2(g)$, $H_2O(g)$

Here μ_B (Bohr magneton) = 9.2732×10^{-21} erg. (gauss)$^{-1}$
$$= 9.2732 \times 10^{-24} \text{ JT}^{-1}$$
and μ_N (Nuclear magneton) = 5.0505×10^{-24} erg. (gauss)$^{-1}$
$$= 5.0505 \times 10^{-27} \text{ JT}^{-1}$$

Molecular rotation is a source of angular momentum and of rotational magnetic moment which is close to the nuclear magneton. They are particularly important for molecules in gas phase and for molecules that do not have any net orbital or electron spin moments. The nuclear moments play an important role in Nuclear Magnetic Resonance and electron moments in Electron Spin Resonance.

14.3 PRECESSION OF A NUCLEUS IN A MAGNETIC FIELD

Nuclei having a (spin) magnetic moment, in presence of an (external) magnetic field behave like a gyroscope in a gravitational field. Consequently, if H_0 represents a steady magnetic field directed along the z-axis, the magnetic moment vector μ (due to gyroscope effect) precesses around the magnetic field with an angular frequency ω_0 given by

$$\omega_0 = \gamma H_0$$

or Larmor frequency $\nu_0 = \gamma \dfrac{H_0}{2\pi}$ (see Fig 14.1).

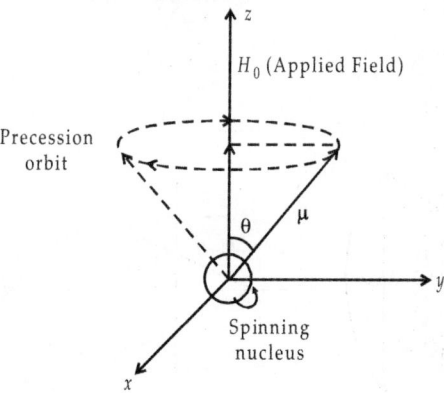

Fig 14.1: Precession of magnetic moment vector μ of a spinning nucleus

The precessions are actually circular movements with respect to the force line and are restricted to a distinct number of angles between the field line and the axis of spin (depending on the values of H_0 and I) as in Fig. 14.2.

14.4 ENERGY LEVELS IN A MAGNETIC FIELD

When a nucleus of magnetic moment μ is placed in a magnetic field B_0 tesla, the potential energy

$$E = -\mathbf{\mu} \cdot \mathbf{B}_0 = -\mu B_0 \cos\theta = -\frac{\mu B_0 m_I}{I}$$

' Since m_I can have $(2I + 1)$ values, we will have $(2I + 1)$ equally spaced energy levels.

In case of a nucleus of spin quantum number $I = ½$, the magnetic moment becomes oriented in one of the two directions with respect to the field (B_0) depending on its magnetic quantum state. The potential energy in these two quantum states is given by

$$E = -\frac{\gamma I \hbar m_I B_0}{I} = -\gamma m \hbar B_0$$

The energy for the lower energy state $(m_I = +½)$ is given by

$$E|_{+1/2} = -\frac{\gamma \hbar B_0}{2}$$

and for the higher energy state $(m_I = -½)$ the energy is

$$E|_{-1/2} = +\frac{\gamma \hbar B_0}{2}$$

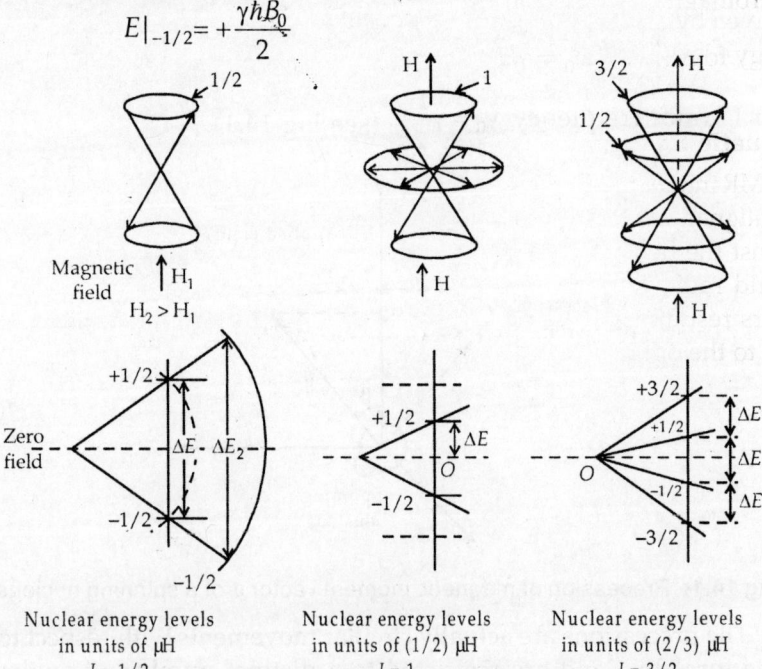

Nuclear energy levels   Nuclear energy levels   Nuclear energy levels
in units of μH   in units of (1/2) μH   in units of (2/3) μH
$I = 1/2$ $I = 1$ $I = 3/2$

Fig. 14.2: Nuclear orientations and energy levels of nuclei (having different spins) in an external magnetic field H

The energy difference ΔE for these two states is given by

$$\Delta E = \frac{\gamma\hbar}{2}B_0 - \left(-\frac{\gamma\hbar}{2}B_0\right) = \gamma\hbar B_0$$

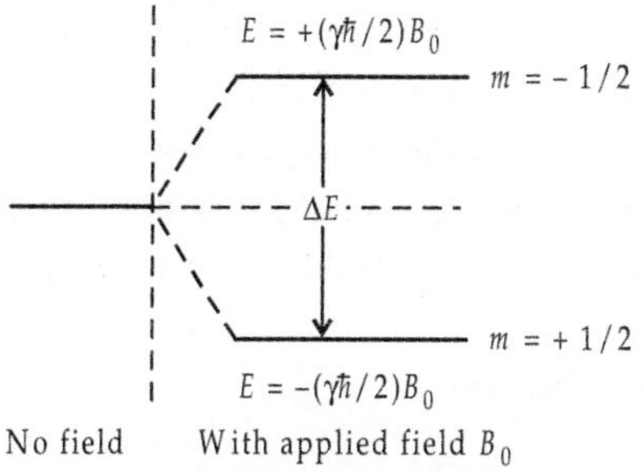

$E = +(\gamma\hbar/2)B_0$

$m = -1/2$

ΔE

$m = +1/2$

$E = -(\gamma\hbar/2)B_0$

No field   With applied field B_0

Fig. 14.3: Energy levels for a nucleus having a spin quantum number of ±1/2

Similar to other types of spectroscopy, transitions between energy states can be brought about by absorption or emission of electromagnetic radiation of a frequency ν_0 that corresponds in energy to ΔE $\left(\text{i.e. } h\nu_0 = \gamma\hbar B_0 / 2\pi \text{ or } \nu_0 = \dfrac{\gamma B_0}{2\pi}\right)$.

Magnetic Resonance

In NMR initially, in a strong magnetic field B_0, the magnetic moment μ is aligned in one of two possible orientations (i.e. either along or against the field). When these spinning nuclei are irradiated by a second radio frequency, weaker field, characteristics absorption occurs resulting in transitions from one alignment in the applied field to the opposite one (Fig. 14.4).

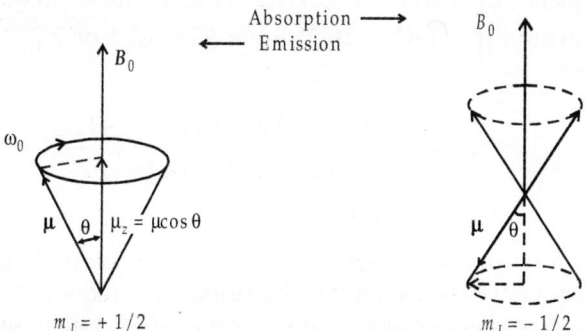

Fig. 14.4: Absorption (or emission) of radiation by a precessing nucleus

In NMR, the amount of energy required to cause a particular nucleus to realign depends upon such factors as field strength, electronic configuration around the particular nucleus, anisotropy, type of molecule and intermolecular interactions. Thus NMR spectroscopy is most widely used by particularly organic chemists to identify and characterise molecules.

The frequency ν_0 represents the resonant frequency that will affect the transitions between the energy levels, with the energy of reorientation of a magnetic dipole equal to $\mu B_0/I$.

The equilibrium population of the nuclear energy levels is given by Boltzmann distribution

$$\frac{n_{\text{upper}}}{n_{\text{lower}}} = e^{-\mu B_0/(IkT)}$$

where k is Boltzmann constant and T is absolute temperature. If only B_0 is applied, moments precess without any phase coherence. A radio frequency magnetic field H_1 applied along the y-axis and represented as an oscillation along the x-axis forces the nuclei to precess in phase. The radio frequency field H_1 has two components clockwise and counterclockwise and only component rotating in the same sense as the nuclear precessing moment would interact with the nucleus. As the frequency is increased, the interaction increases and when the frequency equals that of precessing nuclear frequency, resonance absorption occurs and the nuclei flip from the lower energy level to an upper energy level, i.e. there is a flip over, spins originally precessing with B_0 now precess against B_0.

If the frequency of the radio frequency field is swept through the region of resonance frequency, peak absorption of energy from the radio frequency field will be observed at the resonance frequency.

For a proton μ = 1.41 × 10^{-30} Joule (Gauss)$^{-1}$ or 2.793 nuclear magnetons.

$$\therefore \qquad \nu = \frac{\Delta E}{h} = \frac{\left(1.41 \times 10^{-30}\ \text{J-G}^{-1}\right) \times 141\text{G}}{\left(6.626 \times 10^{-34}\ \text{J-s}\right) \times 1/2}$$

$$\simeq 60 \times 10^6\ \text{s}^{-1}$$

That is, in a magnetic field of ~141 G, the protons will precess with a frequency ~ 60 MHz. It is the resonance frequency required to flip the excess population in the lower energy state to the higher energy state (Fig. 14.5).

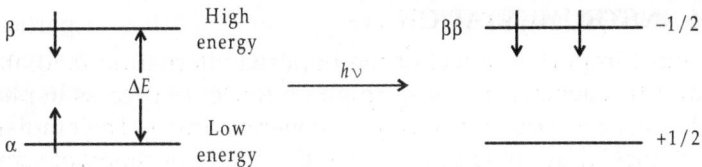

Fig. 14.5: Energetics of excitation in the α to β spin-flip

14.5 NMR SPECTRA

In an NMR technique, the radio frequency is kept constant, for example, at 60 MHz and the strength of magnetic field is varied. At certain value of applied field strength, depending upon the nature of proton or nucleus, the energy required to flip the proton matches the energy of radio frequency radiation. As a result absorption takes place and a signal is observed in the spectrum. Such a signal or peak is called an NMR spectrum.

Nuclei with the $I = 1/2$ give the best resolved spectra (e.g $^1H_1$, ^{13}C, ^{19}F and ^{31}P nuclei). These nuclei act as though they are spherical charge distributions having zero electric quadrupole moment. Nuclei with spin of 1 or more possess nuclear quadrupole moments, the latter measures the electric charge distribution within a nucleus when it possesses non-spherical symmetry. Nuclei possessing nuclear quadrupole moment exhibit smearing out of NMR signal due to shortening of spin life time in a given state.

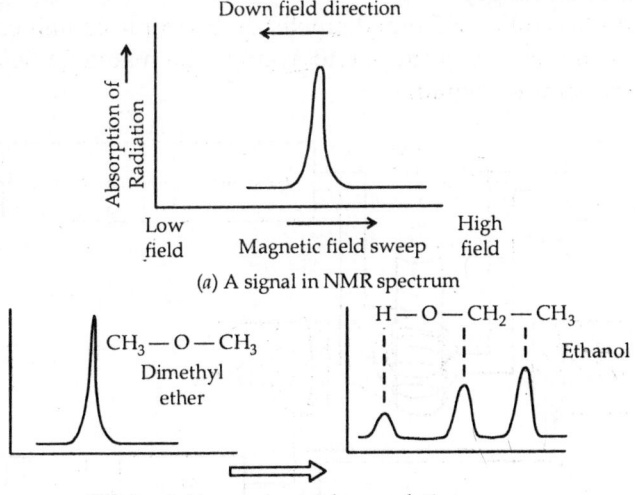

(a) A signal in NMR spectrum

Higher field carried out at low resolution

(b) Proton NMR spectra of C_2H_6O isomers

Fig. 14.6

14.6 INSTRUMENTATION

The most important effect of the imposed alternating field at the Larmor frequency is to cause spinning nuclei to precess in phase, i.e. they act like oscillators forming a coherent source. Their radiation can be picked up by another coil in the neighbourhood of sample positioned with its axis mutually perpendicular to the oscillator coil and the fixed field. The transition between two consecutive energy levels results when the resonance condition given by

$$h\nu = \frac{\mu B_0}{I} = g_N \beta_N B_0 \qquad\qquad ..(1)$$

is satisfied. The spectrum can be obtained experimentally either by varying the external magnetic field (keeping the frequency fixed) or by varying the frequency (keeping the magnetic field fixed). The first approach is generally preferred since it is difficult to vary the frequency at a very high level of stability.

The basic component-requirements of a typical NMR spectrometer are:

   (*i*) An electromagnet giving a powerful, stable and homogenous magnetic field. The field has to be constant over the area of the sample and over the period of time of the experiment.

  (*ii*) A sweep generator which supplies a variable current to the magnet (in order to vary the total applied magnetic field over a small range).

 (*iii*) A glass tube (of 5 mm diameter) as a sample container, spun by an air driven turbine to average the magnetic field over the sample container.

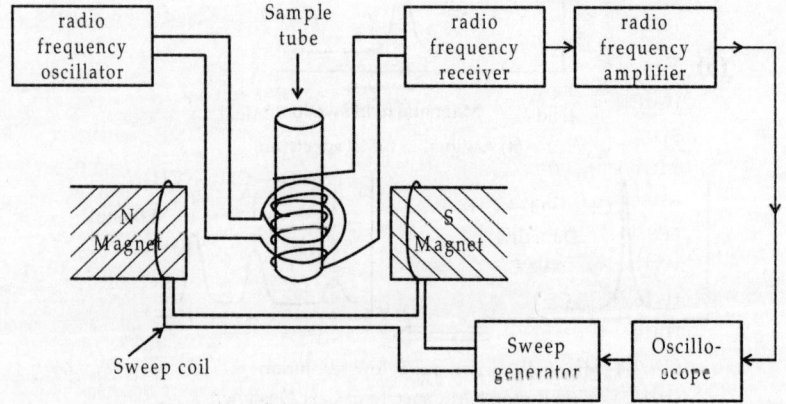

Fig. 14.7: Block diagram of an NMR spectrometer

(*iv*) A radio frequency oscillator connected to a coil (called the transmitter coil) to transmit the energy to the sample. The axis of this coil must be perpendicular to the field.

(*v*) A radio frequency receiver connected to a coil, called the receiver coil, encircling the sample tube. Its axis has to be perpendicular to both the magnetic field and the axis of the transmitter coil.

(*vi*) A read out system, consisting of a radio frequency amplifier, recorder and other accessories to increase the sensitivity, resolution and accuracy.

14.7 INTERPRETATION OF NMR SPECTRA

(*a*) **Some Important Aspects:** The frequency of radio frequency radiation that is absorbed by a given nucleus is strongly affected by its chemical environment, i.e. by nearby electrons and nuclei. Thus even simple molecules provide a wealth of information that can serve to elucidate their chemical structure.

The number of signals (i.e. peaks) signifies how many different kinds of protons are present in the molecule. The positions of the signals tell us about the electron environment of the different types of protons present. The intensities of signals tell us about the different kinds of protons present. The splitting of a signal into several peaks (i.e. hyper fine structure) tells us the number of protons in the adjacent positions (i.e. environment of a proton with respect to other nearby protons).

(*b*) **Environmental Effects:** In ethyl alcohol molecule, (CH_3CH_2OH), the nuclei of C and O atoms have zero nuclear spin, so will not produce any nuclear magnetic spectra. However, H atom has spin 1/2 and it would give a single absorption frequency. But it is evinced that even at low resolution, as many as three absorption peaks are obtained with intensities in the ratio 1 : 2 : 3. On the basis of this ratio, it is logical to attribute the peaks to the hydroxyl, the methylene and the methyl protons respectively. Thus, it is conjectured that different H nuclei exist in the molecule in different environments and consequently each type of H atom interacts differently with the applied magnetic field and exhibits different absorption frequency. Small differences occur in the absorption frequency of the proton (H atom) with differences depending on the group to which the H atom is bonded. This environmental effect is called the *chemical shift*.

The higher resolution spectrum of ethyl alcohol shown in Fig. 14.8 reveals that two of the three proton peaks are split into additional peaks. This secondary environmental effect (which is superimposed upon the chemical shift) has a different cause and is termed as *spin-spin splitting*.

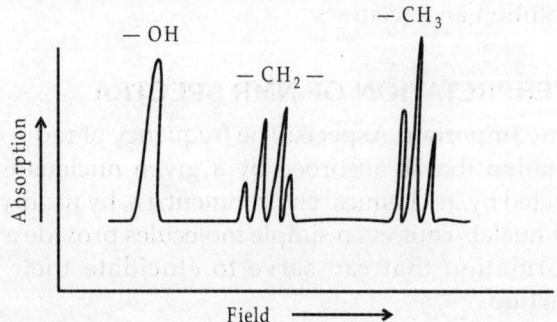

Fig. 14.8: NMR (high resolution) spectrum of ethyl alcohol at a frequency of 60 MHz

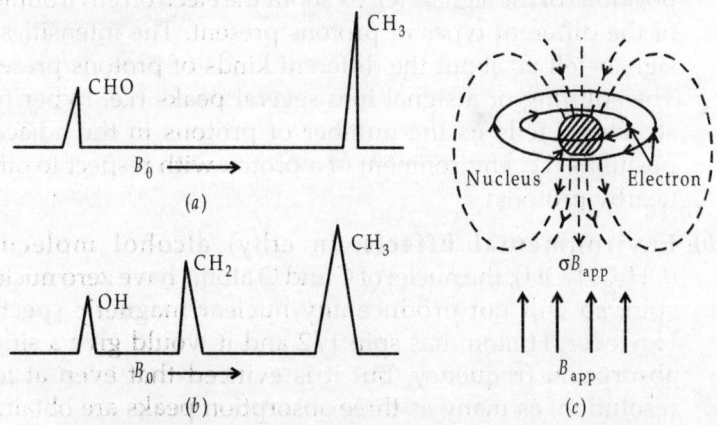

Fig. 14.9: The NMR spectrum of (*a*) acetaldehyde and (*b*) ethanol under low resolution (*c*) diamagnetic shielding of a nucleus*

---

*Diamagnetism is a result of motion induced in bonding electrons by the applied field. This motion (i.e. diamagnetic current) creates a secondary field that opposes the applied field.

(c) **Chemical Shift:** According to the resonance condition, eqn (1) all protons should absorb energy at the same magnetic field. However, this is not so even under low resolution. The spectrum of acetaldehyde (CH_3CHO) shows two lines with intensity ratio 1 : 3 whereas ethanol (CH_3CH_2OH) shows three lines in the intensity ratio 1 : 2 : 3. Moving electrons in a molecule constitute effective currents (within the molecule) that produce a secondary magnetic field which acts oppositely to the applied magnetic field, consequently nuclei are exposed to an effective field that is generally somewhat smaller than the external field. In other words, the nucleus is screened by the surrounding electrons. Thus,

$$B_{eff} = B_{app} - \sigma B_{app} = B_{app}(1 - \sigma)$$

where σ is a dimensionless constant called the *shielding parameter*. The value of σ ($\sim 10^{-5}$) depends on the electron density around the proton. Fig 14.10 illustrates the situation for a shielded spin $1/2$ nucleus.

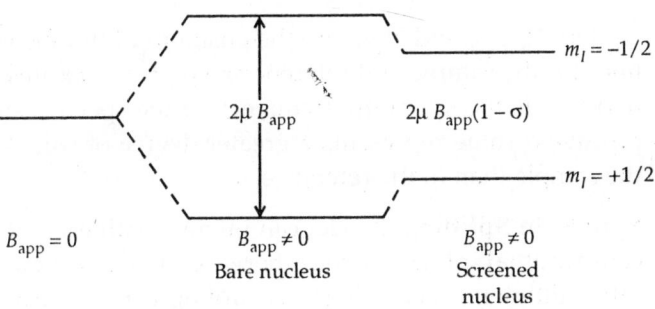

Fig. 14.10: Screening effect for spin 1/2 nucleus in an external magnetic field B_0

Acetaldehyde has two types of protons—CHO and CH_3 protons. The three protons in CH_3 are equivalent. Since O-atom is more electronegative, the electron density around the proton in CHO is less than around the CH_3 protons. Therefore, the shielding (σ) is more for the methyl protons. For a given external field B_0, B_{eff} for CHO proton will be greater than that for CH_3. Consequently, to bring CHO proton into resonance at a fixed frequency, a lesser magnetic field is sufficient. So, the NMR spectrum of CH_3CHO shows two peaks with CHO peak at a lower magnetic field. This situation can be viewed from a different viewpoint, i.e. when the applied field B_0 is kept constant. At a fixed field B_0, the CHO proton finds itself in a greater B_{eff} than the CH_3 protons. Consequently, a higher frequency is required to bring the CHO proton into resonance.

It is desirable to express the position of resonance in field independent units and since NMR spectrometers employing different field strengths are in use, therefore the resonance position is measured with respect to a reference compound. For proton spectra (in non-aqueous media) the reference compound used is tetramethyl silane [$(CH_3)_4Si$], abbreviated TMS whose position is assigned as exactly $\delta = 0$. (TMS contains 12 protons, all are chemically equivalent and so give rise to a single sharp signal). The chemical shift δ is expressed as

$$\delta = \frac{H_{sample} - H_{TMS}}{\nu_1} \times 10^6 \text{ parts per million or (ppm)}$$

Here H_{sample} and H_{TMS} are the positions of the absorption lines for the sample and reference respectively expressed in hertz; ν_1 is the operating frequency of the spectrometer. A positive δ value represents a greater degree of shielding in the sample than in the reference.

(d) **Spin-Spin Splitting:** Nuclei can interact with each other to cause mutual splitting of the otherwise sharp resonance lines into multiplets called spin-spin coupling, e.g. in case of NMR spectra of CH_3CH_2OH, the peak for CH_2 group contains four closely packed absorption lines and CH_2 group has two possible spin orientations. After coupling there would be three possible combinations producing three closely spaced peaks of CH_3.

$$\left(+\frac{1}{2} \ \ +\frac{1}{2} \right) \qquad \begin{pmatrix} +\dfrac{1}{2} & -\dfrac{1}{2} \\[2mm] -\dfrac{1}{2} & +\dfrac{1}{2} \end{pmatrix} \qquad \left(-\frac{1}{2} \ \ -\frac{1}{2} \right) \qquad \begin{array}{c} \overset{\leftarrow}{\rightarrow} \\ \underset{B_{app}}{\overset{\leftarrow}{\rightarrow}} \underset{\rightleftharpoons}{} \ \ \overset{\rightarrow}{\rightarrow} \end{array}$$

| Parallel spins | Opposed spins | Antiparallel spins | Possible spin orientations of CH_2 protons |

The interaction taking place between the various protons causes slight shifts in the magnetic moments which lead to small changes in absorption frequencies. Hence due to influence of CH_2, there is a triplet ($n + 1$ rule) in the fine structure of CH_3 absorption peak in the NMR spectrum (where n = number of neighbouring protons).

In similar manner, CH_2 group is affected by the possible orientations of the three protons in the CH_3 group. When coupled, there would be following possible combinations of the methyl protons:

$$\left(+\frac{1}{2} \quad +\frac{1}{2} \quad +\frac{1}{2}\right) \begin{pmatrix} +\dfrac{1}{2} & -\dfrac{1}{2} & -\dfrac{1}{2} \\[2mm] -\dfrac{1}{2} & -\dfrac{1}{2} & +\dfrac{1}{2} \\[2mm] -\dfrac{1}{2} & +\dfrac{1}{2} & -\dfrac{1}{2} \end{pmatrix} \begin{pmatrix} +\dfrac{1}{2} & +\dfrac{1}{2} & -\dfrac{1}{2} \\[2mm] +\dfrac{1}{2} & -\dfrac{1}{2} & +\dfrac{1}{2} \\[2mm] -\dfrac{1}{2} & +\dfrac{1}{2} & +\dfrac{1}{2} \end{pmatrix}$$

$$\left(-\frac{1}{2} \quad -\frac{1}{2} \quad -\frac{1}{2}\right)$$

Hence the spectrum of CH_2 group would split into four lines with intensities $1 : 3 : 3 : 1$.

Thus we see that the observed behaviour in spin-spin splitting is attributed to the effect that the spins of one set of nuclei (protons) exert upon the resonance behaviour of the other. The multiplets arise because magnetic moments of nuclei interact via the strongly magnetic electrons in the intervening bonds and the strength of the coupling, denoted by J, is given by the spacing of the multiplets and is expressed in hertz.

Applications

A careful study of NMR spectra reveals:

1. The presence of a particular functional group.
2. Relative number of nuclei present in the group.
3. The relative positions of these groups from the multiplicities of the lines.

14.8 RELAXATION PROCESSES IN NMR

Initially, in the strong magnetic field, when a nucleus is exposed to a radiation of a suitable frequency (in NMR), there occurs absorption because of presence of slight excess of lower energy state nuclei. Since this excess is usually small, there is always a danger that the absorption process will equalize the number of nuclei in the upper and the lower states, in which case the absorption signal may decrease and approach zero. This state of the spin system is referred to as *saturation*. In order to avoid saturation, it is necessary that the rate of relaxation of excited nuclei to their lower energy state be greater than the rate at which they absorb the radio frequency photons. In other words, there must be some mechanism in the system which allows the spins occupying the upper state to return to the lower state by transferring the energy from the spin-system to other degrees of freedom. This is referred to as *relaxation*. In NMR studies, non- radiative relaxations are of prime importance. In this process excess spin energy is shared either with the surroundings or with other nuclei and the time taken for a fraction $1/e$ of the excess energy to be dissipated is called the *relaxation time*.

Two different relaxation processes are common for nuclei. In the first, the excess spin energy equilibrates with the surroundings termed the lattice (regardless of whether the sample is a solid, a liquid or a gas). This is known as *spin-lattice relaxation* or *longitudinal relaxation*. Such relaxation comes about by lattice motions, e.g. atomic vibrations in a solid, having approximately the right frequency to interact coherently with nuclear spins. The corresponding relaxation time T_1 varies greatly, being 10^{-2}–10^4 s for solids and 10^{-4}–10 s for liquids.

Secondly, there is a sharing of excess spin energy directly between nuclei via *spin-spin* or *transverse relaxation*, corresponding relaxation time is known as T_2. For solids T_2 is usually very short, $\sim 10^{-4}$ s while for liquids $T_2 \approx T_1$.

Spin-lattice relaxation is strongly affected by the mobility of the lattice. In crystalline solids and viscous liquids, where mobilities are low, T_1 is large. As the mobility increases, the vibrational and rotational frequencies increase and T_1 becomes shorter.

Spin-spin relaxation takes place by interaction between neighbouring nuclei having identical precession rates but different magnetic quantum states. This type of interaction leads to an interchange of quantum states between the two nuclei, i.e. the nucleus in the lower spin state is excited while the excited nucleus relaxes to the lower energy state. No net change in the relative spin state population (and so no decrease in saturation) results, but the average lifetime of a particular excited nucleus is shortened. This causes line broadening.

14.9 SPIN FLIP IN NMR AND THE PRINCIPLE OF RESONANCE

A proton has a magnetic moment given by

$$\vec{\mu} = g_p \frac{e}{2m_p} \vec{I}$$

where g_p is the gyromagnetic constant of proton and $\vec{I}$ is the proton spin. (The measured value of g_p is 5.5). There can be a coupling between the nuclear spin and external magnetic field $\vec{B}$, the energy is $-\vec{\mu} \cdot \vec{B}$. In NMR, (in particular Magnetic Resonance Imaging), we use this energy to explore the presence of hydrogen in an environment such as the human body. The magnetic field exerts a torque on a magnetic dipole and the dynamical equation for the torque is

$$\frac{dI}{dt} = \vec{\mu} \times \vec{B} = + \frac{e g_p}{2m_p} \vec{I} \times \vec{B}$$

If $\vec{B}$ is along the z-axis then the r.h.s. of eqn. has no z-component and

$$\frac{dI_z}{dt} = 0$$

i.e. z-component of proton spin is unchanging. Further,

$$\frac{dI_x}{dt} = \frac{e g_p}{2m_p} B I_y \text{ and } \frac{dI_y}{dt} = -\frac{e g_p}{2m_p} B I_x$$

These equations describe a rotation in a plane perpendicular to the magnetic field with a frequency $\omega = \dfrac{e g_p B}{2m_p}$. (This is termed as *precession*). We now consider the energy changes.

We know that the energy is $-\vec{\mu} \cdot \vec{B}$, so that when the spin is pointing up ($I_z = +1/2$), the energy is $-(e g_p \hbar / 4m_p)B = -(e g_p B/2m_p)$ $(\hbar/2)$ whereas when the spin is pointing down ($I_z = -1/2$), the energy is $+ (e g_p B/2m_p)(\hbar/2)$. Thus the spin up configuration has lower energy than the spin down configuration. As long as $\vec{B}$ is unchanging, I_z remains the same and the energy remains fixed.

Now, in order to change the energy, we supply an additional weak field $\vec{B}'$, that lies entirely in the xy-plane and that rotates about the z-axis with a frequency ω_0. When ω_0 approaches ω, the field $\vec{B}$, rotates along with the proton spin. But then, the spin vector will precess about the direction of $\vec{B}'$, i.e. a precession that involves a change in I_z. We say that a *spin-flip* occurs. (The new field $\vec{B}'$ is weak enough to influence energy only through it role in causing spin flip). Since energy is conserved, when the spin flips, the proton either absorbs or emits e.m. radiation (in the form of a photon) depending on whether the transition is to the higher energy or lower energy state. In either case, the magnitude of photons energy is

$$\frac{e g_p B}{2 m_p} \frac{\hbar}{2} - \left(-\frac{e g_p B}{2 m_p} \frac{\hbar}{2} \right) = \hbar \frac{e g_p B}{2 m_p} \equiv \hbar \omega$$

i.e. these photons have the frequency of the spin precession. *The emission or absorption of such photons is detectable and betrays the presence of hydrogen.* Here "resonance" refers to the fact that the time varying field must have a frequency (ω_0) that closely matches ω in order to produce the spin-flip.

14.10 MAGNETIC RESONANCE IMAGING (MRI)

In practice, a sample whose hydrogen constant is of interest, is placed in a magnetic field $\vec{B}$. The lower of the two energy levels is more populated (as dictated by the Boltzmann factor). An alternating current with a frequency that can be controlled is sent through a wire surrounding the sample, and this produces an additional time varying magnetic field. When the frequency of this field matches $\omega \left(= \dfrac{e g_p B}{2 m_p} \right)$, a spin flip is induced. On the average, the spin flip corresponds to transition from more populated lower level to the less populated upper level, thus photons are absorbed.

The above process offers a good way of measuring g_p, something of interest in itself (a function of hydrogen content). Once we know g_p, we can use the intensity of energy absorption and it spatial dependence to measure, the spatial concentration of hydrogen in the sample. When the "sample" is a patient, the process is known as magnetic resonance imaging (*MRI*) in medical technology. *The measurement of distribution of hydrogen in the body translates to an ability to detect abnormality and to map non bony tissue.* (Bones contain very little hydrogen).

In order to produce a magnetic resonance image, it is important that the B field not be homogeneous. Then, with a fixed photon frequency ω_0, there will be only one place at which resonance will occur, namely where B has the value $(2m_p\omega_0)/(eg_p)$. By moving the patient in a known space varying B-field, the abnormality can be localized.

Questions

14.1 Explain the principle of NMR and obtain the resonance condition.

14.2 List the basic requirements of a typical NMR spectrometer.

14.3 Explain chemical shift with examples.

14.4 Distinguish between spin-lattice and spin-spin relaxations.

14.5 What is quadrupole relaxation?

Electron Spin
Resonance Spectroscopy

Electron Spin Resonance (ESR) or EPR (Electron Paramagnetic Resonance) is a branch of absorption spectroscopy in which radiation of microwave frequency induces transitions between magnetic energy levels of electrons with unpaired spins. The magnetic energy splitting is created by a static magnetic field.

15.1 ELECTRON SPIN BEHAVIOUR (PRINCIPLE OF ESR)

The electron, like the proton, is a charged particle and due to spin has a magnetic field associated with it. It spins much faster than nuclei and thus has a more strong magnetic field. ESR spectroscopy is based on unpaired electron's spin. Unpaired electrons, relatively unusual in occurrence, are present in odd molecules, free radicals, triplet electronic states and transition metal and rare earth ions.

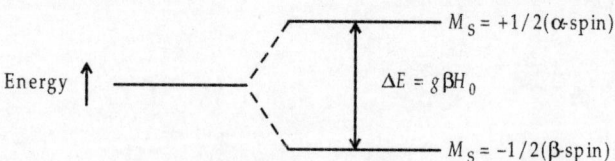

Fig. 15.1: Energy level diagram of an unpaired electron

The spin of an electron gives it an intrinsic angular momentum and is represented by $S\hbar$ where $S = 1/2$ is called the electron spin quantum number. The z-component of the spin angular momentum is expressed by $M_s\hbar$ with $M_s = \pm 1/2$. The magnetic moment associated with electron spin is given by

$$\mu = -g\beta S = -g\beta\sqrt{S(S+1)} \qquad \text{..(1)}$$

where β = Bohr magneton = $\dfrac{e\hbar}{2m_e}$ $\qquad$ ..(2)

$$\equiv 9.274 \times 10^{-26} \text{ JT}^{-1}$$

$$g = \frac{3}{2} + \frac{S(S+1) - L(L+1)}{2J(J+1)} \qquad \text{..(3)}$$

(g is Lande's splitting factor)

If the electron magnetic moment arises from the spin angular momentum only, (e.g. in an organic radical) with $L = 0$ and $S = J = 1/2$, we have $g = 2.0000$ but its experimental value is 2.0023. So,

$$\mu = -2.0023 \times 9.274 \times 10^{-26} \times \sqrt{3}/2$$
$$\cong -16.13 \times 10^{-24} \text{ JT}^{-1} \qquad \text{..(4)}$$

The negative sign in eqn (1) accounts for the fact that the direction of the magnetic moment vector is opposite to the angular momentum vector.

Classically, the energy of interaction of the free electron with an external magnetic field is given by

$$E = -\mu \cdot B$$

The quantum-mechanical hamiltonian is obtained by replacing μ by appropriate operator as

$$H = g\beta S \cdot B \qquad \text{..(5)}$$

If the external field is directed along the Z-axis, then $B_x = B_y = 0$ and eqn (5) becomes

$$H = g\beta M_s B_Z \qquad \text{..(6)}$$

This acts only on the spin variables. Its eigenvalues are given by

$$E_2 = +\frac{1}{2}g\beta B \quad \text{for } M_s = \frac{1}{2}; (\alpha\text{-spin})$$

$$E_1 = -\frac{1}{2}g\beta B \quad \text{for } M_s = -\frac{1}{2}; (\beta\text{-spin}) \qquad \text{..(7)}$$

where $E_2 > E_1$. The lower state has the negative sign and corresponds to the magnetic moment aligned parallel to the field (spin antiparallel to the field). Thus, due to the external field B, the degeneracy is lifted (as shown in Fig. 15.1). A spin transition from $E_1 \rightarrow E_2$ will lead to absorption of frequency ν given by

$$E_2 - E_1 = \Delta E = h\nu = g\beta B$$

or $\qquad \nu = \dfrac{g\beta B}{h}$ $\qquad\qquad$ ..(8)

The *ESR* spectrometer can have two different designs:

(*i*) The frequency ν could be varied at constant magnetic field *B*.

(*ii*) The magnetic field *B* could be varied at constant frequency ν.

However, the latter method is more convenient and thus preferred.

Now, if a sample containing unpaired electrons is in a thermodynamic equilibrium in a magnetic field, then the population ratio is given by, the Boltzmann distribution:

$$\frac{n_1}{n_2} = \exp\left(-\frac{\Delta E}{k_B T}\right) \qquad\qquad ..(9)$$

where n_1 and n_2 are the populations in the upper and lower levels respectively. (For a single spin system, the excess of population of the lower energy level over that in the upper level is extremely small). If radiation is supplied to the sample such that $h\nu \equiv \Delta E = g$ $\mu_B H$) then resonance occurs; electrons in the lower energy level will absorb the radiation and get excited to the higher level. Electrons in the higher level will emit radiation of the same frequency ν and return to the lower level. In order to maintain steady state, there must be some mechanism viz. relaxation to allow excited electrons to lose energy to the lattice otherwise there will be continuous absorption until both levels get equally populated. This condition is known as *saturation*.

Saturation depends on the intensity of the microwave as well as upon the time required for a molecule in the upper level to fall back to the lower level. This time (related to spin-lattice relaxation time) is a measure of the interaction of the unpaired electron with the molecules of the environment, i.e. lattice. The condition of saturation is usually avoided in ESR measurements by working with a low power level of radiation to induce the electron resonance transitions. Equation (3) gives the resonance condition for ESR observation. For free electron in a field of 0.33 tesla, from eqn (3)

$$\nu = \frac{2.0023 \times 9.274 \times 10^{-24} \times 0.33}{6.625 \times 10^{-34}} \simeq 9250 \text{ MHz}$$

This frequency falls in the microwave region. Thus instrumentation used involves radar type components such as waveguides, microwave cavities and klystrons.

15.2 ESR INSTRUMENTATION AND WORKING

The most obvious difference between the instrumentation for ESR and NMR is that because of the high frequencies involved, the power conduction is better by rigid waveguides than by flexible coaxial cables.

The sample cell is inserted through an orifice in the waveguide at a point where the magnetic vector of the electromagnetic wave is undergoing maximum amplitude fluctuation.

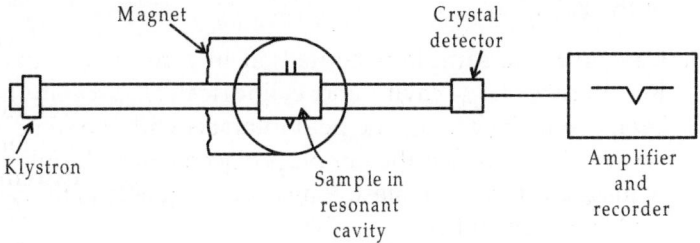

Fig. 15.2

A very simple instrument for carrying out ESR spectroscopy at microwave frequencies and operating in the visible region is shown in Fig. 15.2.

Klystron is the source of radiation. Microwaves are conducted to the sample through a waveguide. The sample tube containing the sample is placed in a microwave cavity between the poles of a magnet operating at ~3300 Gauss. Crystal diode acts as a detector and processes a DC output proportional to the power of radiation falling on it.

The above method, though simple, is not usually used because of its poor sensitivity.

A modern ESR instrument is depicted in Fig. 15.3. It consists of:

1. **Klystron:** Klystron tube acts as the source of radiation. It is stabilized against temperature fluctuations by immersion in an oil bath or by forced air cooling. It is kept at a fixed frequency by an automatic frequency control circuit and provides a power output of ~300 milliwatts.

2. **Waveguide:** It is a hollow rectangular brass tube (0.9×0.4 inches). It is used to convey the microwave radiation to the sample and crystal.

3. **Attenuators:** The power propagated down the waveguide may be continuously decreased by inserting a piece of resistive material into the waveguide. This piece is called variable attenuator and is used in varying the power of the sample.

4. **Isolators:** These are used to prevent the reflection of microwave power back into the radiation source. (It is a strip of ferrite material which allows microwaves to propagate in one direction only). It also helps in stabilizing the frequency of the Klystron.

5. **Cavities:** The sample is contained in a resonance cavity. Rectangular TE_{102} cavity and cylindrical TE_{011} cavity have been used. Since magnetic field interacts with the sample to cause spin resonance, the sample is placed where the intensity of magnetic field is greatest. A measure of quality of the cavity is Q factor given by

$$Q = \frac{\text{Energy stored in cavity}}{\text{Energy lost}}$$

The sensitivity of the spectrometer is directly proportional to the Q value.

6. **Couplers and Matching Screws:** The various components of the microwave assembly may be coupled together by making use of irises or slots of various sizes.

7. **Crystals Detectors and Holders:** A Si-crystal detector which converts the radiation into DC has been used as a detector of microwave radiation. A bridge arrangement has also been used in place of detection technique. Microwave bridges, such as magic-T are commonly used.

8. **Magnets:** An electromagnet capable of producing field of at least 5000 Gauss, is required for ESR.

    The ESR spectrum is recorded by slowly varying the magnetic field through the resonance condition by sweeping the current supplied to the magnet by the power supply. This sweep is usually accomplished by using a variable speed motor drive.

9. **Modulation Coils:** The modulation of the signal at a frequency consistent with good signal to noise ratio in the crystal detector is accomplished by a small alternating variation of the magnetic field. This variation is produced by supplying an AC signal to modulation coils.

10. **Display Devices:** A cathode ray tube is employed. A strip chart or X-Y recorder is used for recording the signal.

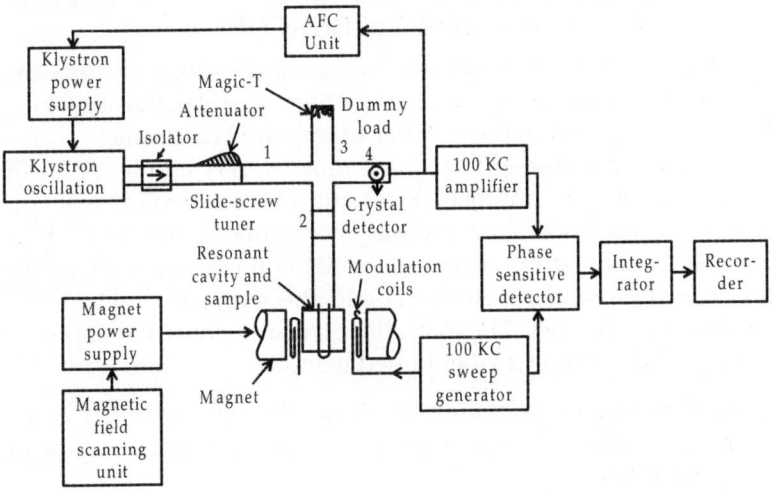

Fig. 15.3: Block diagram of a reflection ESR spectrometer

Working: When the bridge is in a balanced position, microwave power flows only in the arms to the cavity and to the dummy load. There will be no power in the fourth arm. Power in the fourth arm will be there only when the bridge is not balanced. Thus if the balance exists, no signal (initially) appears at the detector. When the sample absorbs, the balancing of the bridge is lost and power appears in the fourth arm.

The width of ESR lines are fairly large and hence the spectrum is usually recorded in the first derivative mode which enables one to fix up the frequency position and estimation of intensity more precisely. Another advantage of using derivative mode is that it gives a well defined line width ΔB.

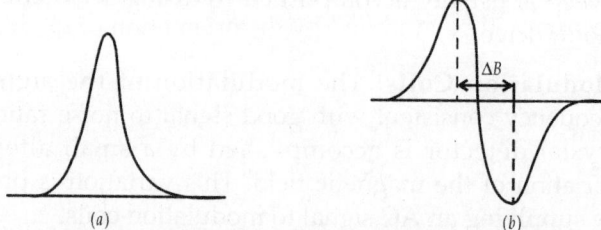

Fig. 15.4: ESR signal (*a*) a single absorption line (*b*) its first derivative

15.3 HYPERFINE INTERACTIONS AND QUALITATIVE ANALYSIS IN ESR SPECTROSCOPY

The detection of electron spin resonance signals proves the presence of unpaired electrons in the sample. The intensity of the absorption line is proportional to the number of unpaired electrons.

The ESR spectroscopy has been most widely employed in the study of chemical, photochemical and electrochemical reactions which proceed via free radical mechanisms because of the occurrence of hyperfine structure, which is the result of interaction between the unpaired electrons and the magnetic nuclei in the paramagnetic species. The hyperfine structure gives two important pieces of information as mentioned below:

1. It provides information about the environment of the molecule and distribution of electron density within the molecule.

2. It allows identification of paramagnetic substance in a number of cases.

The resonance frequency of an electron actually depends upon the magnetic field at the electron and electron actually is affected by the applied field H_0 and any local field due to magnetic fields of nuclei or other effects (H_{local}), i.e.

$$h\nu = g\beta \, (H_0 + H_{local})$$

The effect of magnetic moment of nuclei on the ESR spectrum is known as hyperfine interaction and it is responsible for splitting of the ESR line giving rise to hyperfine structure.

Let us consider the interaction of an unpaired electron, say, with nucleus of a hydrogen atom (i.e. a proton). The proton is a charged spinning particle with a nuclear spin $I = 1/2$. The proton thus has a magnetic moment and the electron will be affected by the magnetic field of the nucleus, as well as the applied magnetic field. The energy of the electron will be modified by the orientation of the magnetic moment of the proton which can be either parallel or antiparallel to the magnetic field.

The relative orientation of the nuclear magnetic moment and the electron's magnetic moment causes a splitting of the original two levels to four levels.

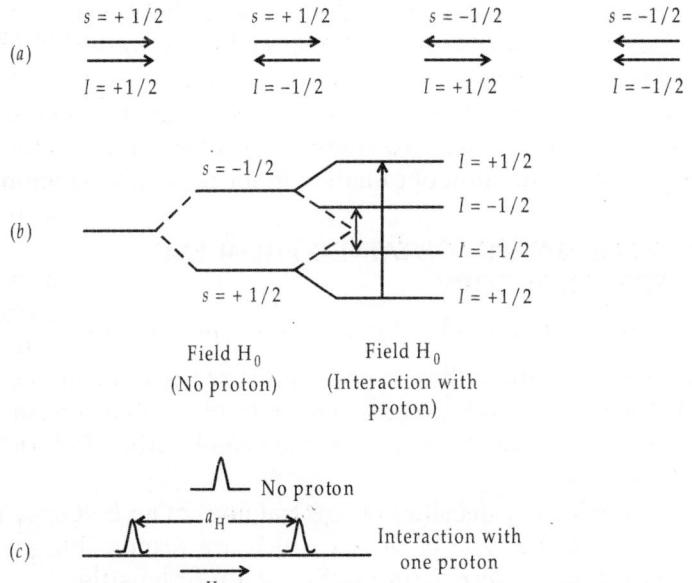

Fig. 15.5: Interaction of unpaired electron with one proton (*a*) possible orientations of electron's and nuclear magnetic moments (*b*) Splitting of energy levels (*c*) Splitting of spectral line

Only two transitions are found between these two states because the nuclear moment remains fixed during electronic transitions so that transitions occur only between $I = +1/2$ or $I = -1/2$ states. The selection rules for the change in magnetic quantum numbers which govern ESR transitions are

$$\Delta M_s = \pm 1, \quad \Delta M_I = 0$$

where M_I is the magnetic quantum number of the nucleus which is $\pm 1/2$ for proton.

Applications: The applications of ESR are limited due to the fact that substance being analyzed contains unpaired electrons.

However, this technique has widely been used in organic chemistry in the analysis of free radicals produced by chemical reactions.

15.4 COMPARISON BETWEEN NMR AND ESR

In NMR spectroscopy, the two different energy states are produced due to the alignment of the *nuclear* magnetic moments relative to the applied field and a transition between these energy states takes place upon the application of a radio-frequency field of appropriate frequency. On the other hand, in ESR spectroscopy, the two different energy states are produced due to the alignment of the *electronic* magnetic moments relative to the applied magnetic field and a transition between these two energy states takes place upon the absorption of a quantum of radiation in the microwave region.

15.5 INFORMATION OBTAINED FROM ESR SPECTROSCOPY

An ESR spectrum provides the following types of information:

    (*i*) It decides about the site of the unpaired electron(s).
   (*ii*) The number of line components decides about the number and type of nuclei present in the neighbourhood of the odd electron.
  (*iii*) The relative intensities of spectral lines in an ESR spectrum confirm the type of nuclei which are responsible for the splitting pattern (summation of the intensities leads to determination of the total number of free electrons in the sample).
  (*iv*) From the ESR spectrum, the value of g can be measured by comparing the position of the line with that of a standard substance of known g value.

Questions

15.1 Explain the principle of ESR.
15.2 Explain the factors responsible for the hyperfine structure in ESR spectra.
15.3 Explain the spin arrangements for four equivalent protons and obtain the degeneracies of the different states.

Fourier Infrared (FIR) Spectroscopy

16.1 INTRODUCTION

It was first developed by astronomers in the early 1950's to study the infrared spectra of distant stars. Nowadays it has developed into a very powerful technique for the detection of very weak signals from the environmental noise. It is based on the Fourier Transform method to resolve a complex wave into its frequency components. The conventional IR spectrometers are not of much use in the far infrared region. FIR spectroscopy has made this energy limited region more accessible. In FIR spectral range non-dispersive spectrometers have a great advantage. As a consequence of high capacity and speed of computers, the non-dispersive or inter-ferometric spectroscopy is also entering the visible spectral range.

The conventional spectroscopy is dispersive or the frequency domain spectroscopy. It records the radiant power $G(\omega)$ as a function of frequency ω. In the FIR, i.e. time domain spectroscopy, the changes in radiant power $f(t)$ are recorded as a function of time t. In a Fourier spectrometer, a time domain plot is converted into a frequency domain spectrum (with the help of computers). Mathematically,

$$G(\omega) = \frac{1}{\sqrt{2\pi}} \int_{-\infty}^{+\infty} f(t) e^{i\omega t} dt$$

$$\text{〰〰〰} \Rightarrow \frac{1}{\nu}$$

Time domain   Frequency domain

and $\qquad f(t) = \dfrac{1}{\sqrt{2\pi}} \int_{-\infty}^{+\infty} G(\omega)e^{-i\omega t}d\omega$

16.2 PRINCIPLE OF FOURIER TRANSFORM SPECTROSCOPY

The method basically employs Michelson interferometer as shown in Fig. 16.1. It consists of a source S, a beam splitter B and two plane mirrors M_1 and M_2, mirror M_1 is fixed while M_2 can be moved to and fro.

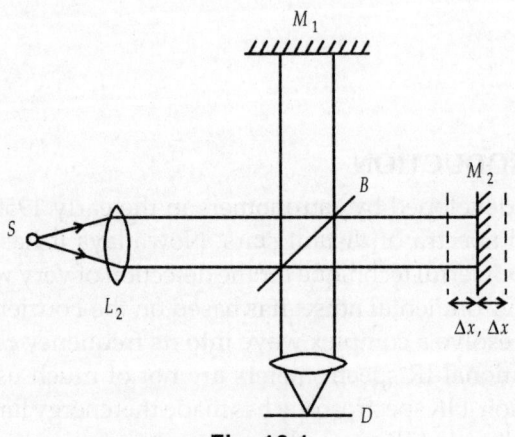

Fig. 16.1

The white light from the source (placed at the focus of lens L_1) is separated into two (equal 50% each) parts by B.

The reflected part is focused onto the detector D after reflection from the stationary mirror M_1 and after a second splitting by the beam splitter B. The transmitted part is also focused onto the detector D after reflection from the mirror M_2 and again splitting by the beam splitter. By suitable movement of the mirror M_2 through distance Δx, interference fringes can be obtained at the detector. The intensity $I(x)$ of the fringes depends on the position x of the mirror M_2. If the incident beam is monochromatic of the form $E(x, t) = E_0\cos(kx - \omega t)$, the field E_D at the detectors is

$$E_D = \frac{1}{2}\Big[E_0 \cos(\omega_0 t - k_0 x) + E_0 \cos\{k_0(x + 2\Delta x) - \omega_0 t\}\Big]$$

where $2\Delta x$ is the optical path difference between the two beams (twice the shift of the mirror M_2). Choosing $x = 0$ and replacing $2\Delta x$ by $2x$ (and k_0 by $2\pi\nu_0$), we have intensity at the detector

$$I(x) = c_0 \, \epsilon_0 \left\langle E_D^2 \right\rangle = \frac{\epsilon_0 \, c_0}{4} E_0^2 \left\{ 1 + \cos(4\pi \nu_0 x) \right\}$$

We can rewrite this using spectral intensity
$$I(\nu) = \epsilon_0 C_0 E_0^2 \, \delta(\nu - \nu_0)/2, \text{ as}$$

$$I(x) = \frac{1}{2} \int_0^\infty I(\nu)\{1 + \cos(4\pi \nu x)\} d\nu$$

We can generalize this equation to an arbitrary intensity spectrum $I(\nu)$ and obtain the following basic relation for Fourier Spectroscopy

$$I'(x) = I(x) - \frac{1}{2} \int_0^\infty I(\nu) d\nu$$

$$= \frac{1}{2} \int_0^\infty I(\nu) \cos(4\pi \nu x) d\nu$$

The function $I'(x)$ contains the whole information about the spectrum $I(\nu)$. In fact $I'(x)$ is the cosine Fourier transform of $I(\nu)$. The observed intensity oscillates around an average intensity $\int I(\nu)d\nu/2 = I_0$ (i.e. half of the total intensity of the beam). Fourier transform of $I'(x)$ yields (for $x = y/2$),

$$\int I'(y/2)\cos(2\pi \nu' y) dy$$

$$= \frac{1}{2} \int_0^\infty I(\nu) d\nu \int \cos(2\pi \nu y)\cos(2\pi \nu' y) dy = \frac{I(\nu')}{2}$$

Since the integration over y gives $\delta(\nu - \nu')$. This equation means that we obtain the spectral components directly from the interferogram (i.e. function I') by Fourier transformation without any spectral dispersion. This is accomplished with the help of computers.

Mirror M_2 is moved smoothly over a period of time (e.g. one second) through about 1 cm while the detector signal—the interferogram—is collected into a multichannel computer.

The detector signal may be monitored, say, every thousandth of a second during the mirror (M_2) traverse, and each piece of information is put serially into one of 1000 different storage points in the computer (Multiplexing). The computer then replaces the 'proper' spectrum piecemeal into the same 1000 locations, ready for plotting out onto paper (i.e. interferogram).

16.3 ADVANTAGES OF FOURIER TRANSFORM SPECTROSCOPY

As compared to conventional dispersive spectroscopy there are three basic advantages viz. speed, the energy advantage and the multiplex advantage.

1. The whole spectrum is since contained in the interferogram, which recorded in the computer within one second, adding the computing and plotting time of say 15 seconds. We find that the overall time to obtain spectrum is very short compared with $\simeq$ the 10 minutes required by conventional methods. The reason is that in the conventional method, each element or observation point of spectrum is examined consecutively whereas in the FIR interferometer, all the elements are examined simultaneously.

2. The energy advantage originates from the fact that during the whole period of measurement, always nearly the total beam intensity hits the detector. Thus, the detection operates on a high signal level, consequently improving the signal to noise ratio.

3. The multiplex advantage is attributed to ability for simultaneous measurement of the full spectrum over the complete period T of detection. In contrast, in dispersive spectroscopy N parts of widths Δv of the spectrum will be measured successively and for each part, only the time T' = T/N is available).

4. In conventional spectroscopy, a very fine slit gives good resolving power, since only a narrow spread of frequencies fall on the detector at any moment. In FT work, parallel beams are used throughout, no slit is required and all the source energy passes through the instrument, therefore, amplifier are less critical and the resolving power is solely governed by the mirror traverse and computer capacity.

5. The resolving power of an FT instrument is constant over the entire spectrum; in a grating or a prism instrument, the resolving power depends on the angle which that component makes with the incident beam, and hence varies with frequency.

6. A great advantage of FT spectroscopy is higher heightness and ability of data precessing. This has enabled to investigate the spectrum of transient species, e.g. unstable molecules in a chemical reaction and for the analysis of environmental samples. Further, high resolution FT spectroscopy has made possible to measure the components in the atmosphere to the level of a few parts per billion.

Questions

16.1 What is Fourier infrared spectroscopy?

16.2 Wha is Fourier transform spectroscopy? Discuss its principle.

16.3 What are advantages of Fourier transform spectroscopy?

17

Photoacoustic Spectroscopy

17.1 INTRODUCTION

Photoacoustic spectroscopy is based upon a light absorption effect that was first investigated in the 1880's by Alexander Graham Bell and others. This spectroscopy was then developed in the early 1970's and it provides a means for obtaining ultraviolet and visible absorption spectra of solids, semisolids or turbid liquids. Spectra for these kinds of samples cannot be acquired by ordinary methods and are often impossible because of light scattering and reflection.

The excitation of a molecule by absorption of radiation initiates a series of events by which the excess energy is dissipated, allowing the molecule to return to its ground state. These events are either radiative (fluorescence/phosphorescence) or non-radiative. PAS employs the non-radiative transitions. Non-radiative transitions can occur from higher vibrational sublevels to the base levels. If the sample is a liquid or a gas, the energy corresponding to these transitions simply increases the thermal agitation of the molecules of the sample. If the sample is a solid, the energy is first converted to increase lattice vibrations and may then be transferred to any gas (or liquid) in contact with the sample.

This energy, directly or indirectly gets converted to heat, so it is generally possible to obtain useful information by measurement of temperature rise. This could have been done by direct measurement of temperature with a thermistor but more successfully is done by the technique of photoacoustic spectroscopy.

17.2 THE PHOTOACOUSTIC EFFECT

This effect is observed when a gas in a closed cell is irradiated with a chopped beam of radiation of a wavelength that is absorbed by the gas. The absorbed radiation causes periodic heating of the gas, which in turn results in regular pressure fluctuations within the chamber. If the chopping rate lies in the acoustic frequency range, these pulses of pressure can be detected by a sensitive microphone.

Thus, in PAS, a pulsating (chopped) beam of radiation is utilized, so, the heating effect produced in the absorbing sample fluctuates at the same frequency. The periodic increase and decrease in vibrational energy propagates through the medium as a sound wave. We know that any wave motion is characterized by an amplitude, a velocity and a frequency. In the present case, the frequency is that at which the radiation is chopped, the velocity is the speed of sound in the medium and the amplitude corresponds to the amount of energy absorbed and converted to heat.

The sonic signal can be measured by a microphone. If the absorbing sample is a gas, the sound wave can be picked up directly by a microphone. If the sample is a solid, the most convenient arrangement is to couple the sound wave in the solid to the microphone by immersing both in an enclosure filled with a gas (usually air). Since there is a substantial loss of energy as the wave crosses the solid-gas boundary, this has to be recovered by electronic amplification.

17.3 INSTRUMENTATION IN PHOTOACOUSTIC SPECTROSCOPY

A block diagram of a PAS spectrometer is shown in Fig. 17.1 and a typical sample cell in Fig. 17.2.

The photoacoustic effect is thus observed, provided the radiation is absorbed by the solid; non-radiative relaxation of the solid causes a periodic heat from the solid to the surrounding gas which results in pressure fluctuations (sonic signals) in the gas, the power of the resulting sound is directly related to the extent of absorption. Radiation reflected (or scattered) by the sample has no effect on the microphone and thus latter property is the most important characteristic of the method.

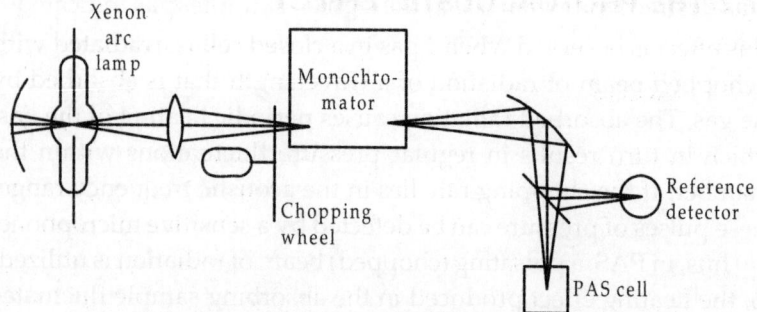

Fig. 17.1: A single beam photoacoustic spectrometer

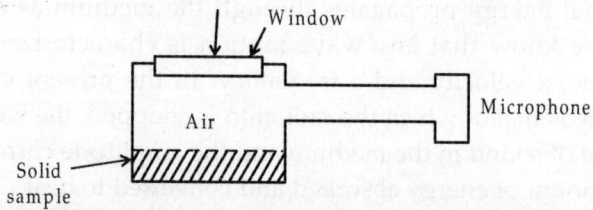

Fig. 17.2: A sample cell for PAS measurement

The pulsed radiation enters through a transparent window and then meets the sample, which is either the gas within the cell or the solid material. The microphone is placed in a side arm to protect it from direct radiation. The dimensions of the cell are so selected that it resonates at the frequency of the chopped radiation.

The chief limitation of the method is due to a saturation phenomenon. This is due to limited rate of diffusion of heat through the sample where a photon gets absorbed to the surface, where a transfer of heat to the surrounding gas is possible. Both optical absorption and thermal diffusion are characterized by constants with the units of cm^{-1}. β is the optical absorption coefficient defined by

$$\frac{P}{P_0} = e^{-\beta l}$$

where l is the path length in the absorbing material and a_s, the coefficient of thermal diffusion defined by

$$a_s = \left(\frac{\pi \nu}{\alpha_s}\right)^{1/2}$$

where ν is the frequency, and the α_s, the thermal diffusivity (in cm^2/s). If $\beta < a_s$, a photoacoustic signal will result that is proportional to β(as desired) but when $\beta \geq a_s$, the signal "saturates" as indicated by c for a hypothetical sample.

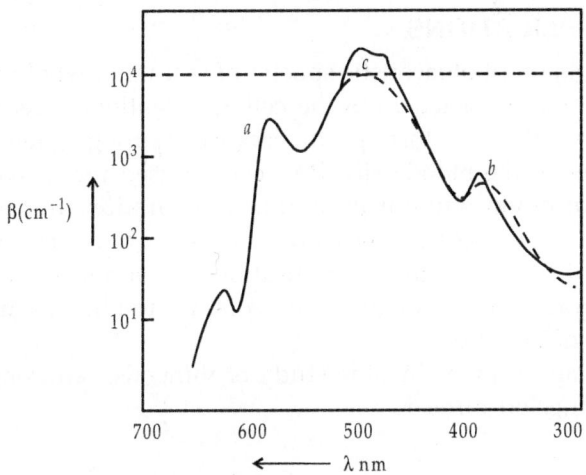

Fig. 17.3: Optical absorption (solid line) and photoacoustic spectrograph (dashed line) of a hypothetical sample

If one were to determine this substance by means of the PAS spectrum, measurements should be made at a or b (and not at c).

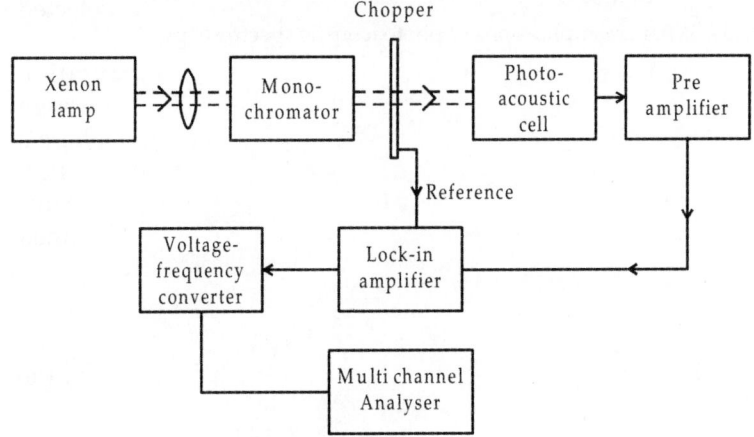

Fig. 17.4: Block diagram of a digitally processed single-beam photoacoustic spectrometer

In Fig. 17.4, the spectrum from the lamp is recorded digitally, followed by the spectrum for the sample. The stored lamp data are then used to correct the output from the sample for variations in lamp output as a function of wavelength.

17.4 APPLICATIONS

We can obtain photoacoustic spectra of smears of whole blood or haemoglobin extracted from the cell. Conventional spectroscopy does not yield satisfactory spectra because of strong light-scattering properties of the blood-cells. PAS can be used for spectroscopic studies of blood without preliminary separation from its large molecules. Photoacoustic measurements in the mid infrared region are useful for qualitative identification of components in solids. Fourier transform techniques are necessary to obtain satisfactory signal to noise ratio.

Other applications include study of minerals, semiconductors, and coatings on surfaces.

Questions

17.1 What is photoacoustic spectroscopy?

17.2 Discuss photoacoustic effect. Describe a single beam photoacoustic spectrometer.

17.3 What are applications of photoacoustic spectroscopy?

Mössbauer Spectroscopy

18.1 INTRODUCTION

Increasing the quantum energy beyond the X-rays leads to the spectral range of γ radiation. This radiation originates from a relaxation of excited and quantized states of the nuclei. The idea of spectroscopy using γ-ray quanta was suggested by Kuhn in 1929. A number of attempts to observe resonance scattering with γ radiation were made but were all without success. The problems of nuclear resonance fluorescence and absorption were complicated due to the large amounts of recoil energies given to the nucleus by the emitted γ-radiations. However, R. Mössbauer reported for the first time recoil free resonance absorption for γ-rays in 1958. This technique is now known as *Mössbauer spectroscopy*.

Mössbauer spectroscopy is the study of γ-ray absorption (or emission) spectra for transitions between nuclear states. For the discovery of this spectroscopic method, Mössbauer was awarded the Nobel Prize in 1961.

18.2 PRINCIPLE OF MÖSSBAUER SPECTROSCOPY (NRF)

Nuclear resonance absorption is in principle expected to occur when gamma radiation emitted in a transition is reabsorbed by another nucleus of the same kind as depicted in Fig. 18.1.

Imagine, now an isolated atom of mass M in an excited nuclear state E_i above the ground state E_f. Let E_0 be the energy difference $E_i - E_f$. If p is the recoil momentum, then kinetic energy of the recoil nucleus

$$E_{re} = \frac{p^2}{2M} \qquad\qquad ..(1)$$

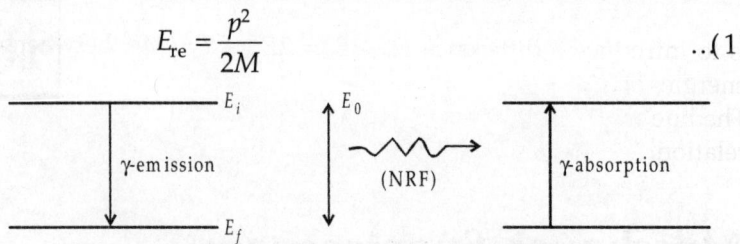

Fig. 18.1: Nuclear resonance fluorescence

The emitted gamma ray will have an energy

$$E_\gamma = E_0 - \frac{p^2}{2M} = E_0 - E_{re} \qquad\qquad ..(2)$$

Therefore, the emitted spectral line will be shifted from the expected position E_0 (as shown in Fig. 18.2).

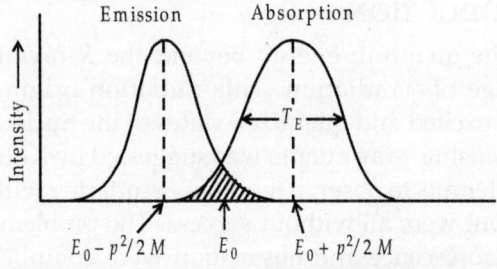

Fig. 18.2: Overlap of emission and absorption lines

By the law of conservation of momentum the recoil momentum of the atom is equal to the momentum of the emitted γ-ray photon (given by the de Broglie relationship), i.e.

$$p = \frac{h}{\lambda} = \frac{E_\gamma}{c} \qquad\qquad (\because c = \nu\lambda) \quad ..(3)$$

where c is the velocity of light

Hence,

$$E_{re} = \frac{p^2}{2M} = \frac{E_\gamma^2}{2Mc^2} = \frac{(E_0 - E_{re})^2}{2Mc^2} \simeq \frac{E_0^2}{2Mc^2} \qquad (\because E_{re} << E_0) \quad ..(4)$$

When the same atom absorbs the γ-ray, its nucleus will have an energy

$$E_\gamma' = E_0 + \frac{p^2}{2M} = E_0 + E_{re} \qquad\qquad ..(5)$$

Thus, the nuclear resonant absorption can only be observed as a successive γ-ray emission and absorption, and the effect of the recoil is to introduce a difference $E_\gamma' - E_\gamma = 2E_{re} \equiv E_0^2/Mc^2$ between the energies of the emitted and absorbed γ-rays in the resonant process. The line width $(\delta\nu/\nu)$ is governed by the energy-time uncertainty relation.

$$\Gamma_E \times \tau = \hbar \text{ or } \Gamma_E = \frac{\hbar}{\tau}; \qquad (\Gamma_E \text{ is energy uncertainty})$$

where τ is the mean lifetime of the nuclei in the excited state and $\hbar = h/2\pi$, h being Planck's constant. The nuclear resonant absorption will have a significant probability if the emission and absorption profiles overlap strongly. Maximum absorption will occur when the recoil energy is zero. This can be achieved when the emitting and absorbing nuclei are found in a crystal lattice. Then the atom (emitter or absorber) will be unable to recoil freely because it is chemically bound to its lattice site. (The crystal mass as a whole will recoil). Equations (2) and (5) contain the reciprocal mass $(1/M)$, which has now become the mass of the crystal (containing $\sim 10^{15}$ atoms).

Thus, the recoil energy of the atom becomes very small, it cannot rupture a chemical bond, rather it becomes the property of the lattice and the lattice as a whole vibrates to dissipate the recoil energy. In other words, there is a finite probability in a solid matrix that a γ-ray photon be emitted or absorbed without recoil or thermal broadening.

18.3 EXPERIMENTAL METHOD

The experimental method is based on using a solid matrix containing the excited nuclei of a given isotope called the *source* next to which is placed a second matrix containing the same *isotope* in ground state, called the absorber. There are a large number of radioactive isotopes produced artificially which can be suitable for the study of Mössbauer effect, e.g. Fe^{57}, Zn^{67} and Sn^{119} have been commonly used. We have seen that the amount of overlap of the energy profiles for the source and absorber decides the extent of resonant absorption, so the basic experimental technique relies on creating *Doppler velocity shift* by some device.

A movement of the source and absorber relative to each other with velocity v produces a Doppler shift in frequency

$$\Delta\nu = \pm\left(\frac{v}{c}\right)\nu$$

or energy shift, $\Delta E = \pm \dfrac{\nu E}{c}$

where c is the velocity of light and ν, the frequency of γ-ray.

The nuclear transition energy involved is usually 10–100 keV mol^{-1} and is far more energetic than chemical bond (10^{-6} keV mol^{-1}) or molecular vibration energy (10^{-7} keV mol^{-1}). The energy time uncertainty relation and overlap condition puts a limit on the value of excited state lifetime of the source material, because natural width of the resonance line depends on the lifetime of the excited state. Consequently, it is desirable that the excited state lifetime of the source should be in the range 10^{-6} s $< \tau < 10^{-10}$ s.

Consider the source ^{57}Co ($I = 7/2$, $e\,Q < 0$) which belongs to ^{57}Fe* through electron capture. This excited nucleus comes to stable ^{57}Fe through emission of γ-rays. The complete energy diagram is shown in Fig. 18.3.

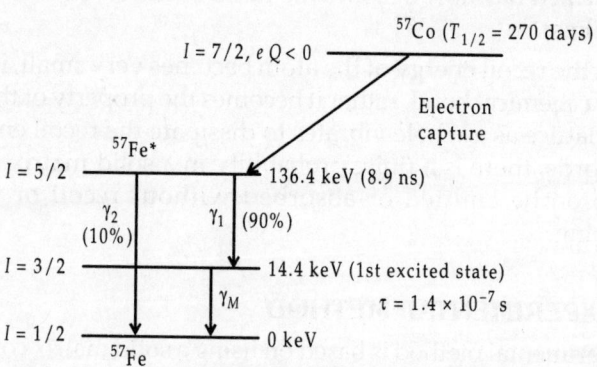

Fig. 18.3: Decay scheme of the excited ^{57}Co to the stable ^{57}Fe and emission of delayed γ-rays

The absorbers are usually stable isotopes of elements with which the Mössbauer effect can be observed. As emitter, one selects the excited state of the same isotope, e.g. ^{57}Co serves as an emitter and ^{57}Fe stable isotope (which absorbs the 14.4 keV γ radiation) serves as good absorber.

A transition from $I = 3/2 \to 1/2$ involving an energy change of $E_0 = 23.07 \times 10^{-16}$ kJ/nuclei is of interest. The frequency of the emitted γ-ray is $\nu = 34.88 \times 10^{17}$ Hz. The absorber placed in the path of the emitted γ-ray leads to the transition $I - 1/2 \to 3/2$. The transitions in any stationary system are not sharp due to the uncertainty relation:

$$\Gamma_E \times \tau = \hbar = 1.054 \times 10^{-34} \text{ Js}$$

where Γ_E is the energy uncertainty and τ is the mean lifetime of the nuclei in the excited state. Since $\tau = 1.4 \times 10^{-7}$ s in the present case, therefore

$$\Gamma_E = \frac{\hbar}{\tau} = 7.5 \times 10^{-28} \text{ J}$$

and $\therefore$ frequency uncertainty

$$\delta\nu = 1.14 \times 10^6 \text{ Hz},$$

and the corresponding line width

$$\delta\nu / \nu = 3.27 \times 10^{-13}$$

This is much smaller spectroscopic widths found in other spectroscopic techniques (viz. for NMR $\approx 10^{-8}$; for $IR \approx 10^{-4}$). So, a very sharp absorption line is observed.

18.4 APPARATUS FOR MÖSSBAUER SPECTROSCOPY

Experimentally, two methods are possible. In one method, the source (^{57}Co) is mounted on a mechanical constant velocity device (e.g. on a screw thread rotation) and the total number of counts is registered in a fixed time by a Geiger Muller counter mounted behind the sample. The same process is repeated with different velocities until the desired velocity is covered. GM counter will show a sudden fall in the count rate when the sample begins to absorb the γ-ray emitted by the source. (A complete spectrum will have to be examined point by point, since for any one source velocity, the Doppler shift is constant).

In the second method, the source is mounted on an oscillating drive which gives it a varying velocity relative to the sample.

The general experimental arrangement is shown in Fig. 18.4 and that of a modern spectrometer is given in Fig. 18.5.

In a modern spectrometer, ^{57}Co is mounted on a loudspeaker coil whose motion is controlled by a signal generator. A sawtooth variation of velocity is given. This provides an oscillatory motion giving the source a varying velocity relative to the sample (zero velocity at the extremities and maximum at the centre).

The absorbing sample is refrigerated. A scintillation counter, or a semiconductor detector may be used for detection. The signal is then fed to a multichannel analyser which collects the results and

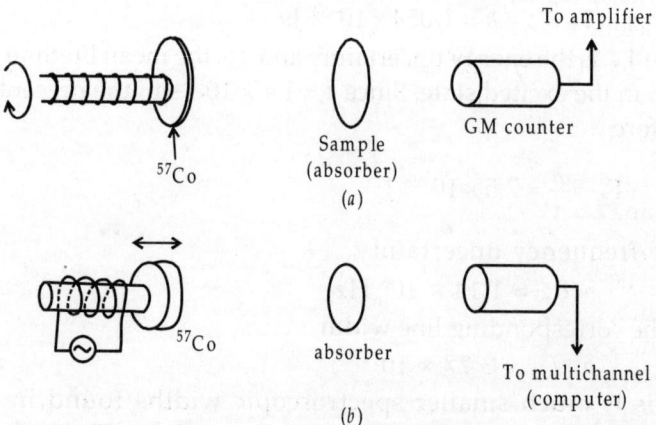

Fig. 18.4: Two experimental arrangements for Mössbauer spectroscopy (a) employing a screw-thread drive (b) using an oscillating drive for the source

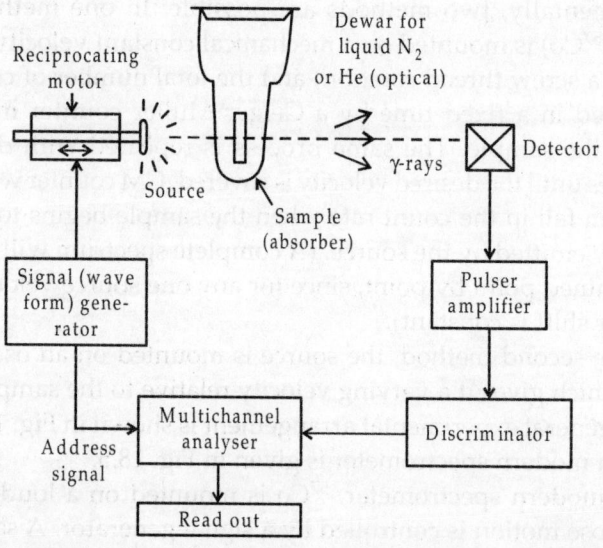

Fig. 18.5: Schematic arrangement of a modern Mössbauer spectrometer

sums it over each cycle. The discriminator rejects the background (non-resonant) radiation. The final Mössbauer spectrum is displayed (as counts/s) as a function of relative velocity between the source and absorber. The absorption by the sample is shown by a fall in counts/s. A relative velocity of 0.1 mm/s is sufficient to produce a marked reduction of the 14.4 keV ^{57}Fe transition.

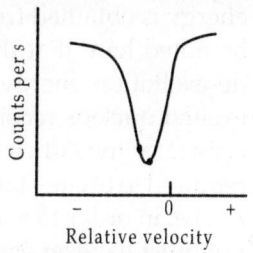

Fig. 18.6: Typical Mössbauer spectrum

18.5 RESONANCE FLUORESCENCE

The Mössbauer effect is known as recoilless gamma resonance fluorescence in which no part of energy is expended in recoil of the nucleus emitting or absorbing the gamma quanta. In resonance fluorescence, a photon of energy E_r and momentum p strikes a target of mass M (initially at rest) and the entire photon-momentum is transferred to the target. The target thus recoils and the energy of recoil (i.e. energy distribution of excited state is as shown in Fig. 18.7 (a). Figure 18.7 (b) shows energy distribution of photons emitted in transition $B \rightarrow A$. This

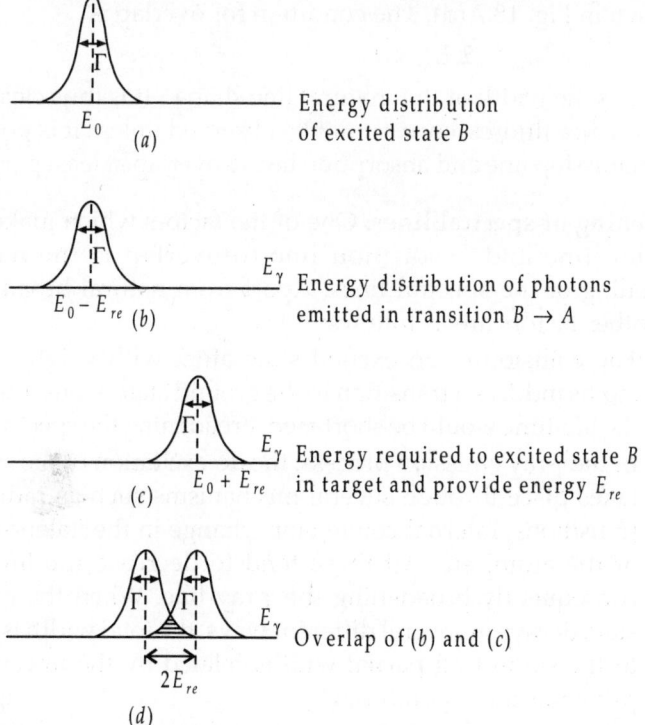

Energy distribution of excited state B

Energy distribution of photons emitted in transition $B \rightarrow A$

Energy required to excited state B in target and provide energy E_{re}

Overlap of (b) and (c)

Fig. 18.7: Energy distribution showing emission of γ ray photons and "natural line shape" of width Γ

energy is obtained from de-excitation of emitting system. It should be noted here that the emitted photon has lesser energy than the de-excitation energy of emitting system due to the fact that the excited nucleus recoils and the momentum is distributed over the crystal by the lattice forces. Figure 18.7 (c) shows energy spectrum required to excite-state B in target and provide centre of mass energy E_{re}, i.e. in order to excite a level of energy E_0, the incoming gamma ray must have an energy $E_0 + E_{re}$ as shown in Fig. 18.7 (c) but only an energy $E_0 - E_{re}$ is available for excitations of internal degrees of freedom.

Resonance fluorescence can occur only if some of the incoming photons possess enough energy to "reach" the state B and at the same time provide energy to recoiling system, i.e. for resonance fluorescence, the absorption line would overlap the emission line. But in actual case only the overlapping part of the spectra, [Figs 18.7 (b) and 18.7 (c)] is responsible for resonance fluorescence as shown in Fig. 18.7 (d). The condition for overlap is

$$2\,E_{re} < \Gamma$$

where Γ is the width of the "natural line shape". It is thus clear that the resonance fluorescence cannot be observed unless it is possible for the emission line and absorption line to overlap at least partially.

Broadening of spectral lines: One of the factors which makes the emission line and absorption line to overlap is the natural broadening of the spectral lines. Apart from natural broadening some other factors are as follows:

(a) The collision of an excited state atom with neighbouring atoms induces a transition to the ground state. Consequently, the life-time would be shortened, broadening the spectral line.

(b) In the γ-ray emission process, the de-excitation of the system takes place through several mechanisms such as radiative-transitions, internal conversion, change in the valence state of the atom, etc. All these tend to decrease the life-time consequently broadening the γ-ray line. When the excited state decays in several different ways, the total width is equal to the sum of all partial widths related by the uncertainty principle correspondingly,

$$\Gamma = \Sigma\,\Gamma_i = h\left(\Sigma\,\frac{1}{T_i}\right)$$

(c) Besides above mentioned broadening processes, there is yet another process called Doppler broadening which arises out of movement of the emitting and absorbing system as a result of temperature. The Doppler broadening has no effect on the value of the absorption integral over the whole spectrum and there is no displacement of the centre of the resonance region. The intensity however decreases rapidly as the distance from the centre of the resonance region is increased.

Cross-Section of Resonance Process

Resonance experiments with γ-rays are usually performed by either measuring the scattered intensity (resonance fluorescence or resonance scattering) or by determining the attenuation of a beam due to resonance absorption. The cross-section for these two processes, for an incident γ-ray of energy E and wavelength λ can be calculated for thin absorber using the expressions:

$$\sigma_{\text{scatt.}}(E) = (2\pi\,\lambda^2)\left(\frac{2\,I_B+1}{2\,I_A+1}\right)\left[\frac{\Gamma_r^2}{4\,(E-E_{re})^2+\Gamma^2}\right]$$

$$\sigma_{\text{abs}}(E) = (2\pi\,\lambda^2)\left(\frac{2\,I_B+1}{2\,I_A+1}\right)\left[\frac{\Gamma\,\Gamma_r^2}{4\,(E-E_{re})^2+\Gamma^2}\right]$$

In these experiments Γ is the total width of the absorption line, Γ_r its γ-ray width and the maximum resonance cross-section $2\pi\,\lambda^2$ $\left(\dfrac{2\,I_B+1}{2\,I_A+1}\right)$ can be designated as σ_0. I_A is the spin of the ground state A and I_B, the spin of the excited state B.

Both $\sigma_{\text{scatt.}}$ and σ_{abs} vary as the square of the wavelength of the incoming radiation, therefore the maximum resonance cross-section for γ rays is

$$\sigma_0 = 2\pi\,\lambda^2\left(\frac{2\,I_B+1}{2\,I_A+1}\right)$$

For γ rays with energies ranging from few KeV to few MeV, σ_0 is 10^6 to 10^{12} times smaller than the corresponding cross-section for optical radiation. For example, for the sodium D_1 line which corresponds to a transition $2p_{3/2} \rightarrow 2s_{1/2}$, the scattering cross section is 4.6×10^6 times the scattering cross-section for 14.4 KeV γ rays from Fe^{37} arising out of a transition from the spin state $3/2$ to the spin state $1/2$. It should however be noted that in the γ ray region, the total absorption line width Γ is different from the γ ray line width due to branching as well as internal conversion.

If from a certain level, we have several transitions giving rise to γ-rays of various energies and if Γ_i is the width of ith γ ray and α_i, the corresponding internal conversion coefficient, then, the total width

$$\Gamma = \Sigma \, \Gamma_i + \Sigma \, \Gamma_i \, \alpha_i$$

If however, the level totally deexcites by emission of one single γ ray then

$$\Gamma = \Gamma_i + \Gamma_i \, \alpha = \Gamma_i \, (1 + \alpha)$$

or

$$\Gamma_i = \frac{\Gamma}{1 + \alpha}$$

Consequently, the scattering cross section and the absorption cross section are modified to

$$\sigma_{scatt} \, (E) = 2\pi\lambda^2 \left(\frac{2I_B + 1}{2I_A + 1} \right) \frac{\Gamma^2}{4 \, (E - E_{re})^2 + \Gamma^2} \frac{1}{(1 + \alpha)^2}$$

$$\sigma_{abs} \, (E) = 2\pi\lambda^2 \left(\frac{2I_B + 1}{2I_A + 1} \right) \frac{\Gamma^2}{4 \, (E - E_{re}) + \Gamma^2} \frac{1}{(1 + \alpha)}$$

It is clear that the scattering cross-section is reduced by $\left(\dfrac{1}{1 + \alpha} \right)^2$

and the absorption cross-section is reduced by $\left(\dfrac{1}{1 + \alpha} \right)$. For example, in case of isotope Fe^{37} emitting 14.4 KeV γ ray with a corresponding internal conversion coefficient of 9, the scattering cross section is reduced by 100 while in case of Ge^{73} emitting 13.5 KeV γ ray, the internal conversion coefficient is 3600 and the scattering cross-section is reduced by 7×10^8. The latter decrease is far greater because of large conversion coefficient. Here, it has been assumed that the emission line is very sharp, in other words, the incident radiation has been assumed to be delta function.

If the incident radiation has an intensity distribution $I(E)$, then, effective scattering and absorption cross-sections are given by

$$\sigma_{eff. \, (scatt.)} = \int 2\pi\lambda^2 \left(\frac{2I_B + 1}{2I_A + 1} \right) \frac{\Gamma^2}{4 \, (E - E_{re})^2 + \Gamma^2} \frac{1}{(1 + \alpha)^2} \, I(E) \, dE$$

$$\sigma_{eff. \, (abs.)} = \int 2\pi\lambda^2 \left(\frac{2I_B + 1}{2I_A + 1} \right) \frac{\Gamma^2}{4 \, (E - E_{re}) + \Gamma^2} \frac{1}{(1 + \alpha)} \, I(E) \, dE$$

The actual lines are much wider because of Doppler broadening and it can be seen by averaging over all possible values of E of the incoming γ-ray energy that effective cross-section is reduced tremendously.

18.6 APPLICATIONS OF MÖSSBAUER SPECTROSCOPY

The most useful feature of the Mössbauer effect is the sharpness of the γ-ray emission (line width $\simeq 10^{-13}$ as compared to 10^{-8} for NMR and 10^{-4} for IR). The width is purely due to nuclear levels and no phonon-excitation is involved. A fractional line width ($\delta\nu/\nu$) of 10^{-13} implies that the energy of the γ-ray emitted is measurable to an accuracy of 1 part in 10^{13}. Such an accuracy in measurement is not achievable even with lasers. Thus, γ-rays provide us the most stable and accurately defined electromagnetic radiation source for use in experiments. Some of the important applications of the Mössbauer effect are as follows:

1. Isomer shift or chemical shift in the Mössbauer spectra reflects differences in the s-electron density around the emitter and absorber nuclei.
2. When the absorber atom is in a reasonably symmetrical environment a single line is observed in the Mössbauer spectrum. When the environment is asymmetrical a feeble internal electric field operates giving rise to a quadrupole splitting. This is used in structure determination.
3. Biochemical Applications: Since the Mössbauer effect produces a powerful probe of the chemical state and the environment of iron atoms, it can be applied to the study of proteins and enzymes.

Questions

18.1 What is Mössbauer spectroscopy? Describe its principle.

18.2 Describe how ^{57}Co can be used for Mössbauer effect?

18.3 Describe a modern spectrometer to study Mössbauer spectroscopy.

18.4 What are applications of Mössbauer spectroscopy?

Lasers

19.1 INTRODUCTION

The word LASER is an acronym for Light Amplification by Stimulated Emission of Radiation. These are light oscillators; just as an electronic oscillator converts DC power to r.f. power, the same thing happens in a laser. A laser is, sometimes, also called an 'optical maser', because it has resulted from an extension of MASER principle (Microwave Amplification by Stimulated Emission of Radiation).

A laser is the result of three phenomena which occur when an electromagnetic wave interacts with a material, viz.

   (*i*) spontaneous emission

   (*ii*) stimulated emission

   (*iii*) absorption.

Spontaneous Emission

Let us consider two energy levels 1 and 2 with energies E_1 and E_2 respectively ($E_1 < E_2$). Let the level 1 be the ground level. Now suppose that an atom of the material is initially in level 2. Since $E_1 < E_2$, the atom will tend to decay to level 1, during this decay, the atom releases energy ($E_2 - E_1$). When this energy released is in the form of electromagnetic wave, the process is called spontaneous emission. The frequency of the radiated wave is

$$\nu = \frac{E_2 - E_1}{h}$$

Thus, the spontaneous emission is characterised by the emission of photon of energy
$h\nu = E_2 - E_1$ when the atom decays from level
2 to level 1. The decay can also occur in the non-
radiative way. In that case, the energy $E_2 - E_1$ is
released in some form other than electro-

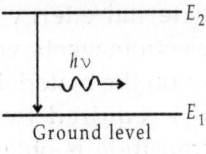

Fig. 19.1

magnetic radiation. Larger the number of atoms in level 2, larger is
the probability for it to decay to level 1.

Stimulated Emission

Suppose that the atom is initially in the level 2. Now, suppose when
an electromagnetic waves is incident on the material, this wave has
the same frequency as the atomic frequency, so there is a finite
probability that this wave will force the
atoms to undergo transition from level 2
to level 1. In this case, the energy $(E_2 - E_1)$
is released in the form of electromagnetic
wave which adds up to the incident wave.
This process is called stimulated emission
and will result in the light amplification.

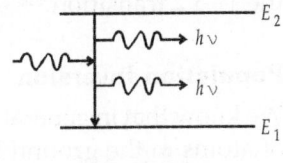

Fig. 19.2

As we know that in the spontaneous emission, the electromagnetic waves emitted by the atom have no fixed phase relation
with that emitted by another atom and so, the emitted wave can be
emitted in any direction. However, in stimulated emission, the
process is forced by the incident electromagnetic wave, the emission
of any photon adds in phase to that of the incoming wave. There is
a phase relationship between the incoming and the emitted wave,
so the induced emission adds the photons travelling in the same
direction as the induced photons. Hence the intensity of the wave
increases as it induces emission in a medium containing excited
electrons. This is the principle behind laser operation. (The incoming
wave also determines the direction of the emitted wave). It is to be
noted here that the idea of stimulated emission was first propounded
theoretically by Einstein in 1917.

Absorption

In this case, we assume the atom to lie in the level 1 initially. If it is
ground level, then the atom remains in this level unless some

external energy is given to it. Suppose an electromagnetic wave of frequency ν be incident upon the material. The energy difference $(E_2 - E_1)$ required by the atom to undergo the transition is obtained from the energy of the incident electromagnetic wave. This process is called absorption.

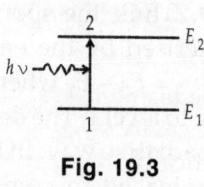

Fig. 19.3

In terms of photons, the above three processes can be described as follows:

In spontaneous emission, the atom decays from level 2 to level 1 through the emission of a photon. In stimulated process, the incident photon stimulates the $2 \rightarrow 1$ transition and then we have two photons—the stimulated photon and the stimulating photon. In the absorption process, the incident photon is simply absorbed to give the $1 \rightarrow 2$ transition.

Population Inversion

We know that in general, under ordinary conditions, the population of atoms in the ground level will be more than in the higher level (i.e., $N_1 > N_2$). However, we would subsequently see that for laser action, we necessarily require a non-equilibrium condition $N_2 > N_1$, for the laser material to act as an amplifier, i.e., population of the higher level greater than the population of the ground level. This condition is called *population inversion*.

The population inversion means that the population difference $(N_2 - N_1) > 0$ is opposite in sign to that which exists under ordinary conditions $(N_2^e - N_1^e < 0)$. The material which exhibits the phenomenon of population inversion is known as *active material*.

If the transition frequency $\nu = (E_2 - E_1)/h$ falls in the microwave region, this type of amplifier is called a MASER-amplifier. But if the transition frequency falls in the optical region, the amplifier is called as LASER-amplifier. The word laser is used for frequency of visible light, in far or infrared, in the ultraviolet or even in the X-ray region.

As we know that an amplifier can be made an oscillator by applying a suitable positive feedback, in the microwave range, this is done by placing the active material in a resonant cavity, having a resonance at the frequency ν_0. But, in the case of lasers, the feedback is generally obtained by placing the active material between two highly reflecting mirrors, which acts like a resonant cavity.

19.2 SCHEMATIC OF A TYPICAL LASER

Laser Cavity

The lasing material is contained in a long narrow container called the cavity. A simple laser cavity consists of two absolutely parallel, optically flat plane mirrors. However, in general, the mirrors can be curved having radii of curvature R_1 and R_2.

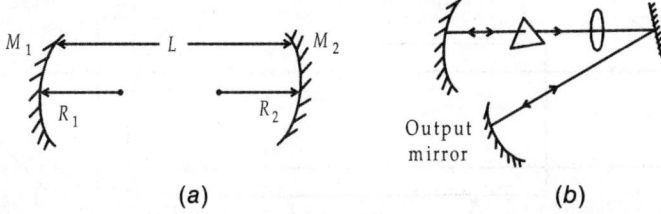

(a)   (b)

Fig. 19.4: (a) a simple laser cavity and (b) a cavity with lenses and prisms

Actually in between the mirrors, we can have lenses and prisms in the propagation path of light. For a laser to operate, we need a laser cavity (for feedback) and an *active medium* which acts as a light amplifier. For example, a typical He-Ne laser, uses Helium-Neon as active medium and electrical discharge as pump. There are different pumps available:

(i) electrical (DC or *r.f.*)
(ii) chemical
(iii) optical
(iv) thermal
(v) nuclear

Most of the common lasers use electrical or optical pumping.

Figure 19.5 shows schematic representation of a typical laser.

In this case, a plane electromagnetic wave travelling in a direction orthogonal to the mirrors bounces back and forth between the two mirrors and gets amplified on each passage through the active material. One of the mirrors is partially transparent, so that a useful output beam can be obtained.

Thus a laser device amplifies and produces a highly directional, high intensity beam that most often has a very pure frequency or wavelength. It comes in sizes ranging from approximately one tenth the diameter of a human hair to the size of a very large building, in powers ranging from 10^{-9} to 10^{20} W and in wavelengths ranging from the microwave to the soft X-ray spectral regions with corresponding frequencies from 10^{11} to 10^{17} Hz respectively. Lasers have pulse energies as high as 10^4 J and pulse duration as short as 6×10^{-5} s. A laser is a specialised light source that can be used when its certain unique properties are required.

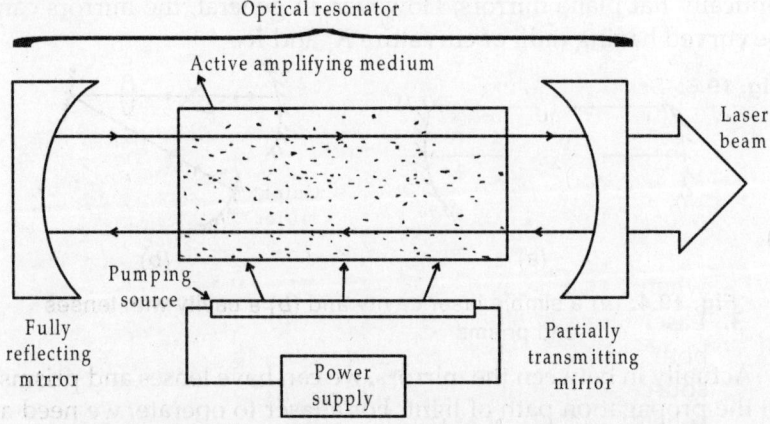

Fig. 19.5: Schematic representation of a typical laser

19.3 UNIQUE PROPERTIES OF LASER LIGHT

The laser light has few characteristics that make it different from other light sources. These are as follows:

1. **Laser Light is Highly Monochromatic:** In contrast to the laser light, light emitted by a tungsten lamp gives no basis for comparison. However, the light from selected lines in a gas discharge tube, can have wavelengths in the visible region that are precise to about 1 part in 10^6. The sharpness of definition of laser light can easily be a thousand times greater, i.e. 1 part in 10^9. This property is due to following two facts: (*i*) only an electromagnetic wave of frequency v_{12} can be amplified. (*ii*) since the mirror arrangement forms a resonant cavity, oscillations can occur only at resonant frequencies of this cavity.

2. **Laser Light is Highly Directional:** As the active material is placed between plane-parallel reflecting surfaces, only electromagnetic wave which propagates along cavity direction or very near to it, is sustained in the cavity and others leak out. Thus high directionality is achieved.

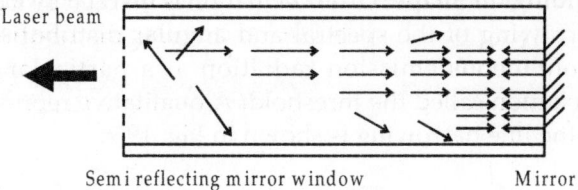

Laser beam

Semi reflecting mirror window   Mirror

Fig. 19.6: Schematic representation of a laser cavity and the build up of
(uni)directioal beam

The beam divergence is given by

$$\alpha_d = \frac{1.22\lambda}{D}$$

where λ is the wavelength of the beam and D is its diameter.

3. **Laser Light is Highly Coherent:** The laser light should be
both space and time coherent. It may be noted here that a
source may have little time coherence but perfect space
coherence. Example of such space coherent source is the light
from a distant star.

(a) **Spatial Coherence:** Suppose the two points P_1 and P_2 at
time t_0 lie in the same wavefront of some given
electromagnetic wave and $E_1(t)$, $E_2(t)$ be the corresponding
electric fields at those points. The two are in same phase
at t_0. Now if this phase difference remains zero at time t,
the coherence between the two points is said to be perfect.
However if it occurs for any other two points on the
wavefront then we say that the wave has perfect spatial
coherence. The directional property of a laser beam is due
to space-coherency. The spatial coherence of a laser beam,
for example, a He-Ne laser can be easily demonstrated with
the help of a diffraction grating and Fraunhofer
interference of a plane wave is obtained in the far field
pattern on a screen.

(b) **Temporal Coherence:** Consider the electric field of the
electromagnetic wave at a point P at time t and $t + \tau$. If the
time delay τ remains same, the phase difference between
the two fields will remain same. But if the phases agree
for all time intervals τ, the light is said to be perfectly time-
coherent. Extreme monochromaticity is the manifestation
of this property of temporal coherence.

4. **Lorentzian Shape:** Laser oscillation is marked by a dramatic narrowing of the spectral and angular distribution of the spontaneous emission radiation at a particular level of excitation called the threshold. A qualitative representation of the line narrowing is shown in Fig. 19.7.

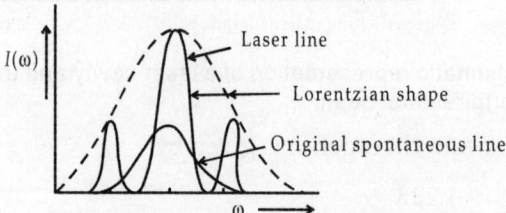

Fig. 19.7: Laser line narrowing

Light from a natural source, consists of a large number of statistically independent wave tracks each of about 10^{-8} second duration. The distribution is Gaussian. In case of laser, the axial mode lying closest to the atomic resonance has the highest gain and thus the laser light concentrates its line width around the resonance resulting in line narrowing. Laser light can thus achieve an enormous spectral purity.

5. **Laser Light can be Sharply Focused:** This property is related to the parallelism of the laser beam. As for light from a star, the size of the focused spot for a laser beam is limited only by diffraction effects and not by the size of the source. Flux densities for focused laser light of 10^{15} W/cm^2 are readily achieved. On the other hand, an oxyacetylene flame has a flux density only of 10^3 W/cm^2.

Apart from the above characteristics, a laser light also has following important peculiarities which differentiate it from thermal light resources.

Laser light is both spatially and temporally coherent, whereas light from other sources is mostly incoherent. The small time bandwidth Δf, and the spatial bandwidth Δf_x, achievable with significant energy by the laser light, are nearly impossible to reproduce by thermal sources. The differences are so dramatic that although both laser light and light from other sources are the same electromagnetic waves, it would be almost correct to say that there is a fundamental difference between them.

6. Intensity for a Fixed Band Width Δf:

A typical He-Ne laser output is 1 mW, whereas a high power pulsed laser output is on the order of 10^{13} W. If the laser output has a wavelength λ, and a spread $\Delta\lambda$, then

$$\frac{\Delta f}{f} = \frac{-\Delta\lambda}{\lambda} \text{ where } f\lambda = c \qquad ..(1)$$

For $\lambda = 0.5$ μm, $\Delta\lambda \cong 10^{-5}$ μm $= 0.1$ Å, we have $\Delta f \sim 1.2 \times 10^9$ Hz. Each photon at $\lambda = 0.5$ μm has energy $\sim 4 \times 10^{19}$ J. Thus typical numbers for photons emitted per second will be 0.25×10^{16} to 0.25×10^{22} photons s^{-1}. The number of photons emitted per unit Hz will be approximately 2×10^4 to 2×10^{10}.

If we now compare the laser light with black body radiation, we have

$$\text{thermal photons/s} - \text{Hz} = \frac{2}{\lambda^2}\left(\frac{1}{e^{hf/kT} - 1}\right)\Delta A \qquad ..(2)$$

where, the emission takes place from an area ΔA and T is the blackbody temperature. The Boltzmann factor for our case is given approximately by $e^{-30,000/T}$ and is in general, very small for ordinary temperatures. For $T = 1000$ K and $\Delta A = 1$ cm^2, we get

$$\text{thermal photons/s} - \text{Hz} < 1. \qquad ..(3)$$

comparing this number with that obtained, even for a 1 mW laser viz. 2×10^5, we see the enormous difference that exists between thermal light and laser light.

7. Radiance:

The angular spread $\Delta\theta$, of a typical laser beyond the Rayleigh distance is approximately given by

$$\Delta\theta \simeq \frac{\lambda}{d} \qquad ..(4)$$

where d is the aperture diameter. Thus

$$\Delta f_x \simeq \frac{1}{d} \qquad ..(5)$$

The far field solid angle $\Delta\Omega$, into which the laser radiation is confined, can be approximated as

$$\Delta\Omega \cong (\Delta\theta)^2 = (\lambda \, \Delta f_x)^2 \approx \frac{\lambda^2}{A} \qquad ..(6)$$

Thus, the radiance of a laser source (1 mW) will be (for $A = 1$ cm^2)

$$R_{\text{laser}} = \frac{10^{-3}W}{\Delta\Omega} = 4 \times 10^5 \text{ W/sr} \qquad ..(7)$$

As the thermal source radiates over a solid angle of 2π steradian, the radiance will be given by

$$R_{\text{thermal}} = \frac{hf}{\lambda^2}\left(\frac{\Delta f}{e^{hf/kT}-1}\right) \qquad ..(8)$$

If we assume for a thermal source, $\Delta\lambda \sim 10^3$ Å and $T = 10^3$ K even then

$$R_{\text{thermal}} \approx 4 \times 10^{-16} \text{ W/sr} \qquad ..(9)$$

This explains why even a 1 mW laser looks "brighter" than a thermal source at T = 10^3 K by twenty orders of magnitude.

8. Brightness

The brightness of a source is given by the power output per steradian of solid angle per Hz of bandwidth. For a laser with output power P, the brightness is given by

$$B_{\text{laser}} = \frac{P}{\left(\frac{\lambda^2}{A}\right)\cdot\Delta f} = \frac{PA}{\lambda^2\cdot\Delta f} \qquad ..(10)$$

Let us assume that we start with a thermal source of temperature $T°$K and filter it spatially as well as temporally to obtain an equivalent brightness for the thermal source equal to the laser source. We obtain

$$B_{\text{thermal}} = \frac{hfA}{\left(e^{hf/kT}-1\right)\lambda^2} = \frac{PA}{\lambda^2\Delta f} \qquad ..(11)$$

To have identical brightness, we need a temperature T given by

$$(e^{hf/kT}-1) = \frac{(hf)\Delta f}{P}$$

or

$$T = \frac{hf}{k}\cdot\frac{1}{ln\left(1+\dfrac{hf\Delta f}{P}\right)} \approx \frac{P}{k\cdot\Delta f} \qquad ..(12)$$

Using $P = 10^{-3}$, $\Delta f \simeq 10^9$, which is equivalent to $\Delta\lambda \sim 10^{-2}$ Å, we obtain

$$T \approx 10^{11} \text{ K} \qquad ..(13)$$

This is also the reason why we should never look at a laser directly even if it is having only 1 mW of power, as it will appear to you as a source with $T \sim 10^{11}$ K.

9. Coherence Length and Coherence Time

The light wavetrain emitted from a source can be characterized by an average time T_0 (or τ) called coherence time. The average length of the wavetrains is called coherence length L_c, given by

$$L_c = c\, T_0 \qquad ..(14)$$

where c is the speed of light,

and
$$\frac{2\pi}{T_0} = \Delta\omega$$

since
$$\Delta\omega = 2\pi\, \Delta v$$

$\therefore$
$$\Delta v = \frac{1}{T_0} \qquad ..(15)$$

Now
$$v = c/\lambda$$

$\therefore$
$$|\Delta v| = +\frac{c}{\lambda^2}\, \Delta\lambda \qquad ..(16)$$

From eqns (14) and (15)

$$\Delta v = \frac{c}{L_c} \qquad ...(17)$$

From eqns (16) and (17)

$$\frac{c}{L_c} = \frac{c}{\lambda^2}\, \Delta\lambda$$

or
$$\Delta\lambda = \frac{\lambda^2}{L_c} \qquad ...(18)$$

$\Delta\lambda$ is called the natural line width. Hence temporal coherence depends upon the values of coherent length L_c and coherent time T_0.

10. Monochromaticity and coherence

The bandwidth Δf (i.e. Δv), of a good stable laser can be less than 1 kHz compared to a thermal source which is of the order of 10^{14} Hz. To appreciate the difference between these two numbers, one can calculate the coherence time and coherence length for both these sources. The coherence length of the laser is $\sim 3 \times 10^5$ m. Thus using a laser source, we can do an interference experiment with path length ~ 300 km, whereas for white light this distance is only 3 μm. This is also the reason the laser light always has speckle pattern and its intensity appears to fluctuate. Any reflection and scattering, even from far away, can interfere and thus cause change in light intensity. For incoherent light, any scatterer beyond ~ 3 μm will not cause any fluctuations of intensity. The coherence time of laser light is of the order of milliseconds, whereas for thermal light, it is $\sim 10^{-14}$ seconds.

19.4 WORKING OF AN OPTICAL RESONATOR (ELEMENTARY IDEA)

Imagine a wave which starts from one of the mirrors and travels towards the other. In passing through the active medium, it gets amplified. The second mirror is made partially reflecting, so, the wave is partially transmitted and the rest reflected back to the first mirror. In travelling to the first mirror, it again gets amplified. Thus, we have waves propagating along both directions which interfere, and for a stable standing wave pattern to be formed, the frequency must satisfy some conditions. If we consider plane wave travelling to and fro, then the total phase change suffered by the wave in one complete round trip must be an integral multiple of 2π so that standing waves are formed in the cavity. Thus if L represents the length of the cavity, then

$$\frac{2\pi}{\lambda} \times \text{path difference} = 2m\pi$$

or $\qquad \dfrac{2\pi}{\lambda}, 2L = 2m\pi, m = 1, 2, 3 \ldots$

where λ is the wavelength of radiation. Since $\lambda = c/\nu$, therefore

$$\nu = m\frac{c}{2L}$$

gives the discrete frequencies of oscillation of the modes. Different values of m lead to different oscillation frequencies which constitute the longitudinal modes of the cavity (i.e. only the longitudinal modes contribute). The frequency difference between adjacent modes is

$$\Delta\nu = \frac{c}{2L}$$

If we consider a resonator made of mirrors of transverse dimensions a and separated by a distance L, then (from physical optics), the wave after reflection at one of the mirrors undergoes diffraction diverges at an angle $\sim\lambda/a$. The angle subtended by one of the mirrors at the other mirror is $\sim a/L$. Hence, for diffraction losses to be low,

$$\lambda/a \ll a/L \quad \text{or} \quad \frac{a^2}{L\lambda} \gg 1$$

The quantity $a^2/(\lambda L)$ is a measure of diffraction losses and is known as *Fresnel number*.

Various Two Mirror Cavities

There are a number of different types of two mirror laser cavities as shown in Fig 19.8. These cavities are distinguished from each other in terms of the relative value of the radius of curvature of the two mirrors in comparison to the separation distance L between the mirrors.

For a confocal curved mirror cavity, the radius of curvature of the mirrors is equal to the spacing between the mirrors. At first glance, all mirror combinations shown in the figure appear to make reasonable laser cavities. However, the following article shows that mirror combinations satisfy a particular stability condition in order that they be a stable resonators.

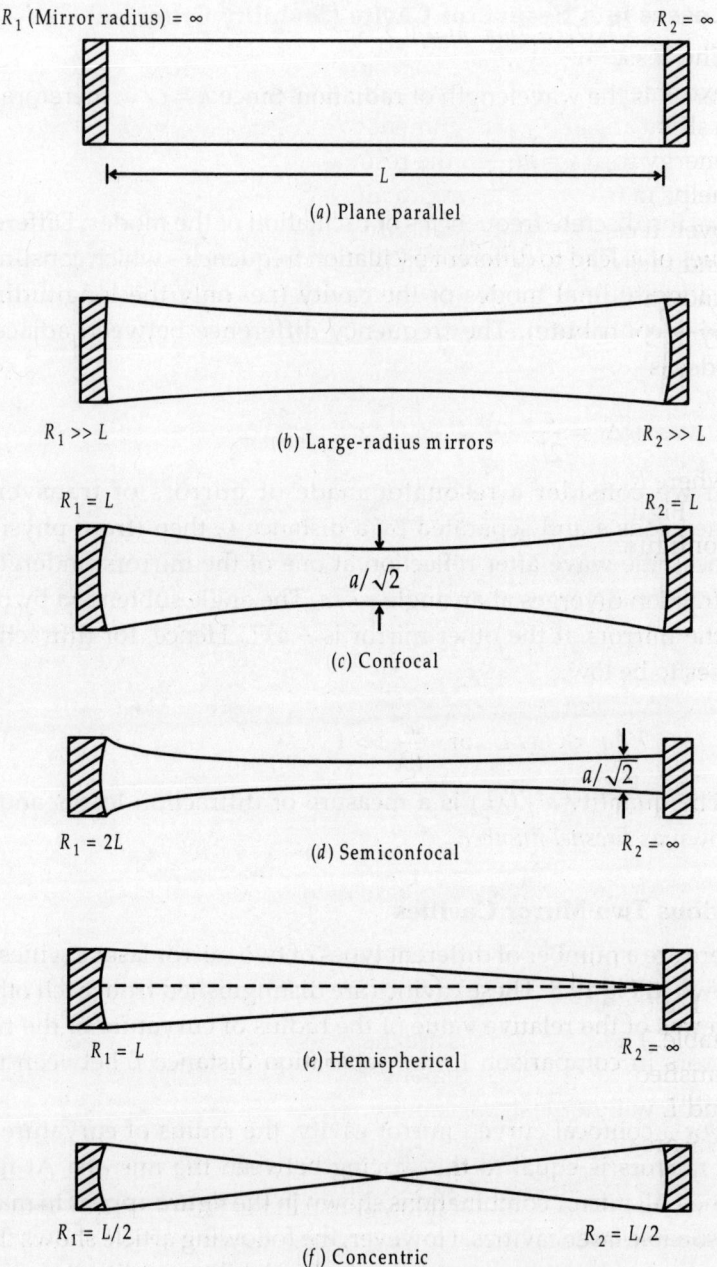

R_1 (Mirror radius) = ∞ $R_2 = ∞$

L

(a) Plane parallel

$R_1 \gg L$ $R_2 \gg L$

(b) Large-radius mirrors

$R_1 = L$ $R_2 = L$

$a/\sqrt{2}$

(c) Confocal

$a/\sqrt{2}$

$R_1 = 2L$ $R_2 = ∞$

(d) Semiconfocal

$R_1 = L$ $R_2 = ∞$

(e) Hemispherical

$R_1 = L/2$ $R_2 = L/2$

(f) Concentric

Fig. 19.8: Various two mirror cavities

Losses in a Resonator Cavity (Stability Criterion)

The losses in a resonator formed by plane parallel mirrors are extremely sensitive to the parallelism of the two mirrors because a slight angular misalignment would cause a large amount of light energy to escape from the resonator. The use of spherical mirrors helps in focusing leading to much less loss due to diffraction spill over. It can be shown that only certain combinations of curvatures and mirror-separations can lead to low diffraction loss. These are known as stable resonators. The stability condition is expressed by

$$0 \le \left(1 - \frac{L}{R_1}\right)\left(1 - \frac{L}{R_2}\right) \le 1 \qquad ..(1)$$

where R_1 and R_2 are the radii of curvature of the mirrors.

Figure 19.9 shows stability criterion for use of resonator configurations.

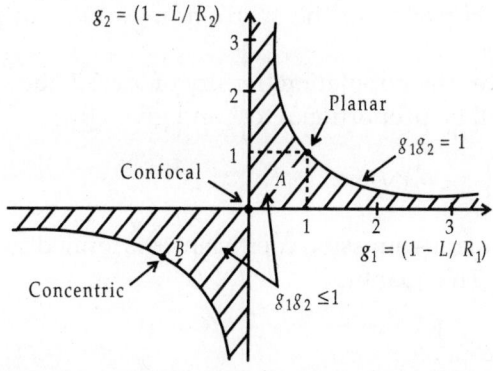

Fig. 19.9: The stability diagram for optical resonators

For the shaded regions, condition (1) is satisfied and the cavity is stable. The clear regions are the regions where this condition is not satisfied, i.e. for such regions, the relationship between R_1 and R_2 and L will not lead to stability. Three particular points in the Fig. 19.9 are of special interest, they are on the verge of instability:

$$R_1 = R_2 = \frac{L}{2} \text{ (symmetric concentric)}$$

$$R_1 = R_2 = L \text{ (confocal)}$$

$$R_1 = R_2 = \infty \text{ (plane parallel)}$$

These three are on the edge of stability curves, and can become extremely lossy for slight deviations. Thus it would be desirable to design those cavities so that the g_1, g_2 parameters move slightly into the stable regions shown by shaded parts of the figure.

19.5 THE EINSTEIN COEFFICIENTS IN RELATION TO SPONTANEOUS AND STIMULATED EMISSION

Let there be two energy levels E_1 and E_2 of a material (out of an innumerable levels possessed by the material). Let us suppose $E_1 < E_2$ and E_2 be the excited level and E_1, the ground level. If the atom is in level 2 then it will have a tendency to decay to level 1. The atom may emit electromagnetic radiation of frequency

$$\nu_{12} = \frac{E_2 - E_1}{h} \qquad ..(2)$$

This process is *spontaneous emission*. The decay can also occur in another way (non-radiative way). In this latter process the energy difference is imparted to the surrounding atoms in the form of kinetic energy.

If N_2 denotes the population density of level 2, the spontaneous decay rate will be proportional to 2 and given by

$$\left(\frac{dN_2}{dt}\right)_{sp} = -AN_2 \qquad ..(3)$$

The spontaneous emission coefficient A is termed as the Einstein A coefficient. The quantity

$$(\tau)_{sp} \equiv \frac{1}{A} \qquad ..(4)$$

is called *spontaneous emission lifetime*

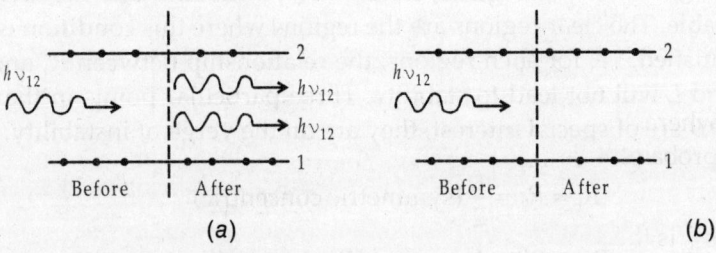

Fig. 19.10: Energy transitions in an atom (a) stimulated emission (b) absorption

Suppose an electromagnetic wave of frequency equal to $\nu_{12} = \dfrac{E_2 - E_1}{h}$ is made to strike the material. This wave has a finite probability of inducing atom to undergo transition from initial level 2 to level 1. In this process, photon of the frequency ν_{12} is released in addition to the incident one. This phenomenon is known as stimulated emission. This is different from spontaneous emission. In case of spontaneous emission, the atom emits electromagnetic wave which has no definite phase relationship with that emitted by another atom. In stimulated emission, the emitted radiation is in phase with the incident radiation. The stimulated emission rate depends on the population density N_2, and is given by

$$\left(\frac{dN_2}{dt} \right) = -W_{21}N_2 \qquad \qquad ..(5)$$

W_{21} is called the stimulated transition probability. (Its reciprocal is called stimulated emission lifetime). In addition to the particular transition, W_{21} depends on the intensity of the incident wave

$$W_{21} = \sigma_{21}F \qquad \qquad ..(6)$$

where F is the incident photon flux and σ_{21} is a quantity having dimensions of area. σ_{21} is known as *stimulated emission cross section* and depends only upon the characteristics of given transition.

Absorption

The Fig 19.10 (*b*) depicts the inverse process viz. absorption. If an electromagnetic wave of frequency $\nu_{12} = \dfrac{E_2 - E_1}{h}$ is incident on the atom, then there is a finite probability that atom makes a transition from ground level (1) to the excited state (2). The rate of absorption is given by

$$\frac{dN_1}{dt} = -W_{12}N_1 \qquad \qquad ..(7)$$

where N_1 is the population density of level 1 and the absorption probability is

$$W_{12} = \sigma_{12}F$$

σ_{12} is the absorption cross section. It is to be noted that

$$\sigma_{12} = \sigma_{21} = \sigma \text{ (the transition cross section)}$$

i.e., probabilities of stimulated emission and absorption are equal. Further the transition probabilities W_{21} and W_{12} are related to Einstein B coefficient (stimulated emission coefficient) as

$$W_{12} = B_{12}\rho(\nu)$$
$$W_{21} = B_{21}\rho(\nu) \qquad ..(8i)$$

Here ρ is electromagnetic energy density and $W_{12} = W_{21}$.

Interrelation between Einstein Coefficients

If absorption and spontaneous emission were the only processes operative, the principle of detailed balance would give

$$\rho(\omega_{21})B_{12}N_1 = A_{21}N_2 \qquad ..(8ii)$$

where ρ is the radiation density.

$$\therefore \qquad \rho(\omega_{21}) = \frac{A_{21}}{B_{12}} \frac{N_2}{N_1} \qquad ..(8iii)$$

In thermal equilibrium, the total number of atoms N will be distributed over the available levels according to the Boltzmann distribution

$$N_i = \frac{Ng_i \exp(-E_i/k_BT)}{\sum_i \exp(-E_i/k_BT)}$$

where g_i is the statistical weight of the level i, T, the absolute temperature, E_i, the energy of ith level and k_B is Boltzmann constant

$$\frac{N_2}{N_1} = \frac{g_2 \exp(-E_2/k_BT)}{g_1 \exp(-E_1/k_BT)}$$

It may be assumed that

$$g_2 = g_1 = 1$$

$$\therefore \quad \frac{N_2}{N_1} = \frac{\exp(-E_2/k_BT)}{\exp(-E_1/k_BT)} = \exp[-(E_2 - E_1)/k_BT] = \exp(-\hbar\omega_{21}/kT)$$

$$..(9)$$

Hence eqn (8iii) changes to

$$\rho(\omega_{21}) = \frac{A_{21}}{B_{12}} \exp(-\hbar\omega_{21}/k_BT) \qquad ..(10)$$

Now, it is an experimental fact, that the black body radiation is given by

$$\rho(\nu) = \frac{8\pi h\nu^3}{c^3} \cdot \frac{1}{e^{h\nu/k_BT} - 1} \qquad ..(11)$$

where $\rho(\nu)$ is the energy density of photons having frequency ν.

Let us consider a medium which has two levels E_2 and E_1. If this medium is placed in a black body enclosure, in equilibrium, the number transmitted from E_2 to E_1 would be equal to that from E_1 to E_2. The number of photons generated per unit time per unit volume would be given by

$$N_2(W_{21} + A_{21}) \qquad\qquad ..(12)$$

where W_{21} is the stimulated emission rate. The term A_{21} is due to spontaneous emission rate.

The photons lost per unit volume will be given by

$$N_1 W_{12} \qquad\qquad ..(13)$$

where W_{12} is the transition rate of spontaneous emission rate from E_1 to E_2.

From (12) and (13) and eqn (8i), we obtain

$$\frac{N_2}{N_1} = \frac{\rho(\nu)B_{12}}{\rho(\nu)B_{21} + A_{21}} \qquad\qquad ..(14)$$

Also from eqn (9) $\dfrac{N_2}{N_1} = e^{-h\nu/k_B T}$ $\qquad\qquad ..(15)$

Thus $\qquad \dfrac{N_2}{N_1} = \dfrac{\rho(\nu)B_{12}}{\rho(\nu)B_{21} + A_{21}} = e^{-h\nu/k_B T}$ $\qquad\qquad ..(16)$

or $\qquad \dfrac{B_{12}e^{h\nu/k_B T} - B_{21}}{e^{h\nu/k_B T} - 1} = \dfrac{c^3}{8\pi h\nu^3}A_{21}$ $\qquad$ (using eqn 15) ..(17)

The left hand side of the above equation is dependent on the temperature T, whereas the right hand side is independent of temperature. In order that this be valid at all temperatures, we must have

$$B_{12} = B_{21} = \frac{c^3}{8\pi h\,\nu^3}A_{21} \qquad\qquad ..(18)$$

This is the famous Einstein relationship between A and B coefficients. Thus obtain for stimulated emission rate

$$W_i = W_{12} \equiv W_{21} = \frac{c^3}{8\pi h\nu^3}A_{21}\,\rho(\nu) \qquad\qquad ..(19)$$

or Rate, $\qquad W_i = \dfrac{c^3}{8\pi h\nu^3}\dfrac{1}{t_{spon}}\rho(\nu)$ $\qquad\qquad ..(20)$

($\because A_{21}$ is inverse to the spontaneous emission rate t_{spon}).

To use the Einstein relationship for the case of lasers, we need to consider finite line width of the energy levels. Let $g(\nu)$ denote the *total line width broadening*. Then eqn (20) has to be modified using eqn (18) due to this line width broadening as follows

$$W_i = \frac{c^3}{8\pi h \nu^3} \frac{1}{t_{spon}} \rho(\nu) g(\nu) \qquad ..(21)$$

19.6 THE SPONTANEOUS EMISSION RATE AND EMISSION INTENSITY

Now, the light from spontaneous emission is noise like with a maxima and is incoherent. Though the light frequency is still given by

$$\nu_0 = \frac{E_2 - E_1}{h}$$

However, the emitted light is actually in form of bursts of decaying exponentials as depicted in Fig. 19.11.

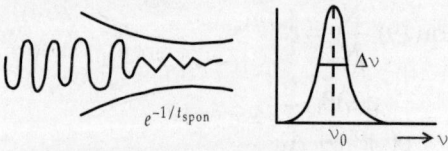

Fig. 19.11: Spontaneous emission

Thus, though the frequency is centred around $\nu = \nu_0$, the spread (in frequency) is found by taking Fourier transform of the decaying light, denoted by

$$E = E_0 e^{i2\pi\nu_0 t} e^{-t/t_{spon}} \qquad ..(22)$$

Its Fourier transform is

$$F[E] = \frac{E_0}{1 + (\omega - \omega_0)^2 t_{spon}^2} ; (\omega_0 = 2\pi\nu_0) \qquad ...(23)$$

The spontaneous emission has a frequency spread given by

$$\Delta\nu = \frac{1}{t_{spon}} \qquad ..(24)$$

The Fourier transform function $g(\nu)$ for the case of homogenous broadening is $1/\Delta\nu$ where $\Delta\nu$ is given but $\Delta\nu \simeq 1/t_{spon}$.

Now, the energy density can easily be related to the intensity of the light beam. Consider the Fig. 19.12.

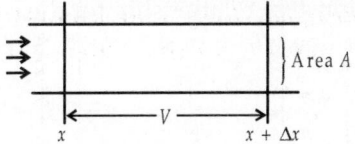

Fig. 19.12: Relationship between photon density and light intensity

The intensity of light is given by the energy per unit area per unit time. As the photons move with a velocity v, in one second, the number of photons moving (crossing) through the area A will be filled up in a cylinder of area A and length V. The volume of this cylinder

$$V = Av \qquad \qquad ..(25)$$

so, the number of photons in this cylinder is given by

$$N = \frac{Av\rho(\nu)}{h\nu} \qquad \qquad ..(26)$$

(since each photon has an energy $h\nu$).

The corresponding energy/area will be the intensity $I(x)$ and is given by

$$I(x) = \frac{Av\rho(\nu)}{A} = v\rho(\nu)$$

or $\qquad \qquad \rho(\nu) = I(x)/v \qquad \qquad ..(27)$

Using this expression, we can rewrite spontaneous emission rate as

$$W_i = \frac{v^2}{8\pi h \nu^3 t_{\text{spon}}} I(\nu)g(\nu) \qquad \qquad ..(28)$$

19.7 POPULATION INVERSION

Under the conditions of thermal equilibrium, we have

$$\frac{N_2}{N_1} = \frac{g_2}{g_1}\exp(-E_2/kT)/\exp(-E_1/kT)$$

and the lower energy level E_1 of the two level atomic system contains more atoms than the upper energy level E_2. This situation prevails at room temperatures and is depicted in Fig. 19.13 (a). However, to achieve stimulated emission and optical amplification it is required to create a non-equilibrium distribution of atoms such that the population of the upper energy level is greater than that of the lower energy level (i.e., $N_2 > N_1$). This condition is known as population inversion and is depicted in Fig. 19.13 (b).

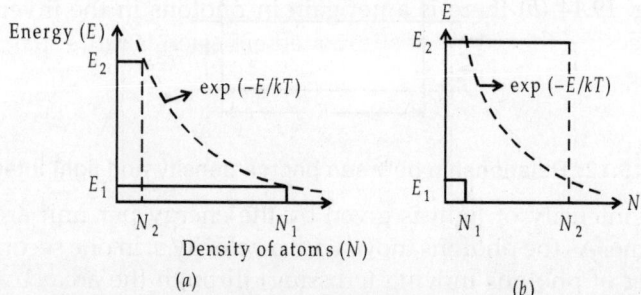

Fig. 19.13: Populations in a two level system (*a*) Boltzmann distribution
for a system in thermal equilibrium (*b*) a non-equilibrium situation
depicting population inversion

Light Amplification by Population Inversion

Population inversion in a laser is brought about by pumping.
Figure 19.14 shows a comparison of the effect of passage of radiation
through a non-inverted and an inverted population. In each case,
the population is shown as being made up of nine molecules of the
lasing medium. In non-inverted population, three molecules are in
the excited state and six are in the ground state. Three of the
incoming photons (Fig. 19.14 (*a*)) are absorbed by the medium
(producing 3 additional excited molecules). However, the radiation
stimulates emission of two photons from excited molecules. Thus
there is a net deficit of one photon.

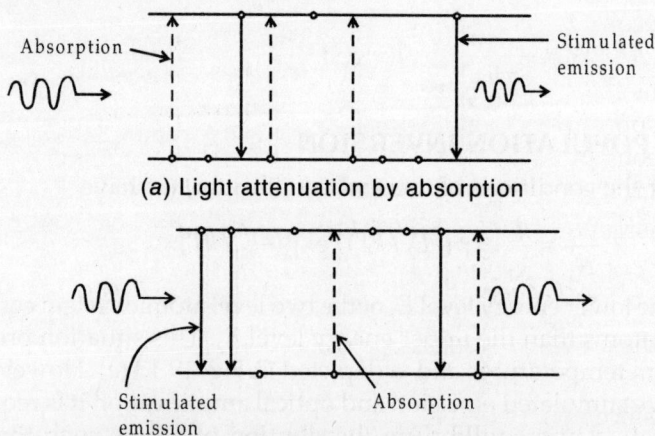

Fig. 19.14: Passage of radiation through (*a*) a non-inverted population
(*b*) an inverted population

In Fig. 19.14 (*b*) there is a net gain in photons in the inverted system because the extent of stimulated emission is more than the absorption.

19.8 LASING GAIN CONSTANT AND INVERSION TEMPERATURE

To obtain an expression for the lasing gain constant, we require to relate the photon's density to the intensity. For this we proceed as follows. We recollect that photons travel with a velocity c (the velocity of light in the particular medium). Let the intensity of the light beam be $I(x)$ at $x = x$ and $I(x + \Delta x)$ at $x = x + \Delta x$ as shown in Fig. 19.15.

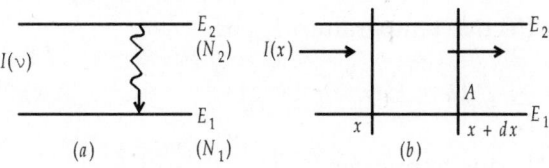

Fig. 19.15: Calculation of light intensity

Let us consider the volume $\Delta x\,A$ as shown in Fig. 19.15 (*b*) where A is the cross-sectional area. We can write

$$I(x + \Delta x) = I(x) + \Delta x \frac{\partial I(x)}{\partial x} \qquad ..(29)$$

The second term on the right hand side arises because in the volume $(A\Delta x)$, some photons are generated. However, the number of photons generated per unit time by the stimulation process (which unlike spontaneous emission) gives coherent output. (The spontaneous emission is like a noise to the system). The stimulated emission is defined by

photons generated = photons emitted – photons absorbed

$$= A\Delta x{\cdot}N_2 W_i - A\Delta x N_1 W_i$$
$$= A\Delta x W_i(N_2 - N_1)$$

Thus the power generated in the volume $A\Delta x$ is given by

$$A\Delta x h\nu\,(N_2 - N_1)W_i$$

(since each photon is having energy $h\nu$).

Hence

$$\frac{\partial I(x)}{\partial x} = h\nu(N_2 - N_1)\frac{c^2 A_{21}g(\nu)}{8\pi h\nu^3}I(x) \qquad ..(30)$$

or $\qquad I(x) = I_0 e^{2\nu x}$ ..(31)

where $\quad \nu = (N_2 - N_1) \dfrac{c^2 g(\nu)}{16\pi\nu^2 t_{spon}}$ ..(32)

Thus, the amplification factor is proportional to $(N_2 - N_1)$. Ordinarily it is given by

$$\frac{N_2}{N_1} = e^{-h\nu/k_B T} \quad \text{or} \quad (N_2 - N_1) = (e^{-h\nu/k_B T} - 1)N_1 \qquad ..(33)$$

In general we do not have amplification unless and until we pump the medium, i.e. energize the medium so that $N_2 > N_1$. Then we can have amplification.

In the situation where there exists population inversion, we often define an effective temperature T_{eff} by

$$\frac{N_2}{N_1} = e^{-h\nu/(k_B T_{eff})} \qquad ..(34)$$

If $N_2 > N_1$, due to pumping,

$$T_{eff} = -\frac{h\nu}{k_B}\left[\ln\frac{N_2}{N_1}\right]^{-1} \qquad ..(35)$$

Thus, theoretically, the population inversion corresponds effectively to a negative temperature.

19.9 LASER CAVITY MODES

When mirrors are placed at the ends of an amplifying medium, they provide an effective increase of the gain length. Further, they also place boundary conditions upon the electromagnetic field (the laser beam) that develops between the two mirrors. These lead to the development of certain modes (longitudinal modes) in the laser cavity. In this context we will subsequently discuss a two mirrored cavity, known as a Fabry-Perot resonator. When the mirrors at the two ends are of finite size, there occur diffraction losses and this consideration leads to the development of transverse modes. We now discuss the Fabry-Perot resonator with no active medium between the mirrors. We will then later consider the effect of placing an amplifying medium in between the mirrors of the Fabry-Perot resonator.

Fabry-Perot Resonator

A Fabry-Perot resonator consists of two parallel partially reflecting surfaces separated by a distance d. Consider a beam of light of amplitude E_0 incident upon the mirror (1) at an angle θ to the normal to this mirror. (Initially we assume that an index of refraction of unity exists in all regions of the analysis, later we will allow for a material with an index of refraction η to exist between the mirrors).

A portion of the light incident upon one surface is reflected, and a portion is transmitted. The reflected portion will be $E_0 r$ and the transmitted portion will be $E_0 t$ where r and t are the reflection and transmission coefficients with values ranging from 0 to 1. (We ignore the time dependence of E_0 for the analysis). The transmitted beam continues until it reaches the second surface where a part of it is reflected and a part is transmitted. The transmitted beam has an amplitude $E_0 t^2$ and the reflected beam has intensity $E_0 tr$. The reflected beam then suffers reflection at the first surface (reflected part $E_0 tr^2$) and a part $E_0 t^2 r$ is transmitted. This continues indefinitely as shown in Fig. 19.16

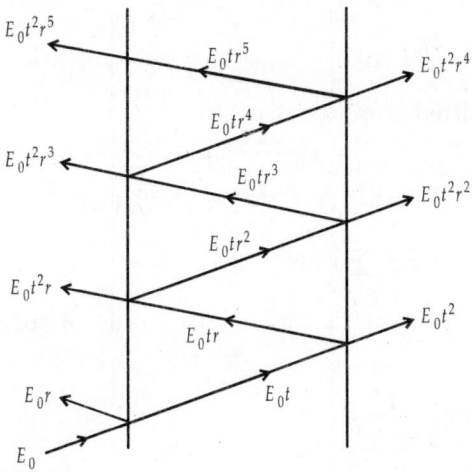

Fig. 19.16: Successive reflections (and transmissions) from two partially reflecting surfaces

Let us consider the extra path length that the reflected ray traverses in one successive reflection between the two surfaces (indicated by dark line in Fig. 19.17).

After the additional path length $a + b$, the ray CD can add in phase to the original ray AB.

$$a = \frac{L}{\cos\theta} ; b = a\cos 2\theta$$

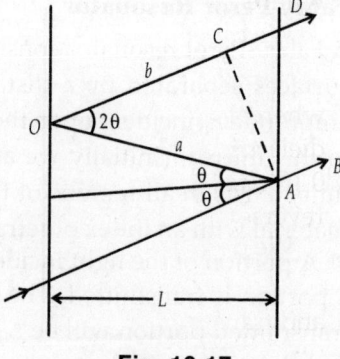

Fig. 19.17

The extra path length (AOC) when it returns to the wavefront location C of wavefront AC is

$$a + b = \frac{L}{\cos\theta}[1 + \cos 2\theta]$$

$$= \frac{L}{\cos\theta}[1 + 2\cos^2\theta - 1]$$

$$= 2L\cos\theta \ (\equiv z) \qquad ..(36)$$

We can represent the wavefront by an electromagnetic plane wave of the form e^{ikz} or $e^{i\phi}$ with phase factor

$$kz \equiv \phi = 2kL\cos\theta$$

or

$$\phi = 2\left(\frac{2\pi}{\lambda}\right)L\cos\theta$$

$$= \frac{4\pi}{\lambda}L\cos\theta \qquad ..(37)$$

The transmitted amplitude is

$$E_t = E_0 t^2 + E_0 t^2 r^2 e^{i\phi} + E_0 t^2 r^4 e^{2i\phi} + ...$$

$$= E_0 t^2 (1 + r^2 e^{i\phi} + r^4 e^{2i\phi} + ..)$$

$$= E_0 t^2 \sum_{n=0}^{\infty} r^{2n} e^{in\phi} \qquad ..(38)$$

This is just a geometric series with a ratio of successive terms $r^2 e^{i\phi}$ hence

$$\sum_{n=0}^{\infty} r^{2n} e^{in\phi} = \frac{1}{1 - r^2 e^{i\phi}} \qquad ..(39)$$

and

$$E_t = \frac{E_0 t^2}{1 - r^2 e^{i\phi}} \qquad ..(40)$$

Since the transmitted intensity is given by $I_t = |E_t|^2$, we have

$$I_t = E_0^2 \frac{t^4}{\left|1 - r^2 e^{i\phi}\right|^2} = I_0 \frac{t^4}{\left|1 - r^2 e^{i\phi}\right|^2} \qquad ..(41)$$

where $I_0 = E_0^2$.

We can allow for a phase change upon reflection such that

$$r = |r| \, e^{i\phi_r/2} \qquad \qquad ..(42)$$

where $\phi_r/2$ is the phase change for one reflection. If there is a dielectric in between the two surfaces, the phase change $\phi_r/2$ is either 0 (for ray going from a denser to a rarer medium) or π (for the reverse). For a metal, the phase change could have any value.

We define the reflectivity or reflectance R as

$$R = |r|^2$$

and transmittance T as

$$T = |t|^2$$

Then, transmitted intensity

$$I_t = I_0 \frac{T^2}{\left|1 - Re^{i\Phi}\right|^2} \qquad \qquad ..(43)$$

where $\Phi = \phi + \phi_r$

The expression in the denominator of eqn (43) can be written as

$$|1 - Re^{i\Phi}|^2 = (1 - Re^{i\Phi})(1 - Re^{-i\Phi})$$

$$= 1 - Re^{i\Phi} - Re^{-i\Phi} + R^2$$

$$= 1 - 2R\cos\Phi + R^2$$

$$= (1 - R)^2 \left[1 + \frac{4R}{(1 - R)^2} \sin^2 \frac{\Phi}{2}\right] \qquad ..(44)$$

If we set

$$F' = \frac{4R}{(1 - R)^2} \qquad \qquad ..(45)$$

then we can write I_t as

$$I_t = I_0 \frac{T^2}{(1 - R)^2} \left[\frac{1}{1 + F'\sin^2(\Phi/2)}\right] \qquad ..(46)$$

The expression

$$\frac{1}{1 + F'\sin^2(\Phi/2)}$$

is referred to as the *Airy function*. If there is no absorption, then $R + T = 1$ and the ratio of transmitted intensity to the incident intensity can be expressed as

$$\frac{I_t}{I_0} = \frac{1}{1 + F'\sin^2(\Phi/2)} \qquad ..(47)$$

which is just the *Airy function*. A plot of I_t/I_0 versus $\Phi/2$ is shown in Fig. 19.18 for various values of R. The function has a maximum value of unity for $\sin(\Phi/2) = 0$ or $\phi/2 = m\pi$, $m = 0, 1, 2$ ...and a minimum for $\sin(\Phi/2) = 1$ or $\phi/2 = (2m + 1)\pi/2$. The minimum depends upon F' but can be very small for large values of R. The maximum values of I_t/I_0 occur at

$$\Phi/2 = m\pi, \, m = 0, 1, 2, \ldots$$

Representing those values of Φ as Φ_m, we have

$$\Phi_m = 2m\pi = \frac{4\pi}{\lambda_0} L \cos\theta + \phi_r \qquad ..(48)$$

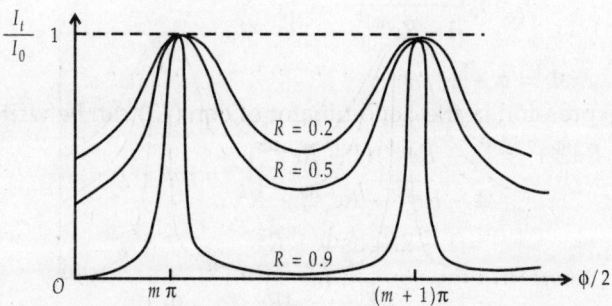

Fig. 19.18: Plot of the transmitted intensity from a Fabry-Perot interferometer versus phase change

Each of the peaks of the Airy function shown in the Fig. 19.18 is identical in shape. We can obtain the full width at half maximum (FWHM) of the Airy function by considering the width of the peak for $m = 0$. For larger values of R (say $R > 0.6$) we can approximate $\sin(\Phi/2)$ by $(\Phi/2)$. The value of Φ at which the *Airy function* reduces to half its maximum value is referred to as Φ'. This is given by

$$\frac{1}{1 + F'(\Phi'/2)^2} = \frac{1}{2}$$

or $F'(\Phi'/2)^2 = 1$

which gives

$$\Phi' = \frac{2}{\sqrt{F'}} \qquad ..(49)$$

The FWHM is just twice of this value

$$\text{FWHM} = 2\Phi' = \frac{4}{\sqrt{F}} \qquad ..(50)$$

Since the separation between the peak values of the function is $\Delta\Phi = 2\pi$, we can obtain an expression F for the ratio of the separation between peaks to the FWHM as

$$F = \frac{\Delta\Phi}{\text{FWHM}} = \frac{2\pi}{\dfrac{4}{\sqrt{F}}} = \frac{\pi\sqrt{F}}{2} = \frac{\pi}{2}\frac{2\sqrt{R}}{(1-R)} = \frac{\pi\sqrt{R}}{(1-R)} \qquad ..(51)$$

This F is referred to as the *finesse*. If the two mirrors have different reflectivities R_1 and R_2, then finesse

$$F = \frac{\pi(R_1 R_2)^{1/4}}{1-(R_1 R_2)^{1/2}} \qquad ..(52)$$

We now consider the simple case in which $\theta = 0$ and $\phi_r = 0$ (for a dielectric). From eqn (48), we can then obtain the following relation between the integer m, the wavelength λ and mirror separation L:

$$m = \frac{2L}{\lambda} \qquad ..(53)$$

The wavelength at which the maxima occur in Fig. 19.18 can be referred to as λ_m^{max}

where $\qquad \lambda_m^{\text{max}} = \dfrac{2L}{m} \qquad\qquad ..(54)$

It is more convenient to write these maxima in terms of the frequencies ν_m^{max} at which the maxima occur by using the relation $\lambda\nu = v = c/\eta$ where c is the velocity of light and η, the refractive index. Thus,

$$\nu_m^{\text{max}} = \frac{mC}{2\eta L} \qquad ..(55)$$

We now take the difference between two successive frequencies $\nu_{m+1}^{\text{max}} - \nu_m^{\text{max}}$ as

$$\nu_{m+1}^{\text{max}} - \nu_m^{\text{max}} = \frac{c}{2\eta L}[(m+1)-m] = \frac{c}{2\eta L} \qquad ..(56)$$

This frequency difference is independent of m. Thus

$$\Delta\nu = \frac{c}{2\eta L} \qquad ..(57)$$

where ηL is the optical path length separating the mirror surfaces. For $\eta = 1$

$$\Delta\nu = \frac{c}{2L} \qquad ..(58)$$

These maxima therefore appear at equal frequency spacings that are independent of the specific value of the frequency or wavelength.

We have thus derived the expression necessary to understand the response of a two-mirrored cavity (resonator) to a beam of light of intensity I_0. Such a cavity is known as a Fabry-Perot interferometer for most optical applications and as a Fabry-Perot etalon if the spacing L between the mirrors is firmly fixed. For laser applications it is generally referred to as a two-mirrored optical cavity or resonator. The separation between maxima in terms of frequency is given by eqn (57), so we can obtain $\Delta\nu_{FWHM}$ for each of these maxima by using

$$\Delta\nu_{FWHM} = \frac{\Delta\nu}{F}$$

and the expression (57). This gives

$$\Delta\nu_{FWHM} = \frac{1}{F} \times \frac{c}{2\eta L} = \frac{c(1-R)}{2\pi\eta L\sqrt{R}} \qquad ..(59)$$

for mirrors of equal reflectivities.

However, for mirrors of unequal reflectivities R_1 and R_2, we have

$$\Delta\nu_{FWHM} = \frac{c[1-(R_1 R_2)^{1/2}]}{2\pi\eta L(R_1 R_2)^{1/4}} \qquad ..(60)$$

Fabry-Perot Cavity Modes (Longitudinal Laser Modes)

We can rewrite

$$L = m\left(\frac{\lambda^{max}}{2}\right) \qquad ..(61)$$

assuming $\eta = 1$ for simplicity.

This equation indicates that each enhancement occurs when an integral multiple of half wavelengths fit into the cavity spacing of length L such that the electric vector of the electromagnetic wave is zero at the reflecting surfaces. Each of these waves that is enhanced by virtue of its exactly fitting into the cavity (of spacing L), is referred to as a *mode*. These modes are the result of interference effects that occur with light interacting due to reflections at two parallel reflecting surfaces. We can rewrite (for $\eta = 1$):

$$\nu = \frac{mc}{2L} \qquad\qquad ..(62)$$

Thus there are essentially an infinite number of frequencies that would fit within such a cavity. In order for the resonance condition (57) or (61) to be applicable over a large mirror surface, the mirror quality or the variation in L at different portions of the reflecting surface would have to be less than $(\lambda/2)/2$ or $\lambda/4$. In actual practice, the surface variation (referred to as the surface figure) must be better than $\lambda/10$ to obtain the condition of high finesse F with lasers.

When a laser gain medium is inserted within a Fabry-Perot cavity, with mirrors at the ends of the medium, a similar set of enhancements (modes) in the form of standing wave patterns equally spaced in frequency will build up within the cavity. When the gain medium is initially turned on, the amplifier begins emitting spontaneous emission at the laser frequency in all directions. Later, since the rays are directed toward the mirrors, they get reflected and return through the amplifier and thus, are enhanced by the gain during each transit through the medium. A highly directional beam eventually evolves in the axial direction. At this point, the beam begins to approximate a plane wave in that it has very low divergence—the case for which we obtained the modes within the Fabry-Perot cavity. The boundary condition

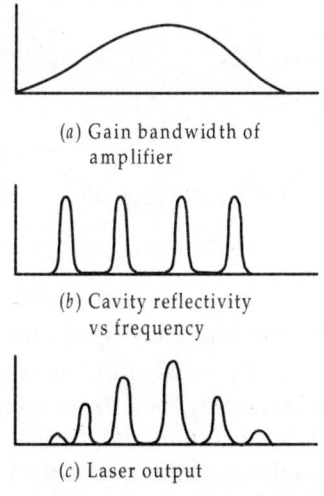

(a) Gain bandwidth of amplifier

(b) Cavity reflectivity vs frequency

(c) Laser output

Fig. 19.19: Resulting laser cavity modes when a gain bandwidth of a laser amplifier is combined with resonances of a two mirror cavity

at the mirrors is $\Phi/2 = n\pi$ which is equivalent to requiring the electric field to be zero at the mirrors. This condition establishes a standing wave pattern of the laser beam within the cavity. The various standing waves, each of a different frequency according to eqn (57) are referred to as *Longitudinal modes*.

Modes of a Rectangular Cavity

Consider a rectangular cavity of dimensions $2a$, $2b$ and d as shown in Fig. 19.20.

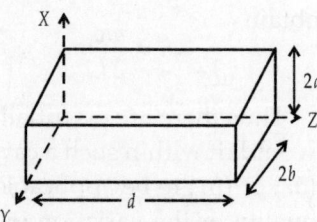

Fig. 19.20

Starting from Maxwell's equations, one can show that the electric and magnetic fields satisfy a wave equation of the form

$$\nabla^2 E - \frac{1}{c^2}\frac{\partial^2 E}{\partial t^2} = 0 \qquad ..(63)$$

where c is the velocity of light in the medium filling the rectangular cavity which is assumed presently to be free space. We can obtain eqn (63) as

$$\nabla \times (\nabla \times E) = -\mu_0 \frac{\partial}{\partial t}(\nabla \times H)$$

$$= -\varepsilon_0 \mu_0 \frac{\partial^2 E}{\partial t^2}$$

Using the identity

$$\nabla \times (\nabla \times E) = \nabla(\nabla \cdot E) - \nabla^2 E$$

$$= -\nabla^2 E$$

We immediately obtain eqn (63).

If the walls of the rectangular cavity are assumed to be perfectly conducting then the tangential component of the electric field must vanish at the walls. Thus if $\hat{e}$ represents the unit vector along normal to the wall then we must have

$$E \times \hat{e} = 0$$

on the walls of the cavity.

Let us consider, cartesian components (say) of the electric vector, which will satisfy

$$\frac{\partial^2 E_x}{\partial x^2} + \frac{\partial^2 E_x}{\partial y^2} + \frac{\partial^2 E_x}{\partial z^2} = \frac{1}{c^2}\frac{\partial \cdot E_x}{\partial t^2}$$

In order to solve this equation, we use the method of separation of variables and write

$$E_x = X(x)\,Y(y)\,Z(z)T(t)$$

Substituting this in eqn (63) and dividing throughout by E_x, we obtain

$$\frac{1}{X}\frac{d^2X}{dx^2} + \frac{1}{Y}\frac{d^2Y}{dy^2} + \frac{1}{Z}\frac{d^2Z}{dz^2} = \frac{1}{c^2T}\frac{d^2T}{dt^2}$$

Each side must equal a constant and separately

$$\left.\begin{aligned}\frac{1}{X}\frac{d^2X}{dx^2} &= -k_x^2 \\[4pt] \frac{1}{Y}\frac{d^2Y}{dy^2} &= -k_y^2 \\[4pt] \frac{1}{Z}\frac{d^2Z}{dz^2} &= -k_z^2\end{aligned}\right\} \qquad ..(64)$$

and $\qquad \dfrac{1}{c^2T}\dfrac{d^2T}{dt^2} = -k^2 \qquad\qquad\qquad ..(65)$

where $\qquad k^2 = k_x^2 + k_y^2 + k_z^2$

Equation (65) indicates that the time dependence is of the form

$$T(t) = Ae^{-i\omega t}$$

where $\omega = ck$ represents the angular frequency of the wave and A is a constant (the eqn (65) corresponds to an outgoing wave). Since E_x is tangential to the planes $y = 0$, $y = 2b$ and $z = 0$ and $z = d$, it has to vanish on these planes and the solutions of eqns (64) would be $\sin k_y y$ and $\sin k_z z$ respectively with

$$k_y = \frac{n\pi}{2b} \quad k_z = \frac{q\pi}{d} \quad n, q = 0, 1, 2, 3, \dots \qquad ..(66)$$

The value zero has been included in n, q values in order to incorporate the trivial solution of E_x vanishing everywhere. Similarly, the x and z-dependence of E_y would be $\sin k_x x$ and $\sin k_z z$ respectively with

$$k_x = \frac{m\pi}{2a} \quad m = 0, 1, 2 \dots \qquad\qquad\qquad ..(67)$$

Finally, the x and y dependence of E_z would be $\sin k_x x$ and $\sin k_y y$ respectively.

Consequently, because of the above forms of the x dependence of E_y and E_z viz. $\partial E_y/\partial y$ and $\partial E_z/\partial z$ would vanish on the surfaces $x = 0$ and $x = 2a$.

Thus on the planes $x = 0$ and $x = 2a$, the equation $\nabla\cdot E = 0$ leads to $\partial E_x/\partial x = 0$.

Hence the x dependence of E_x will be of the form $\cos k_x x$ with k_x given by (67). Note that the case $m = 0$ now corresponds to a non-trivial solution. In a similar manner one may obtain the solutions for E_y and E_z. The complete solution (apart from time dependence) would therefore be given by

$$E_x = E_{0x}\cos k_x x \, \sin k_y y \, \sin k_z z \qquad ..(68)$$

$$E_y = E_{0y}\sin k_x x \, \cos k_y y \, \sin k_z z \qquad ..(69)$$

$$E_z = E_{0z}\sin k_x x \, \sin k_y y \, \cos k_z z \qquad ..(70)$$

where coefficients E_{0x}, E_{0y}, E_{0z} are constants. Maxwell's equation $\nabla\cdot E$ yields (using eqns 68 and 69)

$$E_0 \cdot k = 0 \qquad ..(71)$$

where $\qquad k = k_x\hat{x} + k_y\hat{y} + k_z\hat{z}$

Using eqns (66) and (67), we get

$$\omega^2 = c^2 k^2 = c^2(k_x^2 + k_y^2 + k_z^2)$$

$$= c^2\pi^2\left(\frac{m^2}{4a^2} + \frac{n^2}{4b^2} + \frac{q^2}{d^2}\right)$$

or $\qquad \omega = c\pi\left(\frac{m^2}{4a^2} + \frac{n^2}{4b^2} + \frac{q^2}{d^2}\right)^{1/2} \qquad ..(72)$

which gives the allowed frequencies of oscillation of the field in the cavity. Field configurations given by eqns (68), (69) and (70) represent standing wave patterns in the cavity and are called the modes of oscillations of the cavity. Since the coefficients E_{0x}, E_{0y} and E_{0z} have to satisfy eqn (71), it implies that for given values of m, n and q, only two of the components of E_0 can be chosen independently. Thus a given mode can have two independent states of polarization.

Transverse Laser Cavity Modes

In addition to longitudinal modes, a laser also has certain transverse modes of oscillations. These modes are represented by a particular distribution of the electric field over the cross-section of the beam and characterized with the symbols TEM_{ij} (Transverse Electromagnetic Modes).

Finite size of the mirrors at the two ends of amplifier lead to the formation of transverse modes. Let $U(x, y)$ represent the transverse spatial dependence of radiation distribution on the first mirror, which is the sum of the contributions of radiation from all points leaving the second mirror and arriving at the location (x, y) on the first mirror. It can be shown that $U(x, y)$ can be written as the products of Hermite polynomials and a Gaussian distribution function as

$$U_{pq}(x, y) = H_p\left(\frac{\sqrt{2}x}{\omega}\right) H_q\left(\frac{\sqrt{2}y}{\omega}\right) e^{-(x^2 + y^2)/\omega^2} \qquad ..(73)$$

where ω is a scaling constant and p and q are integers that designate the order of the Hermite polynomials. Every set of (p, q) represents a specific stable distribution of wave amplitude at one of the mirrors or a specific transverse mode of the cavity. We can designate these transverse electromagnetic mode distributions as TEM_{pq}. Different transverse modes are modes having the same longitudinal mode number n but with different values of p and/or q. Since, in a two-mirrored cavity, a standing wave pattern is always produced that leads to longitudinal modes, therefore, every transverse mode will be found to consist of one or more longitudinal modes. Figure 19.21 shows some mode patterns for various transverse laser modes for square mirrors.

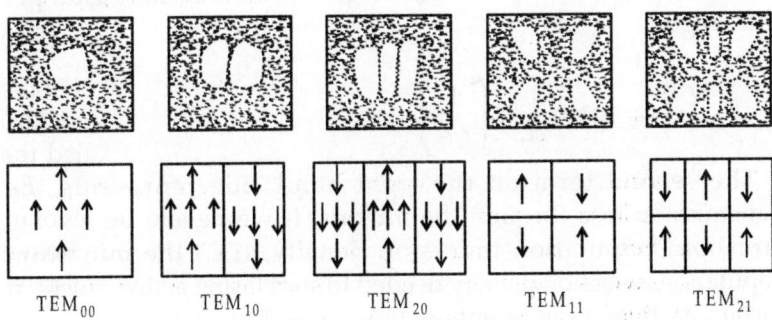

TEM$_{00}$ TEM$_{10}$ TEM$_{20}$ TEM$_{11}$ TEM$_{21}$

Fig. 19.21: Various transverse mode patterns as obtained on screen, for square (plane) mirrors at the two ends of the cavity. In TEM$_{ij}$, the indices i and j count the zeros along the x and y direction respectively

19.10 THRESHOLD CONDITION FOR LASER OSCILLATION AND LASER RATE EQUATIONS

Although population inversion between the energy levels providing the laser transition is necessary for oscillation to be established, it is not alone sufficient for lasing to occur. In addition, a minimum threshold gain within the amplifying medium must be attained to initiate and sustain laser oscillations. This threshold gain may be evaluated by considering the change in energy of a light beam as it passes through the amplifying medium. For convenience, all the losses (except those due to transmission through the mirrors) may be included in a single loss coefficient per unit length α cm^{-1}. It is assumed that the amplifying medium has a length L and completely fills the region between the two mirrors which have reflectivities r_1 and r_2. On each round trip, the beam passes through the medium twice. Hence the fractional loss of the light beam is

$$\text{Fractional loss} = r_1 r_2 \exp(-2\alpha L) \qquad ..(74)$$

Further, it is found that the increase in beam intensity resulting from stimulated emission is exponential. So, if the gain coefficient per unit length produced by stimulated emission is g cm^{-1}, the fractional round trip gain is given by

$$\text{Fractional gain} = \exp(2gL) \qquad ..(75)$$

Hence $\exp(2gL) \times r_1 r_2 \exp(-2\alpha L) = 1$

and $r_1 r_2 \exp 2(g - \alpha) L = 1$

Rearranging this expression, we get for the threshold gain per unit length,

$$g_{\text{th}} = \alpha + \frac{1}{2L} \ln\left(\frac{1}{r_1 r_2}\right) \qquad ..(76)$$

The second term on the right hand side represents the transmission loss through the mirrors. If we denote by N_T, the threshold population inversion density, (i.e., the minimum population inversion density needed to start lasing action and then sustain it), then it can be shown to be given by

$$N_T = (N_2 - N_1)_{\text{th}} = \frac{16\pi \nu^2 t_{\text{spon}}}{c^2 g(\nu)}\left[\alpha - \frac{1}{2L}\ln(r_1 r_2)\right] \qquad ..(77)$$

The cavity losses are characterized by defining the cavity decay constant t_c. This is associated with decay of an electric field as it bounces back and forth between the mirrors and the lasing medium is not pumped, i.e.

$$E(t) = E_0 e^{-t/tc} \qquad ..(78)$$

If T denotes the time taken by the electric field to make a round trip through the cavity, then

$$T = \frac{2L}{V} \qquad ..(79)$$

Thus if we observe the output at any mirror, then we will find the decaying pulses of light coming out of the mirror at every T seconds.

The equation (77) shows that threshold is reached when the population inversion $(N_2 - N_1)$ reaches a threshold value known as *critical inversion*. Once this is reached, the oscillation will build up from the spontaneous emission. This is the basis of the laser.

Non-radiative Decay

An atom can also undergo a non-radiative decay from level 2 to level 1. In this case, the energy difference $(E_2 - E_1)$ is given to surrounding molecules in the form of translational, vibrational or rotational energy.

The phenomenon can be described through a characteristic lifetime known as the non-radiative lifetime τ_{nr}.

As a result of simultaneous occurrence of both radiative and non-radiative decay, the time variation of the upper state population N_2 can be written as

$$\frac{dN_2}{dt} = -\left(\frac{N_2}{\tau_{sp}} + \frac{N_2}{\tau_{nr}} \right) \equiv -\frac{N_2}{\tau} \qquad ..(80)$$

The overall lifetime τ is given by

$$\frac{1}{\tau} = \frac{1}{\tau_{sp}} + \frac{1}{\tau_{nr}} \qquad ..(81)$$

Saturation

Let us study the time evolution of the population of a two level system in the presence of an incident monochromatic electromagnetic wave of intensity I and frequency ω. Let us consider a monochromatic line. We suppose that the total population N_t does not vary with time while the populations N_1 and N_2 of levels 1 and 2 vary with time.

When the atoms are transferred from one level to another, there will always be a band of energy levels. In the stimulated transition, the photons will be emitted. Even though we have supplied energy, still there may not be additional photons. This means that though the atoms have been transferred from level 2 to level 1, still it has not contributed any photon to the pre-exciting photons. This has been termed as *non-radiative decay*. Let τ_{nr} denote the lifetime of the non-radiative decay and τ_{st}, the lifetime of the stimulated emission by the radiation. Hence, the atom transfer from one level to another can take place in two ways. The rate of change of population will be consisting of radiative and non-radiative processes and so will be given by

$$\frac{dN_2}{dt} = \left(\frac{N_2}{\tau_{nr}} + \frac{N_2}{\tau_{st}} \right) \equiv \frac{N_2}{\tau}$$

or
$$\frac{dN_2}{dt} = -\frac{N_2}{\tau}$$

Negative sign has been placed to take into account decrease of population. If N_1 and N_2 denote the populations of levels 1 and 2 respectively, then total population

$$N_t = N_1 + N_2 \qquad\qquad ..(82)$$

If $N_1 > N_2$, it leads to absorption

If $N_1 < N_2$, it leads to population inversion

and if $\qquad N_2 = N_1$, it leads to saturation.
    In fact

$$\frac{dN_2}{dt} = -W(N_2 - N_1) - \frac{N_2}{\tau} \qquad\qquad ..(83)$$

where WN_1 = Probability of absorption

and WN_2 = Probability of transition, per unit time.

The equation (83) gives the rate of change of N_2 with time as the difference between the rate WN_1 of stimulated transitions upward to the level 2 and the rate of downward transitions to level 1, and (N_2/τ) accounts for lifetimes of radiative and non- radiative decays.

If we put

$$N_1 - N_2 = \Delta N$$

then $\quad \dfrac{d}{dt}(\Delta N) = \dfrac{dN_1}{dt} - \dfrac{dN_2}{dt}$ ..(84)

In eqn (82), N_1 and N_2 change with time but N_t is constant in time. Therefore, $\dfrac{d}{dt}(N_1 + N_2)$ will remain constant in time. So,

$$\dfrac{d}{dt}(N_1 + N_2) = 0 = \dfrac{dN_1}{dt} + \dfrac{dN_2}{dt}$$

or $\quad \dfrac{dN_1}{dt} = -\dfrac{dN_2}{dt}$ ..(85)

From eqns (84) and (85), we get

$$\dfrac{d}{dt}(\Delta N) = -\dfrac{2dN_2}{dt}$$ ..(86)

Eqn (83) can be written as,

$$\dfrac{dN_2}{dt} = -W(\Delta N) - \dfrac{N_2}{\tau}$$

$$= -\dfrac{1}{2}\left(-2W\,\Delta N + \dfrac{2N_2}{\tau}\right)$$

Adding and subtracting N_1/τ, we get

$$\dfrac{dN_2}{dt} = -\dfrac{1}{2}\left[-2W\,\Delta N + \dfrac{2N_2}{\tau} + \dfrac{N_1}{\tau} - \dfrac{N_1}{\tau}\right]$$

$$= -\dfrac{1}{2}\left[-2W\,\Delta N + \dfrac{N_1 + N_2}{\tau} + \dfrac{N_2 - N_1}{\tau}\right]$$

$$= -\dfrac{1}{2}\left[-2W\,\Delta N + \dfrac{N_t}{\tau} - \dfrac{\Delta N}{\tau}\right]$$

or $\quad -2\dfrac{dN_2}{dt} = -\left[2W(\Delta N) + \dfrac{\Delta N}{\tau}\right] + \dfrac{N_t}{\tau}$ ..(87)

From eqns (86) and (87), we get

$$\dfrac{d(\Delta N)}{dt} = -\Delta N\left(2W + \dfrac{1}{\tau}\right) + \dfrac{N_t}{\tau}$$ ..(88)

Hence, we see that the rate of change of population depends upon

(i) W, i.e. depends upon the intensity I of incident light or strength of pump.

(ii) τ, i.e. depends upon the lifetime of the pump.

(iii) N_t, i.e. depends upon the total population of the levels.

When $\dfrac{d}{dt}(\Delta N) = 0$, i.e. at steady state, we have

$$\Delta N = \frac{N_t}{1 + 2W\tau} \qquad\qquad ..(89)$$

When the intensity I increases, W increases, hence ΔN will decrease and in the limit $I \to \infty$, $\Delta N \to 0$, i.e. $N_1 \simeq N_2 \approx \dfrac{N_t}{2}$. In this case, the population of the two levels become equal. This phenomenon is called as *saturation*.

Now, in order to maintain the population difference, the system must absorb the energy from the pump. The power absorbed per unit volume by the system to maintain the population difference is given by

$$\frac{dP}{dV} = \Delta N \cdot W(\omega\hbar) \qquad\qquad ..(90)$$

(= (population difference/volume) × probability × energy)

$$= W\omega\hbar \frac{N_t}{1 + 2W\tau} \qquad \text{(using exp 89)} \qquad ...(91)$$

Hence, at saturation, when $W\tau \gg 1$,

$$\left(\frac{dP}{dV}\right)_s = \frac{(\hbar\omega)N_t}{2\tau} \qquad\qquad ...(92)$$

This equation shows that the power $(dP/dV)_s$ which must be absorbed to keep the system in saturation is equal to the power lost by the material due to the decay of the upper states.

19.11 THREE LEVEL LASER RATE EQUATIONS

Let us consider a laser operating in a three level scheme and has only one pump band.

Let the populations of the three levels be N_1, N_2 and N_3.

Suppose that the laser is oscillating in only one cavity, and let q be the number of photons in the cavity. Now, when the photon reaches from N_1 to N_3 then lifetime is such that it immediately goes back to the level 2, i.e. $N_3 = 0$. Hence total number of atoms is

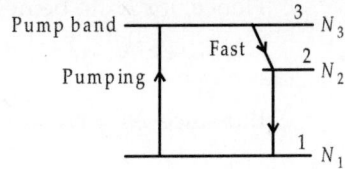

Fig. 19.22: Three level laser

$$N_t = N_1 + N_2 \qquad ..(93)$$

When the photon goes from N_2 to N_1, the N_1 gains atoms due to absorption and non-radiative transition while N_2 loses photons by stimulated radiation emission.

The number of atoms contributing to N_2 due to pumping is $W_p N_1$; the number of atoms due to absorption from N_1 to N_2 is WN_1 and the number of atoms due to stimulated emission from N_2 to N_1 is $(-WN_2)$; the number of atoms due to radiative losses is $(-N_2/\tau)$, hence the rate equation will be

$$\dot{N}_2 = \frac{dN_2}{dt} = W_P N_1 + WN_1 - WN_2 - \frac{N_2}{\tau} \qquad ..(94)$$

But since, in a laser, the number of photons is changing and it depends upon the enclosing volume V_2, thus the change in number of photons due to absorption is $V_a W(N_2 - N_1)$ and the losses in the cavity are q/τ_c where τ_c is time characterizing the photon losses in the cavity. Hence

$$\dot{q} = \frac{dq}{dt} = V_a W(N_2 - N_1) - \frac{q}{\tau_c} \qquad ..(95)$$

Now, let us define the stimulated transition rate per photon as B, then eqns (94) and (95) give

$$\dot{N}_2 = W_P N_1 - Bq(N_2 - N_1) - \frac{N_2}{\tau} \qquad ..(96)$$

and
$$\dot{q} = V_a Bq(N_2 - N_1) - \frac{q}{\tau_c} \qquad ..(97)$$

We now observe that

(i) If the decay rate of level 3 is much faster than the pump rate, then level 3 can be considered to be empty. Suppose τ_3 is the lifetime of level 3, then

$$N_3 = W_P \tau_3 N_1 \qquad ..(98)$$

Hence, for N_3 to be negligible, we must have

$N_3 \ll N_1$ or N_2.

But since $N_1 \approx N_2 \approx \dfrac{N_t}{2}$, hence we get

$$\tau_3 \ll \frac{1}{W_P} \qquad\qquad ..(99)$$

(ii) In eqn (97), we have neglected the term for spontaneous emission. Now if we put $q = 0$, then eqn (97) gives $\dot{q} = 0$. So the laser action will not start because laser action is actually started by spontaneous emission. When spontaneous emission is taken into account, the term $V_a B_q N_2$ in eqn (97) becomes $V_a B(q + 1)N_2$. Hence everything behaves as if we should add an extra photon to the term describing stimulated emission.

Now, if we put the population difference as

$\Delta = N_2 - N_1$

$$\dot{\Delta} = \dot{N}_2 - \dot{N}_1 \qquad\qquad ..(100)$$

and since, we have $N_t = N_1 + N_2$, but N_t is independent of time, so

$$\dot{N}_1 + \dot{N}_2 = \dot{N}_t = 0 \qquad\qquad ..(101)$$

$\therefore \qquad \dot{\Delta} = 2\dot{N}_2 = -2\dot{N}_1 \qquad\qquad ..(102)$

and since $N_1 = N_t - N_2$

$\therefore \quad N_1 + N_1 = N_t - N_2 + N_1$

or $\quad 2N_1 = N_t - (N_2 - N_1)$

or $\quad 2N_1 = N_t - \Delta \qquad\qquad ..(103)$

Hence putting $\dot{N}_2 = \dfrac{\dot{\Delta}}{2}$, (from eqn 102) $\qquad ..(104)$

$N_2 - N_1 = \Delta$

and $\quad N_1 = \dfrac{1}{2}(N_t - \Lambda) \qquad\qquad ..(105)$

Hence, from eqns (96), (97) and (105), we get

$$\dot{q} = V_a Bq\Delta - \frac{q}{\tau_c} \qquad \qquad ..(106)$$

$$\frac{\dot{\Delta}}{2} = W_P \frac{(N_t - \Delta)}{2} - Bq\Delta - \frac{(N_t + \Delta)}{2\tau} \qquad \qquad ..(107)$$

$$(\because N_t + \Delta = N_1 + N_2 + N_2 - N_1 = 2N_2)$$

or $\quad \dot{\Delta} = W_P(N_t - \Delta) - 2Bq\Delta - \dfrac{(N_t + \Delta)}{\tau} \qquad \qquad ..(108)$

and $\quad N_t - (N_1 + N_2) = 0 \qquad \qquad ..(109)$

The three equations (106), (108) and (109) define the static and dynamic behaviour of laser. Now, in order to have laser action, we must have population inversion and we must supply some critical energy. Suppose q = constant; then $\dot{q} = 0$, and there will be no laser action. Hence in order to have laser action, we must have $\dot{q} > 0$ or

$$V_a B\Delta - \frac{1}{\tau_c} > 0$$

Hence the critical population difference can be defined as

$$\Delta_c = \frac{1}{V_a B \tau_c} \qquad \qquad ..(110)$$

Then, the laser action will take place if

$$\Delta > \Delta_c$$

Now let us evaluate the critical pumping rate W_{cp}. The first requirement for laser action is that population difference must be greater than the critical population difference (i.e. $\Delta > \Delta_c$), secondly $q = 0$, thirdly $\dot{\Delta} = 0$ because $\dot{\Delta}$ is the rate of change of population from N_2 to N_1. Substituting

$W_P = W_{cp}$, $\dot{\Delta} = 0$, $q = 0$ and $\Delta = \Delta_c$ in eqn (108), we get

$$0 = W_{cp}(N_t - \Delta_c) - 0 - \frac{(N_t + \Delta_c)}{\tau}$$

or $\quad W_{cp} = \left(\dfrac{N_t + \Delta_c}{N_t - \Delta_c} \right) \dfrac{1}{\tau} \qquad \qquad ..(111)$

This gives the critical pumping rate. This equation can be understood readily, noticing that the critical population of the upper state as $N_t = N_1 + N_2$ and of lower state is $\Delta = N_2 - N_1$ then

$$N_{2c} = (N_t + \Delta_c)/\tau \quad \text{and} \quad N_{1c} = \frac{N_t - \Delta_c}{\tau}$$

Then eqn (111) can be written as

$$W_{cp} = \frac{N_{2c}}{N_{1c}} \frac{1}{\tau}$$

This shows a balance between pump transitions and spontaneous transitions. But for practical purposes $N_c << N_t$ so eqn (111) gives

$$W_{cp} \approx \frac{1}{\tau}$$

Critical Energy of Laser

We assume that:

(*i*) W_p is independent of time

(*ii*) Pumping rate is greater than critical pumping rate, i.e. $W_p > W_{cp}$.

In such cases the photon number q will grow from the initial value determined by spontaneous emission and ultimately a steady state is reached. Let the number of photons at steady state be q_0. This equilibrium value and the corresponding population difference Δ_0 are obtained by setting

$$\dot{\Delta} = 0 = \dot{q}$$

in eqns (106) and (108), which gives

$$\Delta_0 = \frac{1}{V_a B \tau_c} = \Delta_c \qquad \qquad ..(112)$$

and

$$q_0 = \frac{V_a \tau_c}{2}\left[W_p(N_t - \Delta_0) - \frac{(N_t + \Delta_0)}{\tau}\right] \qquad ..(113)$$

The equations (112) and (113) define the continuous wave (CW) behaviour of lasers. In eqn (113) we have replaced Δ_0 by $1/(V_a B \tau_c)$ (from eqn (107)).

Now, since all the photons generated in the cavity never contribute to the laser, a certain fraction is always lost due to cavity losses. Hence, putting $\Delta_0 = \Delta_c$ in eqn (111), we get

$$(N_t - \Delta_0) = \frac{N_t + \Delta_0}{\tau W_{cp}} \qquad ..(114)$$

Hence eqn (113) gives

$$q_0 = \frac{V_0 \tau_c}{2\tau}(N_t + \Delta_0)\left[\frac{W_p}{W_{cp}} - 1\right] \qquad ..(115)$$

Thus at $W_c = W_{cp}$, q_0 number of photons is zero. Thus for laser action $W_p > W_{cp}$ must hold good. Thus, the energy output of the system at equilibrium ε is

$$\varepsilon = \hbar \omega q_0$$

(i.e. energy of a photon × number of photons)

or $$\varepsilon = \hbar \omega \frac{V_0 \tau_c}{2\tau}(N_t + \Delta_0)\left(\frac{W_p}{W_{cp}} - 1\right) \qquad ..(116)$$

where τ_c is cavity loss.

19.12 FOUR LEVEL LASER RATE EQUATIONS

The atoms when raised to level N_3, make transition from level 3 to level 2. Then, between N_1 and N_2 laser action takes place, i.e. population inversion is set up between N_1 and N_2. For N_3 and N_1, we have fast decay, hence due to this fast decay $N_1 \approx N_3 = 0$ as compared to N_0 and N_2. Now, when N_2 is very much greater than

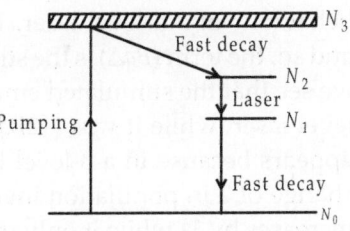

Fig. 19.23: A four level laser

N_1, then there is a population inversion. The energy gap between N_0 and N_1 is such that the atoms cannot go from N_0 to N_1 by thermal conduction, i.e. $kT << \varepsilon_0$. Hence the rate equation can be written as

$$N_0 + N_1 + N_2 + N_3 = N_t \qquad (\because N_1 \approx N_3 = 0)$$

or $$N_0 + N_2 = N_t \qquad ..(117)$$

and the rate of change of population is

$$\frac{dN_2}{dt} = \dot{N}_2 = W_p N_0 - Bq(N_2 - N_1) - \frac{N_2}{\tau}$$

or $\qquad \dot{N}_2 = W_{\text{p}}N_0 - BqN_2 - \dfrac{N_2}{\tau}$ $\qquad\qquad$ ..(118)

The rate of change of number of photons is

$$\frac{dq}{dt} = \dot{q} = BV_{\text{a}}(N_2 - N_1)q - \frac{q}{\tau_{\text{c}}}$$

or $\qquad\qquad \dot{q} = BqV_{\text{a}}N_2 - \dfrac{q}{\tau_{\text{c}}}$ $\qquad\qquad$ ..(119)

Now, defining $\Delta = N_2 - N_1$, we have

$$\dot{\Delta} = \dot{N}_2 \qquad\qquad\qquad (\because N_1 = 0)$$

So, eqns (118) and (119) give

$$\dot{\Delta} = W_{\text{p}}N_0 - Bq\Delta - \frac{\Delta}{\tau}$$

or $\qquad\qquad \dot{\Delta} = W_{\text{p}}(N_t - \Delta) - Bq\Delta - \dfrac{\Delta}{\tau}$ $\qquad$ ..(120)

and $\qquad\qquad \dot{q} = \left(BV_{\text{a}}\Delta - \dfrac{1}{\tau_{\text{c}}} \right)q$ $\qquad\qquad$ ..(121)

The eqns (117), (120) and (121) define the static and dynamic behaviour of a 4-level laser. Here B is the stimulated probability and so, the term $(Bq\Delta)$ is the stimulated photon emission term. Here we see that the stimulated emission term is given by $(-Bq\Delta)$ for a 4-level laser, while it was $(-2Bq\Delta)$ in a 3-level laser. This factor of 2 appears because in a 3-level laser, the photon emission implies a change of 2 in population inversion (i.e. N_2 decreases by 1 and N_1 increases by 1) while it only implies a population change of 1 in 4-level laser. When N_2 again decreases by 1, N_1 remains approximately unchanged on account of fast 1 to 0 decay. Hence rate of change of population inversion is faster in this case.

19.13 SINGLE MODE OSCILLATION OR THE TIME DEPENDENT BEHAVIOUR OF A THREE LEVEL LASER

As we know that three level rate equations are

$$\dot{\Delta} = \frac{d}{dt}\Delta(0) = W_{\text{p}}(N_t - \Delta) - 2Bq\Delta - \frac{\Delta_t + \Delta}{\tau} \qquad ..(122)$$

and $\qquad \dot{q} = \dfrac{dq(t)}{dt} = \left(BV_a\Delta - \dfrac{1}{\tau_0} \right) q$ ..(123)

The solution of these equations depends upon the variation or the type of the pump to be applied. Let us take the variation of $W_p(t)$ as

$$\begin{cases} W_p(t) = 0 \text{ for } t < 0 \\ \qquad = W_p(0) \text{ constant of } t > 0 \end{cases} \qquad ..(124)$$

The source is such that the time variation is not much off from the equilibrium condition. Let q_0 and Δ_0 be the equilibrium values of q (number of photons) and Δ. Then, we can make the approximation

$$\Delta = \Delta_0 + \delta\Delta$$
$$q = q_0 + \delta q \qquad ..(125)$$

From eqn (122), we can write

$$\delta\dot{\Delta} = W_p(0 - \delta\Delta) - 2B[(\Delta_0 + \delta\Delta)]\delta q + (q_0 + \delta q)\delta\Delta] - \frac{\delta\Delta}{\tau}$$

or $\qquad \delta\dot{\Delta} = -\delta\Delta W_p - 2B\left\{\Delta_0\delta q + q_0\delta\Delta\right\} - \dfrac{\delta\Delta}{\tau}$

or $\qquad \delta\dot{\Delta} = -\delta\Delta\left(W_p + \dfrac{1}{\tau} \right) - 2B(\Delta_0\delta q + q_0\delta\Delta)$ ..(126)

Similarly for $\delta\dot{q}$, we get

$$\delta\dot{q} = BV_a q_0\delta\Delta + \left(BV_a\Delta_0 - \frac{1}{\tau_0} \right)\delta q$$

or $\qquad \delta\dot{q} = BV_a q_0\delta\Delta$ (because 2nd term is zero) ..(127)

Differentiating again eqn (127), we get

$$\delta\ddot{q} = BV_a q_0\delta\dot{\Delta}$$

$\Rightarrow \qquad \delta\dot{\Delta} = \dfrac{1}{BV_a q_0}\delta\ddot{q}$ ..(128)

Equating eqns (126) and (128), we get

$$\frac{1}{BV_a q_0}\delta\ddot{q} = -\delta\Delta\left(W_p + \frac{1}{\tau} \right) - 2B(\Delta_0\delta q + q_0\delta\Delta)$$

or $\delta\ddot{q} + BV_a q_0 \delta\Delta\left(W_p + \dfrac{1}{\tau}\right) + 2B^2 V_a q_0 \Delta_0 \delta q + 2B^2 q_0 V_a q_0 \delta\Delta = 0$

Substituting $\delta\dot{q} = BV_a(\delta\Delta)q_0$ from eqn (127) in the above eqn, we get

$$\delta\ddot{q} + \delta\dot{q}\left(W_p + \dfrac{1}{\tau} + 2Bq_0\right) + 2B^2 V_a q_0 \Delta_0 \delta q = 0$$

Putting $W_p + \dfrac{1}{\tau} + 2Bq_0 = \dfrac{1}{t_0}$

and $2B^2 V_a q_0 \Delta_0 = \omega^2$

we get

$$\delta\ddot{q} + \dfrac{\delta\dot{q}}{t} + \omega^2 \delta q = 0 \qquad\qquad ..(129)$$

The solution of this equation is

$$\delta q = C e^{-t/t_0} \sin(\omega t + \phi) \qquad\qquad ..(130)$$

where C and ϕ are constants and can be determined from initial conditions (eqn 125). Hence eqn (130) gives information about the number of photons. Substituting eqn (130) in eqn (127) we get

$$\delta\Delta = \dfrac{\omega C}{Bq_0 V_a} e^{-t/t_0} \cos(\omega t + \phi) \qquad\qquad ..(131)$$

Hence, from eqns (130) and (131) we conclude that the change in the photon number (δq) and the change in population inversion ($\delta\Delta$) have a phase difference of $\pi/2$ ($\delta\Delta$ lags behind δq by 90°).

The above discussed transient behaviour also applies to the case when the laser is already operating in the steady state and a sudden perturbation is applied. In this case eqns (130) and (131) are still valid and show that the perturbations introduced at $t = 0$ decay with time as damped sinusoidal waves. The steady state solutions correspond to a stable equilibrium.

When the perturbation is not small as compared to the equilibrium value, the results calculated by the calculator are as shown in Fig. 19.24.

The initial conditions are $N(0) = -N_t$ and $q_0 = q_i$ where q_i is a small integer and is needed only to allow the laser action to start. The Fig. 19.24 clearly shows the time delay between $q(t)$ and $\Delta(t)$. At the start of laser action, the time behaviour of $q(t)$ and hence that of the output light consists of a succession of

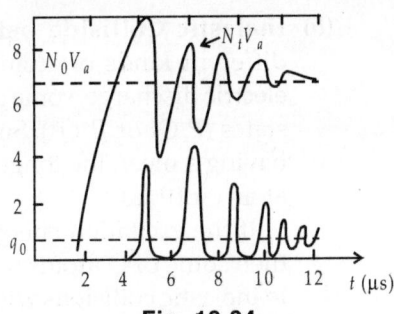

Fig. 19.24

light pulses (laser spikes) which are regularly spaced in time and decreasing in amplitude. This is referred to as the *regular spiking case*. We also see that for small oscillations about the equilibrium position, the time behaviour of $\Delta(t)$ and $q(t)$ is in agreement with eqns (130) and (131).

19.14 METHODS OF ACHIEVING POPULATION INVERSION

There are primarily four types of methods for achieving population inversion:

(*i*) Optical pumping
(*ii*) Electrical pumping
(*iii*) Chemical method
(*iv*) Semiconductor laser.

(*i*) **Optical Pumping:** When the atoms are exposed to electromagnetic radiation (photons) of frequency ν_{12}, then there is selective absorption due to which the atoms are raised to excited state, e.g. Ruby or Neodymium laser and liquid state (dye) lasers. It is particularly suitable for solid state lasers.

(*ii*) **Electrical Pumping**

(*a*) **Excitation by Electrons:** This method is used in gas lasers. Electrons are released from the atoms due to intense electric discharge through gas. The electrons are then accelerated to high velocities due to the strong electric field inside the discharge tube. When these electrons collide with the neutral gas atoms, a fraction of atoms are raised to excited state.

(b) **Inelastic Collision between Atoms:** If a gas has two different kinds of atoms (say P and Q) then during an electric discharge, some of these atoms are raised to excited states ($P\ Q^*$ or P^*Q^*). Suppose Q^* is in a metastable state having a mean life longer than the mean life of an excited atom ($\approx 10^{-8}$s).

If the excitation energy of P^* and Q^* are nearly equal, then some of Q atoms may be raised to excited states due to inelastic collisions with P^*. $P^* + Y \to P + Y^*$. As a result, the number of Q^* increases. This type of collision is known as collision of second kind. This method is used in He-Ne laser.

(iii) **Excitation by Chemical Method:** Sometimes an atom or a molecule which is a product of a chemical reaction is produced in an excited state, e.g. H and F combine to produce HF molecule in the excited state and the number of such (excited) molecules may be considerably larger than those in the ground state (separate atoms), causing thereby population inversion. This method is used in excimer laser.

(iv) **Semiconductor Diode Laser:** The lasing material in these lasers consists of a p-n homojunction (i.e. of a single crystal) made of heavily doped ($\sim 10^{18}$ cm^{-3}) n and p type semiconductor material such as GaAs. With a forward bias, the electrons concentrate on the p-side and the holes, on the n-side. The recombination of the electrons and holes in the junction region gives rise to laser action. If the current density at the junction is high enough, there may be population inversion between the electrons and hole levels. Thus stimulated emission may take place if the optical gain in semiconductor cavity is higher than the junction layer (recombination) loss.

19.15 RUBY LASER

It is a three-level laser. Theodore Maiman, in 1960, demonstrated the first successful laser action by using a crystal of ruby as a laser material. Ruby is crystalline Al_2O_3 containing a small (0.05%) percentage of Cr^{3+} ions (density 1.62×10^{19}/cm^3). This doping is done by adding small amounts of Cr_2O_3 in the melt of highly purified Al_2O_3. The normal green of Cr^{3+} is modified to red by the distortion of the local crystal field stemming from the replacement of an Al^{3+} ion by a slightly larger Cr^{3+} ion. The bulk of the material Al_2O_3 does not participate in the laser action, it rather acts as a host, but the Cr^{3+} ions form the active material. Ruby is a three-level laser. Maiman's first laser was in the form shown in Fig. 19.25.

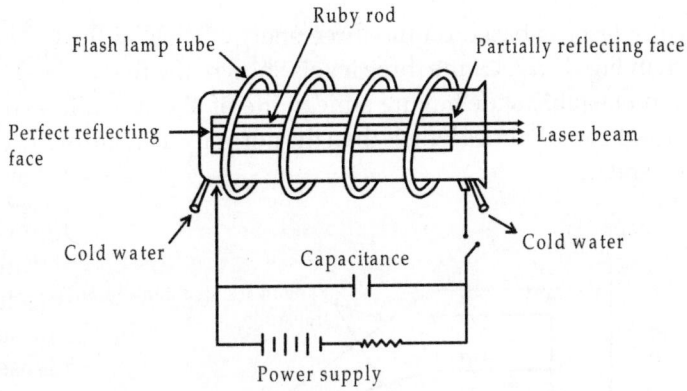

Fig. 19.25: A typical flash lamp pumped pulsed ruby laser

The ruby was in the form of a cylinder ~5 cm in length and ~0.3 cm in diameter. The ends of the ruby rod are grounded and polished plane and parallel (the ends are flat to better than $\lambda/10$ and parallel to within 2 seconds of an arc). One end is silvered to give a completely reflecting surface and the other end is translucent. The rod is kept inside a glass tube which is surrounded by a tube in the form of a coil which is filled with xenon gas. The ends of this flash tube are connected to a power supply to produce flashes of light. Since a larger proportion of energy used in producing flashes goes away as heat energy. Cold water is allowed to circulate continuously in the glass tube.

Instead of flash lamp, pumping scheme, one may use other pumping schemes such as that shown in Fig. 19.26 in which the pump lamp and the laser rod are placed along the two foci of an elliptical

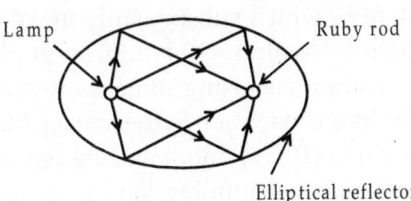

Fig. 19.26: Elliptical pump cavity in which the lamp and the ruby rod are placed at the two foci of the elliptical cylindrical reflector

cylindrical reflector. It is well known that the elliptical reflector focuses the light emerging from one focus into the other focus of the ellipse, leading to an efficient focusing of pump light on the laser rod.

Operation

The ruby laser is based on the three energy levels in the Cr^{3+} ions shown in Fig. 19.27. Of the three levels shown, the first excited state E_1 is a metastable state having a mean life of about 3 milliseconds, which is about 10^5 times longer than the normal mean life of a excited atomic state.

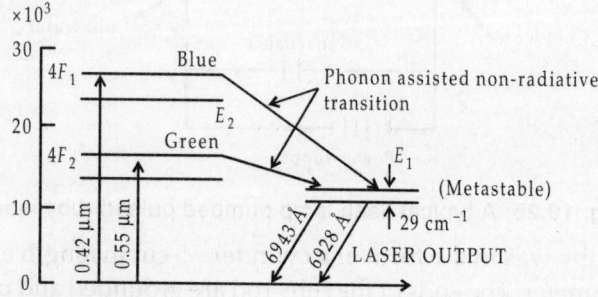

Fig. 19.27: Energy levels of Cr^{3+} in ruby laser

The higher excited state E_2 has a mean life of $\sim 10^{-8}$ s. Thus there are two main pump bands $4F_1$ and $4F_2$ centred around 0.42 μm (blue) and 0.55 μm (green) respectively. Each pump band is about 0.1 μm wide.

When the xenon flash lamp is switched on, it provides light flashes of wavelength 5500 Å. This excites the Cr^{3+} ions to the excited state E_2 which subsequently make transitions to the lower excited state E_1 (within $\sim 10^{-8}$ s) through phonon assisted (non- radiative) transition via losing some energy in collision with the lattice. Since the ions can remain in the metastable state, for comparatively longer periods (10^{-3} s), more and more ions become trapped in this state making their number very large as compared to the ground state (provided pumping excitation from Xe flash lamp is sufficiently intense). In order that induced emission be greater than induced absorption, more than half the Cr^{3+} ions in the ruby rod must be in the metastable state. The metastable state E_1 slowly decays in $\sim 10^{-}$

by emitting a sharp doublet with components 6943 Å and 6928 Å (the laser output beam). The rod's length must be precisely an integral number of half wavelength so that radiation trapped in it forms an optical standing wave.

The population inversion is very difficult to sustain continuously and in practice the ruby laser is pulsed. Typical pulses might consist of 2 J pulses persisting for 10 ns, corresponding to a power of 0.2 GW. Even during the short period of a few tens of microseconds in which the laser is oscillating, the output is a highly irregular function of time with the intensity having random amplitude fluctuations of varying duration (Fig. 19.28.) This is called laser spiking.

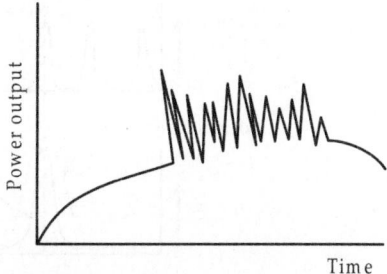

Fig. 19.28: Laser spiking in ruby laser beam

Ruby laser is one of the important practical lasers. The absorption bands of ruby are very well matched with the emission spectra of practically available flash lamps so that an efficient use of the pump can be made. It also has a favourable combination of a long lifetime and a narrow line width. It is also attractive from an application point of view since its output lies in the visible region where photographic emulsions and photodetectors are much more sensitive than in the infrared region. Thus ruby laser finds application in holography and laser ranging. They are also used in measuring such plasma properties as electron density and temperature. Ruby lasers are also used to remove tattos and also skin lesions resulting from excess melanin.

19.16 EFFECT OF LASER CAVITY ON THE EMITTED FREQUENCY

Figure 19.29 shows the effect of laser cavity on the output. For a laser cavity having a mirror separation L, the resonant frequencies are $\nu = mc/(2L)$ where m is an integer and c is the speed of light. For a frequency ν' corresponding to $m' = m + 1$, we find

$$\Delta\nu = \nu' - \nu = \frac{c}{2L}$$

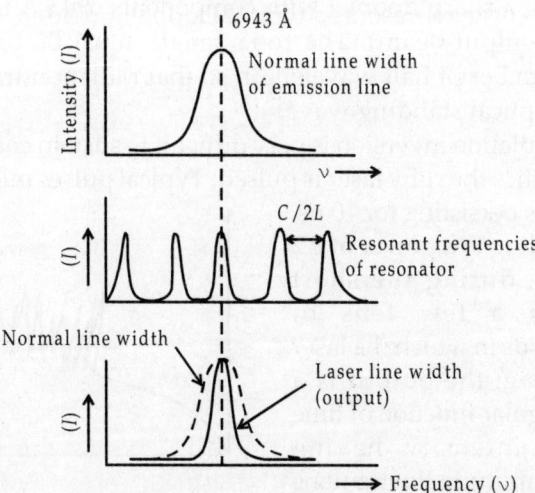

Fig. 19.29: Effect of the laser cavity on frequency

19.17 VARIOUS LINE WIDTHS ASSOCIATED WITH A LASER

Fig. 19.30 shows the various line widths associated with a laser. The broad solid curve represents the spectral width due to Doppler broadening of the laser medium. For example, if we consider the He-Ne laser operating at 6328 Å, the Doppler broadened line width is typically about 1300 MHz. Inside the broad curve are shown the cavity modes as sharp peaks. The frequency separation between two adjacent cavity modes is $c/2L$ which for a typical laser cavity 60 cm long corresponds to 250 MHz, this is much less than the Doppler width.

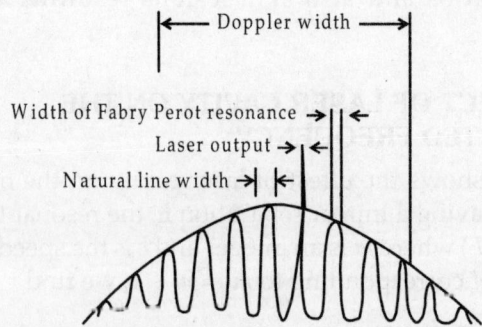

Fig. 19.30: The line widths associated with a laser

The solid curve represents a typical Doppler broadened spectral line. The narrow peaks inside this curve are the cavity modes. The sharp line represents the output of the laser.

The cavity modes are also broadened due to the various losses in the cavity. Thus, for a 60 cm long cavity specified by a fractional loss per round trip of 4×10^{-2}, the width of the cavity mode is about 1.5 MHz. When the losses in the cavity are compensated for by the active medium placed inside the cavity, the resultant emission becomes extremely narrow (i.e. the laser output).

19.18 LASER SPIKING

Transient Laser and Regular Spiking

In a laser, if the pumping rate is constant, we have a continuous laser. However, if we want an oscillating laser, i.e. time dependent laser, then we have to find as to how the number of photons is changing with time. This is connected with pumping rate, and as such a laser is called as the *transient laser*. In order to study a transient laser, let us consider the step-pump pulse and we consider an example: A laser has mirrors 10 cm apart. If the natural width of the emission life for the laser transition is $\Delta \nu_0 = 10^8$ Hz, can the laser have more than one frequency?

Solution: The separation between the frequency modes is

$$\Delta \nu = \frac{c}{2L} = \frac{3 \times 10^8 \, \text{m/s}}{2 \times 0.10 \, \text{m}} = 1.5 \times 10^9 \, \text{Hz}$$

Since the line width is smaller than the separation of the cavity modes, there can be only one mode (or frequency) for this laser.

Spiking in Ruby Laser

Refer to Fig. 19.28. This phenomenon of spiking can be expressed as follows:

When the pump is suddenly switched on to a value much above the threshold, the population inversion builds up and crosses the threshold value, as a consequence, the photon number builds up rapidly to a value much higher than the steady state value. Since the photon number is higher than the steady state value, the rate at which the upper level depletes because of stimulated emission is much higher than the pump rate. Consequently, the inversion becomes below threshold and the laser action ceases. Thus the laser emission stops for a few microseconds, within which time the flash lamp again pumps the ground state atoms to the upper level, and laser oscillations begin again. This process is repeated until the flash lamp power falls below the threshold value and the lasing action stops finally.

19.19 THE HELIUM-NEON LASER

The first successful gas laser was demonstrated by Javan, Bennett and Harriott in 1961. Since then, many different gas lasers using gases of many kinds and mixtures have been put into operation. Because it is less expensive, unusually stable and the first laser to be operative in continuous wave (CW) mode, the He-Ne gas laser is widely used in optics and physics laboratories the world over.

The schematic set up is shown in Fig. 19.31 and the energy levels for He and Ne are given in Table 19.1.

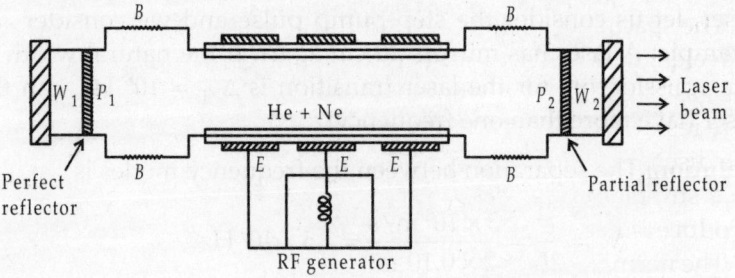

Fig. 19.31: The He-Ne gas laser arrangement

A discharge tube is filled with Ne and He gas. The He pressure is 1 mm and Ne pressure is 0.1 mm. Hence He is in a large quantity (9 parts of He and 1 part of Ne) and so, He will act as host and Ne as an activator. P_1 and P_2 are the optically plane and parallel reflector plates. B_1 and B_2 are flexible pillows to adjust the distance between P_1 and P_2. W_1 and W_2 are the windows. $E's$ are the electrodes. Here we use an *r.f.* generator.

It is a four level laser in which population inversion is achieved by electric discharge. He is there to provide more efficient excitation only and actual laser emissions take place through the Neon levels.

TABLE 19.1: The energy levels and their designations for He and Ne

Element	Electron Configuration	Level Designation	Element	Electron Configuration	Level Designation
He	$1s^2$	$1S_0$			6
He	$1s\,2s$	$3S_1$	Ne	$2p^5\,3p$	7
		$1S_0$			8
Ne	$2p^6$	$1S_0$			9
					10
Ne	$2p^53s$	$^3P_2$ $^3P_1$ $3P_0$ $^1P_1$	Ne	$2p^54s$	$^3P_2$ $^3P_1$ $^3P_0$ $^1P_1$
Ne	$2p^53p$	1 2 3 4 5	Ne	$2p^55s$	$^3P_2$ $^3P_1$ $^3P_0$ $^1P_1$

The spacing of the mirrors is equal to an integral number of half wavelength of the laser light. An electric discharge is produced in the gas by electrodes connected to a high frequency electric source. As a result, collisions between ionised atoms and electrons carrying the discharge current excite the atoms to various energy states and at a sufficiently high voltage, an orange-red glow discharge is produced. The energy level diagram is shown in Fig. 19.32.

The normal state of helium is a $^1S_0$ level arising from two valence electrons in $1s$ orbit. The excitation of either one of these electrons to the $2s$ orbit finds the atom in a $^1S_0$ or a $^3S_1$ state, both quite metastable since transitions to the normal state are forbidden by selection rules. (He atom excited to the $1s\,2s$ state cannot return to the ground state by emitting a 20.61 eV photon because both the states have zero total angular momentum and a photon must carry away at least one $\hbar$ of angular momentum). However, the He atom can lose energy by energy exchange collisions with Ne-atoms initially in the ground state. A $1s\,2s$ He atom, with its initial energy 20.61 eV and a little (0.05 eV) additional (kinetic) energy (provided by kinetic energy of atoms) can collide with a Ne atom in the ground state, exciting it to the $5s$ excited state at 20.66 eV and leaving the He atom in the $^1s_2$ ground state. Consequently, the population in the 5s state is substantially enhanced. Thus He atoms help to achieve a population inversion in the Ne atoms. Stimulated emission from 20.66 eV metastable state then results in the emission of highly coherent laser light at 6328 Å, due to transition to a lower excited state at 18.70 eV. This state ($2p^5 3p$) decays by spontaneous emission. The remaining excitation energy is lost in collisions with the tube walls through radiation less transitions.

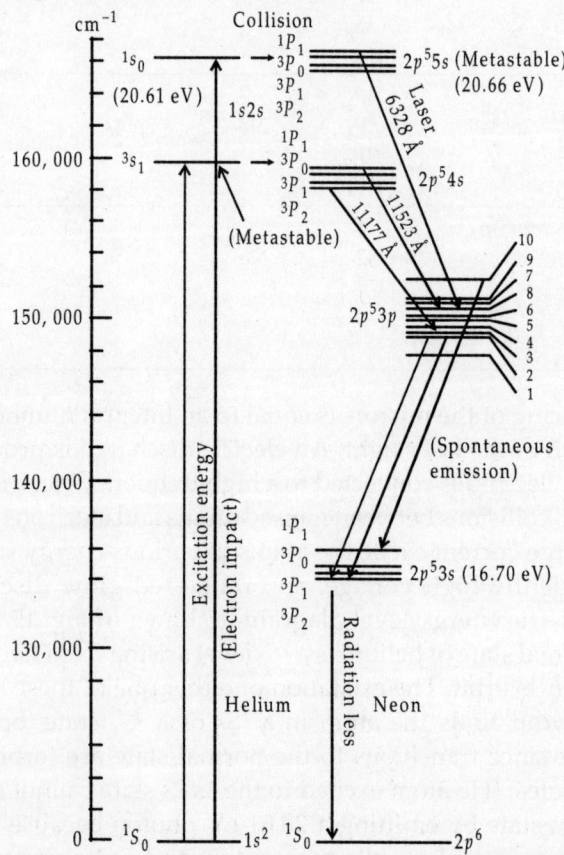

Fig. 19.32: Energy level diagram for He-Ne laser

Since the electron impacts exciting the He and Ne atoms occur all the time, it is a continuous wave (CW) laser. Further since the laser transition does not terminate at the ground state, the power needed for excitation is less than that in a 3 level laser.

Most of the amplified radiation emerging are in the near infrared region of the spectrum, between 10,000 and 35,000 Å, the most intense amplified wavelength in the visible spectrum being the red line at 6328 Å.

19.20 THE Nd-YAG LASER

It is a four level laser and consists of Nd^{3+} ions at low concentration ($\sim$1.38 $\times$ 10^{20}/cm^3) in *Yttrium Aluminium Garnet* (YAG which is specifically $Y_3 Al_5 O_{12}$), hence known as Nd-YAG laser. A cheaper medium instead of YAG is glass, but glass is a poorer thermal conductor than YAG and then the laser must be pulsed. Glass is amorphous, whereas YAG is crystalline, so for YAG line width is much smaller implying much lower thresholds for laser oscillations. For continuous or very high pulse repitition rate operation the Nd-YAG laser will be preferred over Nd-glass whereas for high energy pulsed operation, Nd-glass may be preferred.

The energy level scheme is shown in Fig. 19.33.

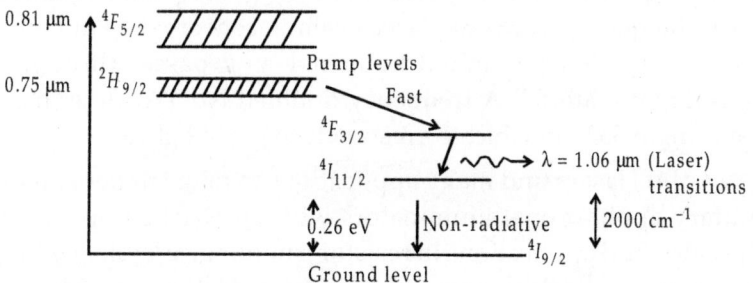

Fig. 19.33: Energy levels of Nd^{3+} in Nd-YAG laser

The energy levels of Nd^{3+} ion take part in laser emission. The levels shown in the figure are all associated with the inner $4f$ shell which is screened from external fields by the $5d$ and $6s$ outer shells as a result of which the lower levels are very narrow. The Nd^{3+} ion surrounds itself with several oxygen atoms that largely shield it from its surroundings.

The main pump bands for excitation of Nd^{3+} ions are in the 0.81 μm and 0.75 μm wavelength regions and pumping is done using Kr arc lamp. Since $F \rightarrow I$ transitions are forbidden in the dipole approximation hence F levels are metastable levels and fluorescence occurs from $^4F_{3/2}$ level to four multiplets. Since the energy difference between the lower laser level ($^4I_{11/2}$) and the ground level is ~ 0.26 eV, the ratio of its population to that of the ground state (at room temperature) is $e^{-\Delta E/k_B T} \approx e^{-9} \ll 1$, hence the lower laser level is almost unpopulated and inversion is easy to achieve. Consequently, any population in the level $^4F_{3/2}$ would give rise to an inversion.

The probability of transition to the level $^4I_{11/2}$, however, is an order of magnitude higher than that to the other members of the fluorescence multiplets. The laser transition $^4F_{3/2} \rightarrow {}^4I_{11/2}$ corresponds to λ = 1.06 μm and has a Lorentzian line shape. The lifetime of the upper level is 0.23×10^{-3} s and the spontaneous lifetime corresponding to the laser transition is 550 μs. The emission line corresponds to homogenous broadening and has a width Δv = 1.2 $\times 10^{11}$ Hz which corresponds to $\Delta\lambda$ ~ 4.5 Å. The transition is at a number of wavelengths in the infrared and transition at 1.064 μm is very efficient and a substantial power output is obtainable. The power is great enough for *frequency doubling*. Frequency doubling is a technique in which the laser beam is converted to radiation with a multiple of its initial frequency as it passes through a transparent material. A frequency doubled Nd-YAG laser has a wavelength 532 nm which corresponds to green light.

Nd-YAG lasers find many applications in range finders (many military applications) illuminators with Q-switched operation. They also find applications in resistor trimming, micromachining operations as well as spot welding, hole drilling, etc. Medical applications include many types of surgery, such as membrane cutting (in cataract surgery) and gallbladder surgery. They are also used in large laser systems for studies of (plasma's) inertial confinement fusion. Frequency multiplication into the green and ultraviolet makes these lasers good pumping sources for pumping tunable dye lasers and other types of laser probes and diagnostics.

19.21 EXCIMER LASER

There are some diatomic molecules which do not exist as molecules in their ground state, i.e. they are unstable in their ground state (so, it is referred to by saying that they can only exist as *monomers* in this state). The potential energy curve of their ground state does not possess a minimum. However, for the excited states, these molecules have a potential energy curve which possesses a minimum. Examples of such molecules are Xe_2^*, Kr_2^*, $(XeCl)^*$, F_2^*, $(KrF)^*$, $(Hg\ Br)^*$. Such molecules are called *excimers*, a contraction of the words *excited dimers*, first suggested by B. Stevens and E. Hutton (*Nature*, **186**, 1045 (1960), to distinguish them from the normal type dimers which exist in the ground state.

The energy level diagram of an excimer molecule (the XeF molecule) is shown in Fig. 19.34.

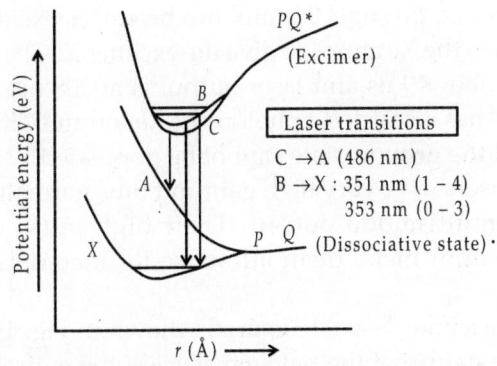

Fig. 19.34: Molecular energy levels in a XeF excimer laser

These molecules have stable excited energy levels (energy curve having a potential well) but the ground state is repulsive (as indicated by *A* level) or very loosely attractive (as indicated by X level). Thus these molecules do not normally exist in nature, i.e. the lower state does not effectively exist. For these combinations of atoms, the molecules can be formed in their excited state (population inversion) only by various special excitation techniques. After the excited states are formed, the population rapidly decays to the ground state in a time typically ranging from 1 to 5 ns and subsequently, the two atoms fly apart. Thus the molecules exist only for a duration corresponding to the lifetime of the excited states. The ground state is always empty, so this facilitates population inversion.

Excimer lasers typically emit in the ultraviolet region and some operate in the visible spectrum (see Table 19.2).

TABLE. 19.2

Excimer Laser	Laser Transition (nm)
XeF	353
XeCl	308
KrF	248
F_2	153

For XeF (Fig. 19.34) $B \rightarrow X$ and $C \rightarrow X$ are ultraviolet transitions whereas $C \rightarrow A$ is visible transition.

There also have been used three atoms to make excimer lasers, e.g. a mixture of Xe, Cl and Ne (which acts as a buffer gas). An electric discharge through the mixture produces excited Cl atoms which attach to the Xe atoms to give the excimer XeCl*. The excimer survives for about 9 ns and laser output is at 308 nm (in UV). As soon as XeCl* has discarded a photon, the Xenon and Chlorine atoms separate and the ground state cannot be populated.

Excimer lasers have very high gain and thus normally produce a high order multi-mode output. Their high pulse energy and ultraviolet output make them attractive for materials processing application.

A typical excimer laser structure is shown in Fig. 19.35. Due to the corrosive nature of the halogen species, the entire structure is made of stainless steel with polyvinyl and teflon components.

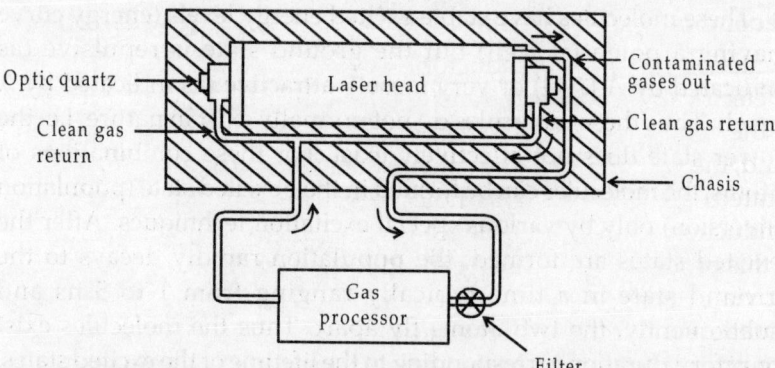

Fig. 19.35: Typical excimer laser (After *Laser Fundamentals*, W. T. Silfvast, Cambridge University Press, (1998))

A transverse discharge through long electrodes is used. The long electrodes are flat metal pieces that have a rounded shape so as to ensure that when voltage is applied, it results in a uniform electric field and uniform excitation. A pre-ionization pulse is also needed to provide the initial electron "seeding" in the region between the electrodes.

The pre-ionization pulse is typically produced by a row of miniature ultraviolet spark discharges. These sparks emit enough UV radiation to produce ionization within the gain region, thereby increasing the electrical conductivity of the gaseous medium.

19.22 SEMICONDUCTOR DIODE LASERS

In contrast to other types of lasers (transitions between discrete atomic energy levels), transitions in semiconductor lasers are associated with the electron states in the valence and conduction bands. Use of semiconductors offers an advantage of the comparative ease to adjust their characteristics by chemical addition of dopants. Laser radiation can be obtained by pumping at the junction of a semiconductor p-n diode. The method of pumping varies with the type of semiconductor used (depending on the band gap). An important difference between a solid state laser and a semiconductor laser is that while in the former, the bulk of the material is the host and the active material is as low as one per cent, in the latter, the entire material is participating in the laser action. Efficiency as high as 70% can be achieved.

Recombination Radiation

Excitation of the semiconductor material by either optical method or electrical method, can transfer electrons from the valence band to the conduction band. When electrons move to the conduction band, the vacant sites in the valence band (where the electrons were located) are referred to as holes. Electrons can decay from the conduction band to the valence band by recombining with holes. In this process, the electrons can either radiate the energy or give it up via collisions with the lattice which induce a form of lattice vibration called phonon relaxation in the material. Which of these two processes takes place depends on whether the semiconductor is a direct or indirect band gap material.

Direct and Indirect Band Gap Semiconductors

The top of valence band and bottom of the conduction band have a wave vector dependence which has a significance in deciding the gain factor. In direct band gap material, the lowest conduction band minimum and the highest valence band maximum correspond to the same crystal momentum or wave vector in the Brillouin Zone but in indirect band gap semiconductor, the above two are at different wave vectors. The optical transition in direct band gap semiconductors is a first order process, since momentum is automatically conserved and the optical gain is high. In the case of indirect semiconductor, the transition involves phonons also and collisional phonon relaxation in the material produces acoustic waves and unwanted heating of the material. In this process the quantum yield of radiation is low, i.e. optical transitions are weaker and the available gain is smaller.

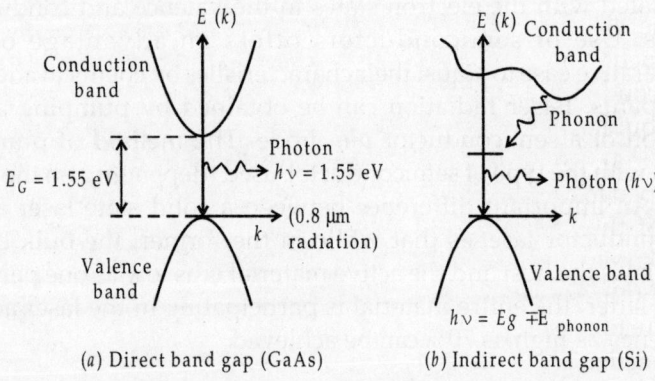

(a) Direct band gap (GaAs)   (b) Indirect band gap (Si)

Fig. 19.36: (a) Direct band gap and (b) Indirect band gap semiconductor materials

Condition for Laser Action in a Semiconductor

In case of metals/semiconductor solids, the probability of occupation $F(E)$ of any energy state E is given by Fermi Dirac statistics

$$(E) = \frac{1}{1 + \exp\left(\dfrac{E - F_0}{kT}\right)} \qquad ..(132)$$

where F_0 is known as Fermi level for the system. F_0 represents the boundary between fully occupied and completely empty levels at $T = 0$ K. When there is a perfect thermal equilibrium between the conduction and valence band, single Fermi level gives the distribution of electrons over the whole range of energies of both bands (Fig. 19.37 (a)).

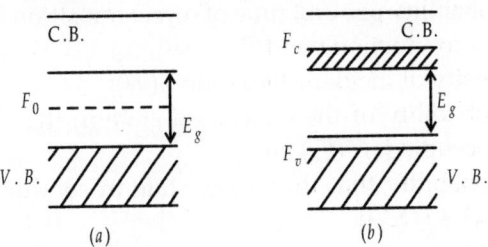

Fig. 19.37: Energy levels of a semiconductor (a) T = 0 K (b) ambient temperature. Some electrons excited to the conduction band

The electrons may be raised from the valence band to the conduction band by (say) optical excitation with energy $E > E_g$. The semiconductor in this condition is not in equilibrium. Nevertheless, the equilibrium is reached within each band in a very short time ($\sim 10^{-13}$ s) in which the electrons in conduction band will have dropped to the lowest level in that band; and any electrons near the top of the valence band will also have dropped to the lowest unoccupied level, leaving the top of valence band full of holes (Fig 19.37 (b)). Thus there is population inversion between bottom of conduction band and top of valence band which may give rise to laser action. (Valence band is empty of electrons down to an energy F_v and the CB is full upto F_c). F_c and F_v are termed quasi-Fermi levels. Although overall thermal equilibrium does not exist, the carriers within a given energy band will be in thermal equilibrium. The occupation probability of a state in the conduction band is given by

$$f_c(E) = \frac{1}{1 + \exp\left(\dfrac{E - F_c}{kT}\right)} \qquad ..(133)$$

Similarly for valence band

$$f_v(E) = \frac{1}{1 + \exp\left(\dfrac{E - F_v}{kT}\right)} \qquad ..(134)$$

At equilibrium $F_c = F_v = F_0$

When a beam of light is incident on such a semiconductor, the number (N_a) of quanta absorbed per unit time will be proportional to

(i) the probability per unit time of direct transition from valence band to conduction band B_{vc}
(ii) the density of incident radiation $\rho(W)$
(iii) the probability of the concerned state in the valence band being occupied: $f_v(E_v)$ and
(iv) the probability that the upper state in conduction band is empty $[1 - f_c(E_c)]$

That is

$$N_a = A B_{vc} f_v(E_v)[1 - f_c(E_c)]\rho(W) \qquad ..(135)$$

Similarly, the number of quanta (N_e) emitted per unit time by stimulated emission is

$$N_e = A B_{cv} f_c(E_c)[1 - f_v(E_v)]\rho(W) \qquad ..(136)$$

The constant of proportionality A in the above expressions includes the density of states of the two bands. For a net amplification, emission must be greater than absorption, i.e. $N_e > N_a$. Assuming $B_{vc} = B_{cv}$, this condition can be written as

$$f_c(E_c)[1 - f_v(E_v)] > f_v(E_v)[1 - f_c(E_c)] \qquad ..(137)$$

Substituting for $f_c(E_c)$ and $f_v(E_v)$ from equations (133) and (134) and simplifying, we get

$$F_c - F_v > E_c - E_v = h\nu$$

This is the necessary condition for stimulated emission to be dominant over absorption in an intrinsic semiconductor. If the semiconductor contains impurities and the impurity energy states are either the initial or final states, it is necessary only to use the quasi-fermi levels for the impurities with the proper degeneracy for the states.

p-n Junction Laser

A simple way of achieving population inversion is to use a semiconductor in the form of a *p-n* junction diode heavily doped with donors (on *n*-side) and acceptors (on *p*-side). Figure 19.38 depicts the energy level diagram.

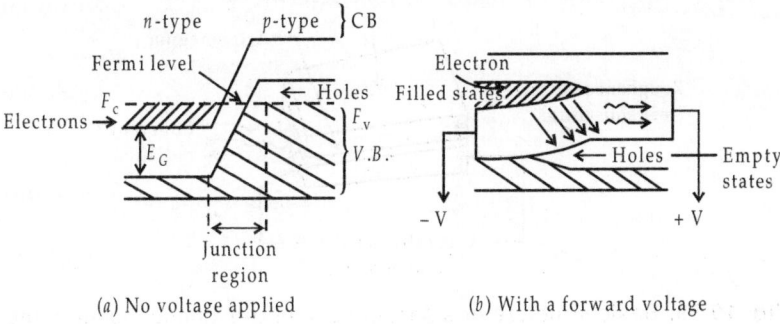

Fig. 19.38: Energy levels of a p-n junction, (a) equilibrium with no voltage applied, (b) with a forward voltage, there results recombination emission

When no voltage is applied, electrons from the n-side flow to the p-side until an electrical potential barrier V_B is built up due to presence of immobile ions (electrons and holes have recombined in the junction region—the depletion layer). This prevents further flow of charges (Fig 19.38 (a)). Due to a forward voltage (Fig 19.38 (b)) we have injected electrons on n-side and injected holes on p-side, both in high concentrations. A high hole concentration means a large number of empty sites into which electrons can fall, i.e. a population inversion in the depletion region. Hence, laser action will take place in this region when the current flow through the diode exceeds a certain threshold value. Similarly population inversion has also occurred on the n-side due to filled levels near the bottom of the conduction band. The recombination of electrons and holes can only take place in the junction region due to lowering of energy gap in presence of forward voltage resulting in an ease for carrier flow and consequent recombination. This results in a generation of coherent (laser) radiation. On the bulk p-side and n-side on both sides, the recombination energy is dissipated through non-radiative processes and so, the laser action is confined to the very thin planar junction region.

Semiconductor Homojunction Laser Structure

Semiconductor lasers were first fabricated in 1962 using GaAs in the form of a diffused p-n homojunction. Figure 19.39 shows the basic structure of a p-n junction laser.

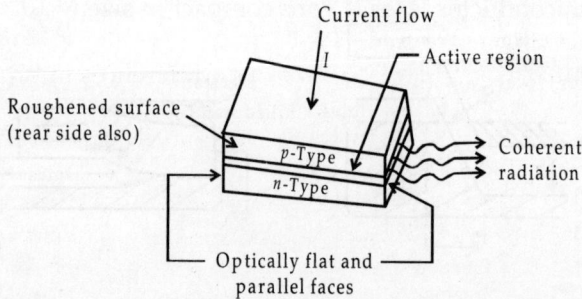

Fig. 19.39: Basic structure of a semiconductor *p-n* junction laser in the form of a Fabry-Perot cavity

The junction is formed by diffusing an acceptor such as Zn into wafers of *n*-type GaAs. The wafers are then cut into small chips with cleaved or polished parallel ends perpendicular to the plane of junction. Because the refractive index of a semiconductor is quite large (~3.6), the reflectivity at the crystal air interface is as high as 32%. The surfaces may be coated to increase the reflectivity. The two remaining sides of the diode are roughened to eliminate lasing in directions other than the main one. The structure is called Fabry-Perot cavity.

For a homostructure GaAs *p-n* junction the threshold current density (J_{th}) increases with temperature. A typical value of J_{th} is ~500 A/(mm)^2 at room temperature. Such large current density imposes serious difficulties in continuous operation of these lasers. Only pulsed operation is possible. The threshold current density can be reduced by using heterojunctions instead of homojunction. A typical single heterojunction laser consisted of a *p-n* junction between GaAs and GaAlAs. At the heterojunction there is carrier confinement as well as optical confinement. The room temperature threshold current density is reduced to ~100 A/(mm)^2.

Advantages of Semiconductor Laser over Other Lasers

Semiconductor lasers differ from other lasers in the following important aspects:

1. In conventional lasers, the quantum transitions occur between discrete energy levels, whereas in semiconductor lasers, the transitions are associated with the band properties of the material.

2. A semiconductor laser is very compact in size ($\simeq 10^{-4}$ cm^3). Because the active region is very narrow (~ or < 1 μm) the divergence of the beam is considerably larger than in a conventional laser.

3. The spatial and spectral characteristics of a semiconductor laser are strongly influenced by the properties of the junction medium (such as band gap and refractive index variation).

4. For the *p-n* junction laser, the laser action is produced by simply passing a forward current through the diode itself. The result is a very efficient overall system that can be modulated easily by modulating the current. Since semiconductor lasers have very short photon lifetimes, modulation at high frequencies can be achieved.

Disadvantages of Semiconductor Lasers

In spite of several advantages, there are unfortunately some drawbacks. The lifetime of these lasers is rather short unless they are operated at low temperatures and in a pulsed mode. The optical transitions in the semiconductors are rather broad of the order of 2 nm yielding a line shape function with a width of the order of about 30 cm^{-1} for a wavelength of 840 nm.

19.23 SEMICONDUCTOR HETEROJUNCTION INJECTION LASER

In the preceding article we have considered the *p-n* junction laser fabricated from a single crystal semiconductor material. This is known as a homojunction. However, we may have improved radiative properties by the use of heterojunctions, i.e. *p-n* junction fabricated from two different semiconductor materials having different energy band gaps. Devices fabricated with heterojunctions are said to have heterostructure. Homojunction lasers can operate only at very low temperatures. In order to make room temperature lasers, it became necessary to develop heterostructures.

Heterojunctions may be either isotype (*p-p* or *n-n*) or anisotype (*p-n*). The isotype heterojunction provides a potential barrier within the structure which is useful for the confinement of minority carriers to a small active region. This is known as carrier confinement. It effectively reduces the carrier diffusion length and thus the volume within the structure where radiative recombination may take place.

This technique is used for the fabrication of injection lasers. Use of anisotype heterojunctions having sufficiently large band gap differences improves the injection efficiency of either electrons or holes. Both types of heterojunction (i.e. isotype/anisotype) provide a dielectric step due to the different refractive indices at either side of the junction. This is used to provide radiation confinement in active region (optical wave guide effect) in injection laser. The confinement efficiency depends on the magnitude of the step which in turn depends on difference in band gap energies and the wavelength of radiation.

Heterostructures are used to provide potential barriers in injection lasers. When a double heterojunction structure (DH), e.g. *p-p-n* is implemented, the resulting carrier and optical confinement reduce the threshold currents necessary for lasing by a factor of about 100. The layer structure and an energy band diagram for a DH injection laser are illustrated in Fig 19.40.

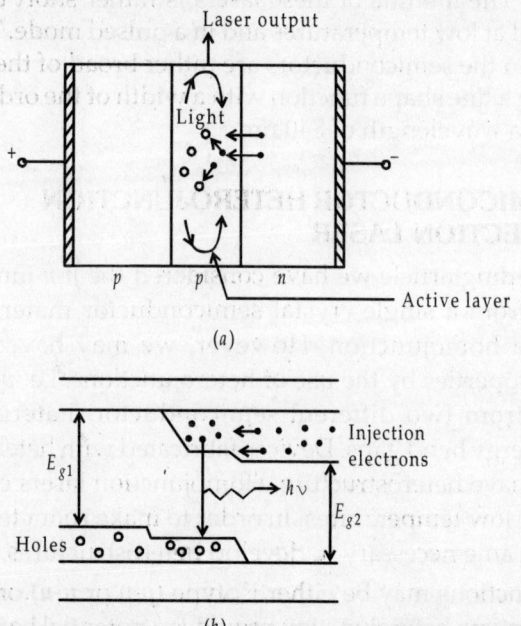

Fig. 19.40: The double heterojunction injection laser (*a*) the layer structure with an applied forward bias (*b*) energy band diagram with a *p-p* heterojunction on the left and a *p-n* heterojunction on the right

There is a heterojunction on either side of the active layer. When a forward voltage corresponding to the band gap energy of the active layer is applied, a large number of electrons are injected from n side to p side (active layer) and holes to the active layer from the left p-end. Consequently, laser oscillations commence. The injected carriers are confined to the active layer by the energy barriers provided by the heterojunctions which are made within the diffusion length of the injected carriers. Further a refractive index step at the heterojunction provides a radiation confinement to the active layer (because, in effect, the active layer behaves as a dielectric waveguide to confine the electroluminiscence within the active region). The Fig. 19.41 illustrates a broad area GaAs/AlGaAs double heterojunction injection laser.

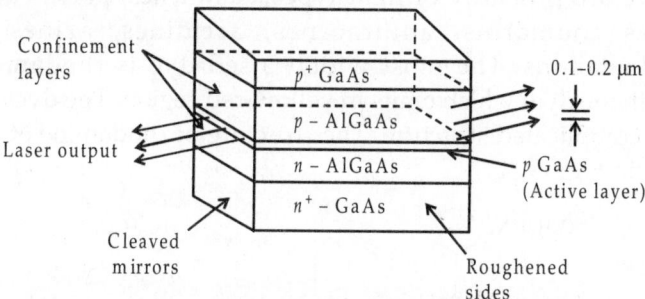

Fig. 19.41: A broad area GaAs/AlGaAs double heterojunction injection laser

Lasing takes place across the whole width of the device and so two sides of the device are roughened to reduce unwanted emission in these directions. Further, to provide optical confinement in the horizontal plane, stripe geometry is used. The stripe is formed by the creation of high resistance areas on either side of the active layer by using the proton bombardment-technique. The forward voltage is applied to the p^+–GaAs and n^+–GaAs which are placed at the two sides of the confinement layers.

Because the charge carriers in heterojunction lasers are confined to a much smaller region than in homojunction lasers, the heat deposition is much lower while the threshold current density (and thus the recombination radiation) are still at a sufficiently high level. Double heterojunction structures provide more control over the size of the active region and provide additional index of refraction variations providing better guiding of optical wave when operated as laser.

19.24 DYE LASERS

Initially the disadvantage of laser sources was their restriction to the emission of a few single and narrow lines. In the course of development of lasers, a few lasers were devised which avoid this problem and allow wavelength tuning at least in a narrow spectral range. These systems are known as dyelasers.

Dye lasers have become important radiation sources in analytical chemistry because they are continuously tunable over a range of 20 to 50 nm. An important and useful type of dye laser is organic dye laser (available in liquid form). In this laser an organic dye dissolved in a suitable liquid such as ethyl alcohol, methyl alcohol, toluene or acetone, etc. is used as the active medium. The dyes that are most effective are primarily of light types: Xanthenes, polymethines, oxaines coumarins, anthracenes, acridines, azines and phthaloziamins. The most widely used dye is rhodamine 6 G(Xanthene dye) which emits in yellow-red region. The dyes have a very complicated structure. The structure of rhodamine 6G is as shown in Fig. 19.42.

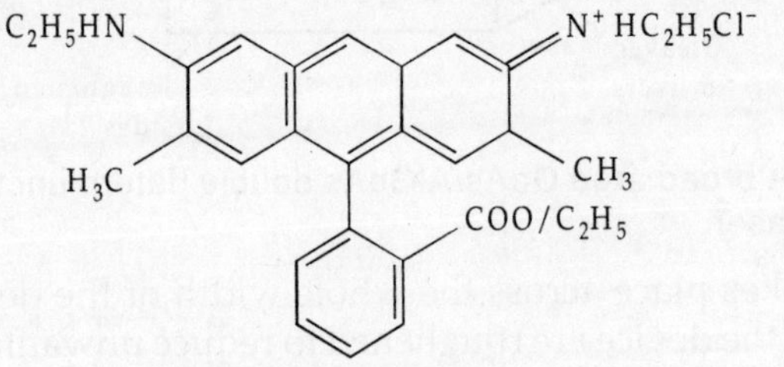

Fig. 19.42: Rhodamine 6G

The active materials in dye lasers are solutions of above organic compounds capable of fluorescing in the ultraviolet, visible or infrared regions.

A peculiar feature of the dye structure is presence of double bonds separated by a single bond; this is known as *conjugated structure*.

There are two possible ways of using an organic dye solution as an active medium. One may use the fluorescence (singlet-singlet) emission or the phosphorescence (triplet-triplet) emission.

The levels taking part in the lasing transition are the different vibrational sublevels of different electronic states of the dye molecule. The electronic states are split into rotational states and a typical ground level S_0 corresponds to an electronic level containing a large number of sublevels constituted by various vibrational and rotational states. The dense collection of the rotational states leads to a near continuum of energy states. That is why we observe the characteristic broad absorption and emission spectrum of a dye molecule. Further only the lowest rotational levels are populated in thermal equilibrium so that a population inversion can be achieved by excitation into higher rotational states. Figure 19.43 shows the luminescence of rhodamine 6 G after excitation from the S_0 ground state to an S_1 excited state. This luminescence is about 10 nm wide.

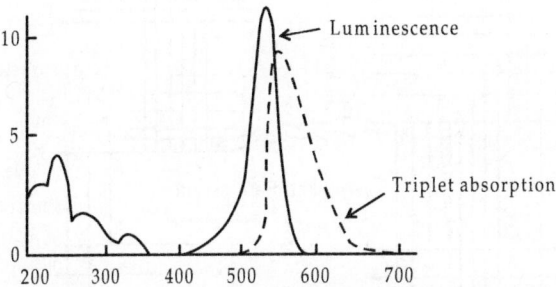

Fig. 19.43: Luminescence and triplet absorption for rhodamine 6G after excitation from the S_0 ground state to an S_1 excited state (A higher vibrational-rotational level)

Because of the thermal redistribution in the level S_1 (in about 10^{-11} s) most of the dye molecules drop to the lowest vibrational level of S_1. Then radiation is emitted when molecules decay from this lowest level to any higher lying sublevel of S_0. (This causes fluorescence). The peak level of the emitted fluorescence spectrum is observed to be higher than that of the absorption spectrum.

Molecules from the state S_1 can also make a non-radiative relaxation to the triplet level T_1. This is referred to as *intersystem crossing*. This state introduces an additional destructive absorption process into the system. It overlaps partly the luminescence and so, decreases the overall gain. This problem is solved by adding a triplet quencher to the dye which transfers the triplet state back to the S_1 state by collisions. Alternatively, the dye can be excited in a rapidly flowing jet stream where the molecules leave the active volume before the triplet state is generated. Figure 19.44 shows the energy level diagram of an organic dye and Fig. 19.45 depicts schematic arrangement of a dye laser.

In Fig. 19.45, the pump light is introduced into the resonator cavity by the pump mirror and excites the dye which is injected by a jet with a speed of about 10 m/s. The cavity is folded and consists of a reflector, a folding mirror and an output coupler. The wavelength for resonance is tuned by a Lyot filter.

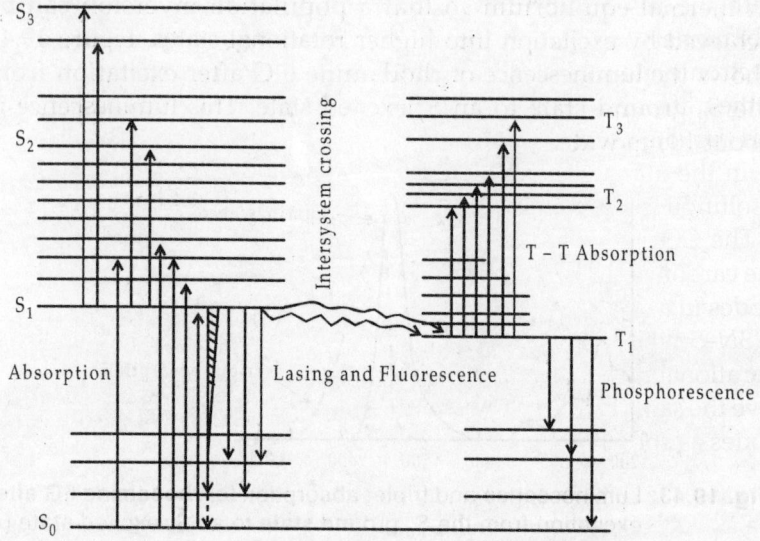

Fig. 19.44: Energy level diagram of an organic dye molecule illustrating singlet-singlet and triplet-triplet absorptions

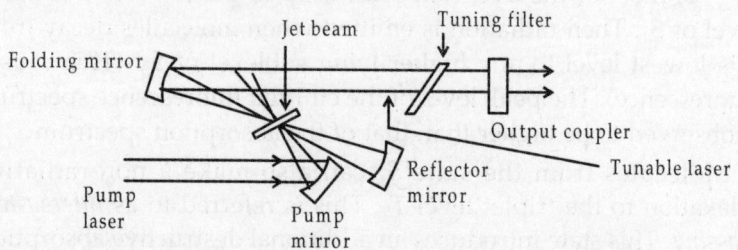

Fig. 19.45: Schematic arrangement of a dye laser

19.25 THE CARBON DIOXIDE LASER

In case of ruby laser or He–Ne laser, we are concerned with electronic energy levels of an atom or an ion. CO_2-laser is a molecular laser (It was first developed in 1964, and was the first molecular laser to be operated).

Like atoms, molecules can also be excited to higher energy levels and the distribution of electrons determines the electronic states of the molecule. Besides the electronic levels, the molecules have rotational as well as vibrational energy levels (due to rotation about the centre of mass, and vibration, of atoms). The energies associated with the rotational levels are generally small compared with vibrational energy levels which in turn are smaller compared to electronic energy levels. In the carbon dioxide laser, one makes use of the transition involving the rotational levels of a vibrational level and the rotational level of a lower vibrational level. The transitions lie in the infra-red region(*) and can be observed only with high resolution instruments.

The CO_2 molecule is a linear, triatomic molecule consisting of one carbon and two oxygen atoms. Now, the number of vibration-modes in a molecule having N atoms is $3N-6$ when it is non-linear or $3N-5$ when it is linear. Thus CO_2 molecule has four possible vibrational modes. Out of the four, two modes are degenerate (i.e. have the same energy). Thus, we are left with only three vibrational modes as shown in Fig. 19.46.

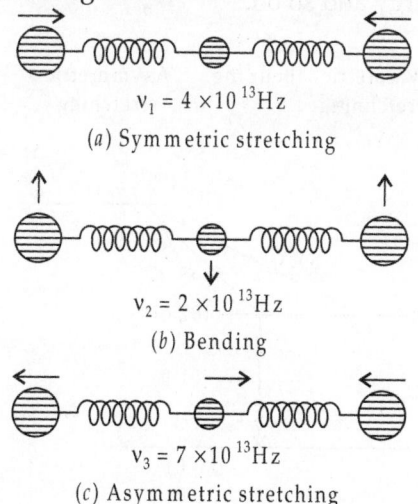

$$v_1 = 4 \times 10^{13} Hz$$

(a) Symmetric stretching

$$v_2 = 2 \times 10^{13} Hz$$

(b) Bending

$$v_3 = 7 \times 10^{13} Hz$$

(c) Asymmetric stretching

Fig.19.46: Vibration modes of CO_2 molecule

(*) The energy difference between various electron levels correspond to visible and ultraviolet regions and energy difference among various vibrational levels involve infra-red region and transitions corresponding to rotational levels lies in microwave or far infra-red region.

These modes are (a) symmetric stretching, (b) bending and (c) asymmetric stretching (the bending mode is doubly degenerate). The frequencies of the three vibrations are as follows.

symmetric stretching, $\nu_1 = 1337$ cm^{-1} $(h\nu_1 = 0.166$ eV$)$

bending stretching, $\nu_2 = 667$ cm^{-1} $(h\nu_2 = 0.083$ eV$)$

asymmetric stretching $\nu_3 = 2349$ cm^{-1} $(h\nu_3 = 0.291$ eV$)$

According to quantum mechanics, the vibration in any mode can take on only discrete energy values which are multiples of some fundamental energy value. The frequency of transition involved is called the fundamental frequency and the multiples are called overtones. At any moment, the CO_2 molecule will be vibrating in a linear combination of the fundamental modes. The energy states of these modes are denoted by vibrational numbers represented in the form (v_1, v_2, v_3). The quantum numbers (v_1, v_2, v_3) represent the amount of energy, or the number of energy quanta associated with each mode. For example, the number (010) represents that the molecule is in the bending mode with one unit of energy and the number (020) represents the molecule in the bending mode with two units of energy and so on.

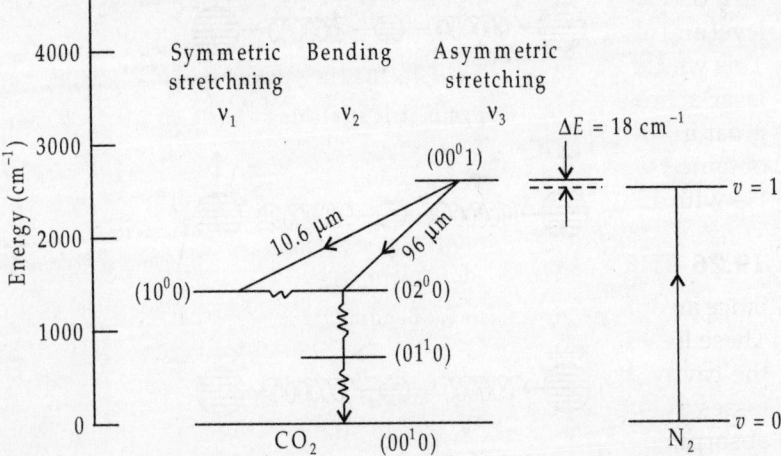

Fig. 19.47: The lowest vibrational levels of the electronic ground state of a CO_2 molecule and a N_2 molecule. For simplicity, rotational levels are not indicated.

The laser uses a mixture of CO_2 and N_2 and Helium. The Fig. 19.47 shows the energy level diagram of CO_2 and N. When a discharge is passed in a tube containing CO_2, electron impacts excite the molecules to higher electronic and vibration rotational levels. The electronic collision cross-section for the excitation to the level (001) is very large. (The level is also populated by radiationless transitions from upper excited levels). The resonant transfer of energy from N_2 added to the (CO_2) gas increases the pumping efficiency. The lifetime of $v'' = 1$ state of N_2 is quite high (as it does not passess a permanent dipole moment) ~ 0.1 sec at 1 torr. Population inversion could be achieved between the level (001), (020) and (100) by the above process. The upper laser level is (001). The lower laser level is provided by (020) and (100) levels. The transition corresponding to 001 – 100 leads to the laser line 10.6 μm and transition corresponding to 001 – 020 gives a laser line at 9.6 μm.

The population inversion to the 001 level is attained through collision of the CO_2 molecule with electrons produced in the plasma by electric discharge and with the excited molecules of nitrogen. The inclusion of He helps in the following way. The levels (100) and (020) and (010) are very close in energy (as can be seen in the figure) and reach thermal equilibrium in a short time. It is necessary that decay from the lowest level (010) to the ground state is very fast, otherwise there would be accumulation of molecules in this level and also in the other two levels because they are in equilibrium. This would spoil population-inversion the condition necessary for laser-action. It has been found that the presence of helium has a great influence on the life-time of these levels. The best result are obtained with a mixture of CO_2 with 1.5 torr, N_2 with 1.5 torr and He with 12 torr. This laser has wide industrial applications.

19.26 THE Q-FACTOR

Since an optical resonator is an open cavity, all modes suffer losses. These losses arise due to finite sizes of the mirrors (at both ends of the cavity, there is a diffraction split-over). In addition, there are losses due to finite reflectivities of the mirrors, and scattering and absorption by the medium filling the resonator cavity. The various losses in the cavity can be related to the passive lifetime t_c which is the time in which energy in the cavity reduces by a factor $1/e$.

The losses in the cavity can also be described by what is known as the quality (or Q) factor. This is defined by

$$Q = \omega_0 \frac{\text{energy stored in the mode}}{\text{energy lost per unit time}} \tag{138}$$

where ω_0 is the oscillation frequency of a particular mode. If $W(t)$ represents the energy in the mode at time t then from eqn (138), we obtain

$$Q = \omega_0 \frac{W(t)}{-dW/dt} \qquad ..(139)$$

or

$$\frac{dW}{dt} = -(\omega_0/Q)W(t) \qquad ..(140)$$

whose solution is

$$W(t) = W(0)e^{-\omega_0 t/Q} \qquad ..(141)$$

Thus if t_c represents the cavity lifetime, then

$$t_c = Q/\omega_0 = \frac{Q}{2\pi\nu_0} \qquad ..(142)$$

We can write for the field associated with the mode

$$E(t) = E_0 e^{i\omega_0 t} e^{-\omega_0 t/2Q} \qquad ..(143)$$

The frequency spectrum of this wave train can be shown to be given by

$$|E(\nu)|^2 = \frac{E_0^2}{4\pi^2} \frac{1}{(\nu-\nu_0)^2 + \nu_0^2/(4Q^2)} \qquad ..(144)$$

which represents a Lorentzian (just as discussed in spontaneous emission). The FWHM of the spectrum is

$$\Delta\nu_{\text{FWHM}} = \frac{\nu_0}{Q} \qquad ..(145)$$

i.e. the line width depends inversely on the quality factor. The higher the quality factor, (i.e. the longer the cavity lifetime) the smaller will be the FWHM. It is a measure of the sharpness of the frequency transmission under consideration.

19.27 Q-SWITCHING

A laser can generate light for as long as the population inversion is maintained. When heat is easily dissipated, the laser may act continuously, for the population of the upper level can be replenished by pumping. In some cases, practical conditions govern whether or not continuous pumping is feasible, while in other cases (especially when overheating is a problem) the laser can be operated only in pulses, perhaps of microsecond or millisecond duration, so that medium has a chance to cool or the lower state discards its population.

It is sometimes desirable to have pulses of radiation rather than a continuous output, with a lot of power concentrated into a brief pulse. One way of achieving pulses is by *Q-switching*. (The modification of the resonance characteristics of the laser cavity). The name is derived from the *Q*-factor which measures the quality of a resonance cavity and in the process the *Q* of the cavity is switched from a small value to a large value.

The aim of *Q*-switching is to achieve a healthy population inversion in the absence of a resonant cavity, then to plunge the population-inverted medium into a cavity, and hence to obtain a sudden pulse of radiation. The switching may be achieved by damaging the resonance characteristics of the cavity in some way while the pumping pulse is active, and then suddenly restoring them. *Q*-switching can give pulses of about ten nanoseconds duration.

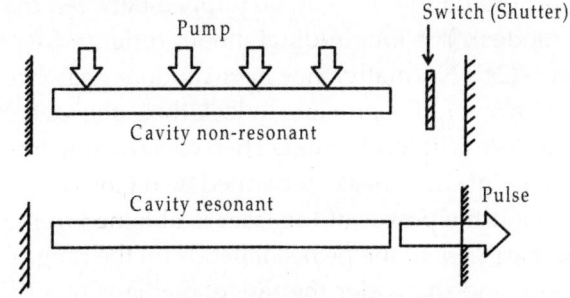

Fig. 19.48: The principle of *Q*-switching

The excited state is populated when the cavity is non-resonant. Then the resonance characteristics are suddenly restored and the stimulated emission emerges as a giant pulse. Figure 19.47 shows schematically the time variation of the cavity loss, cavity *Q*, population inversion and the output power. The figure shows that an intense pulse is generated with the peak intensity appearing when the population inversion in the cavity is equal to the threshold value.

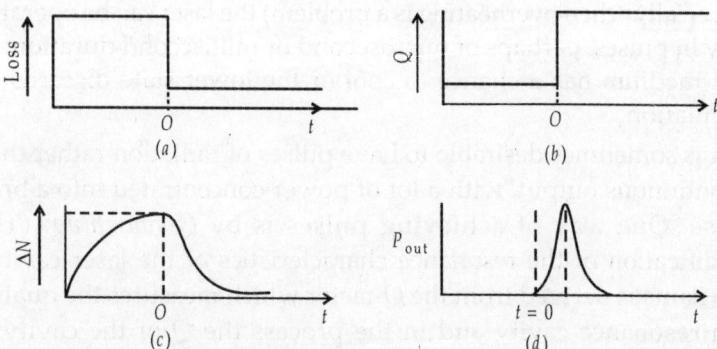

Fig. 19.49: Schematic representation show the various quantities, viz. (a) loss (b) Q, (c) population inversion ΔN and (d) laser output power vary with time when a laser is Q-switched

19.28 MODE LOCKING

The technique of mode locking can produce pulses of picosecond duration. We know that a laser radiates at a number of different frequencies, depending on the precise details of the resonance characteristics of the cavity and in particular, on the number of half-wavelengths of radiation that can be trapped between the mirrors (the cavity modes). The longitudinal modes differ in frequency by multiples of $c/(2L)$. Normally, they have random phases relative to each other. However, it is possible to lock their phases together so that they interfere with each other. Then constructive interference occurs at a series of sharp peaks separated by regions of destructive interference, and the power of the laser is obtained in picosecond bursts. The sharpness of the peaks depends on the range of modes superimposed, and the wider the range, the narrower the pulses. The pulses are separated by an interval, equal to the time it takes for light to make a round trip inside the cavity.

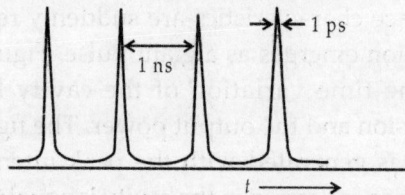

Fig. 19.50: The output of a mode-locked laser

The Effect of Mode Locking

It can be proved mathematically, as follows, that if N modes with frequencies differing by $c/2L$ are superimposed, they give rise to a series of peaks separated by $2L/c$.

Now, each wave has the form

$$\psi_n = E_0 \exp\left[2\pi i\left(\nu + \frac{nc}{2L}\right)\right]t \qquad ..(146)$$

where ν is the lowest frequency and $n = 0, 1, 2, .., N - 1$. The total amplitude of the wave is

$$\psi = \sum_n \psi_n = E_0 \exp\left[2\pi i\nu t\right]\sum_0^{N-1} e^{i\pi nct/L} \qquad ..(147)$$

The summation in this expression is the geometric progression

$$\sum_0^{N-1} e^{i\pi nct/L} = 1 + e^{i\pi ct/L} + e^{2i\pi ct/L} + ... + e^{(N-1)i\pi ct/L}$$

$$= \frac{\sin(N\pi ct/2L)}{\sin(\pi ct/2L)} \times e^{(N-1)i\pi ct/2L} \qquad ..(148)$$

The intensity of the radiation is equal to the square modulus $|\psi^*\psi|$, so

$$I \equiv \psi^*\psi = E_0^2 \frac{\sin^2(N\pi ct/2L)}{\sin^2(\pi ct/2L)} \qquad ...(149)$$

This represents a series of peaks with maxima separated by $t = 2L/c$, the round trip transit time of the light in the cavity.

19.29 NON-LINEAR OPTICAL EFFECTS WITH LASERS

When an electromagnetic wave passes through a dielectric medium, due to electromagnetic field of the radiation, there results polarization of the molecules comprising the medium. For ordinary radiation, polarization $|P|$ is given by

$$|P| = \alpha E \qquad ..(150)$$

where E is the magnitude of the electric field of the radiation and α is the proportionality constant known as the polarizability of the medium. Optical phenomena satisfying this relation are said to be *linear*.

At high intensities of laser radiation, the above relationship breaks down, particularly when $|E|$ overcomes the binding energy of the electrons. Under these situations, non-linear optical effects are observed for which the following relation holds true:

$$P = \alpha E + \beta E^2 + \gamma E^3 + \dots \qquad ..(151)$$

where the magnitudes of the constants are in the order $\alpha > \beta > \gamma$. At moderately high radiation intensities, only first two terms are sufficient to describe the degree of polarization and eqn (151) can be rewritten in terms of angular frequency ω and the amplitude of the field strength E_0 as

$$P = \alpha E_0 \sin \omega t + \beta E_0^2 \sin^2 \omega t \qquad ..(152)$$

Substituting $\sin^2 \omega t = \dfrac{1}{2}(1 - \cos 2\omega t)$, we get

$$P = \alpha E_0 \sin \omega t + \frac{\beta E_0^2}{2}(1 - \cos 2\omega t) \qquad ..(153)$$

The first term in this expression is the normal linear term dominant at low radiation intensities. At high intensities, the second term becomes significant and results in radiation that has a frequency twice that of the incident radiation. This process of frequency doubling is now widely used to produce laser frequencies of shorter wavelengths. For example, the 1064 nm near infrared radiation from a Nd-YAG laser can be frequently doubled to produce a 30% yield of green radiation at 532 nm by passing the radiation through a crystal of potassium dihydrogen phosphate.

Non-linear radiation from laser sources is nowadays beginning to find application in various types of spectroscopy, particularly in Raman spectroscopy.

19.30 APPLICATIONS OF LASERS

With the advent of lasers, the lasers have had a major impact on fundamental and applied areas of research. Consequently, it has found widespread applications in various scientific disciplines such as Physics, Chemistry, Biology, Medicine, etc. and technology and industry. The unique features of the light from lasers are its coherence and monochromaticity, high power levels and highly collimated beams. Due to developments in laser technology, the concepts and techniques of electronics and microwaves are extended to the infrared and optical regions. These developments have led to various novel scientific experiments and technological applications of considerable importance.

1. **Communication by Laser:** The use of coherent light for communication is an important application. The advantage of laser beams over microwaves for communication uses are the higher directionality for comparable antenna size and their tremendous bandwidths. The four techniques commonly used for communication over a long distance are as follows:

   (i) **Coaxial Cable System:** The cable consists of a copper tube about 3/8 inch in diameter with a single copper-wire conductor in the centre. The cables carry radio waves with frequencies from 5×10^5 to 20×10^6 cycles per second. Amplifying equipment is located every two to four miles along the cable.

   (ii) **Microwave-Radio Relay:** The relay towers are located some 20 to 30 miles apart. The system employs microwave radiation in the frequency band between 1–10 billion cycles per second.

   (iii) **Wave-Guide:** A simple hollow tube about two inches in diameter is used as a waveguide to transmit millimetre waves with frequencies between 30 billion to 90 billion cycles per second.

   (iv) **Artificial Earth Satellite:** This broad band communication operates within the microwave-radio band. The main principle involved in these long distance communication systems is the principle of "multiplexing"—the simultaneous transmission of different messages over the same pathways.

   Nowadays attention has been directed to the possibility of using glass fibres as the medium for transmission of light waves employing the phenomenon of total internal reflection. Optical communication systems based on lasers and low-loss optical fibres are now being developed rapidly. Optical fibres carry not only time variable waveforms but also spatial signals. A two dimensional picture can be conveyed to an appropriate place by means of optical fibres. Further, laser communication lines based on fibre-optics have been extremely useful in the modern computer networks.

2. Medical Applications: It was in the early sixties that experiments of exploratory nature were performed in several ophthalmological laboratories to treat certain diseases of the eye. Nowadays lasers are used in ophthalmology to reattach a detached retina. The green beam of the Ar^+ laser is focused on the desired spot on the retina. The beam passes through the lens of the eye and the vitrious chamber without being absorbed, but is strongly absorbed by the red blood cells of the retina and the resulting thermal effects lead to a tiny spot-weld thereby reattachment of the retina. The duration of the laser pulse is ~ 0.01 s and this being very short, the operation is virtually painless.

Laser radiation is often used in controlling haemorrhage. Argon ion and CO_2 lasers are common sources in the treatment of liver and lungs and for elimination of tumours developing in the skin tissues.

Laser technology has made many surgical procedures less painful for the patients. Laser's exquisite precision, (spatial and temporal) and precision of delivery through fibre-optic endoscopes has been the principal attraction for surgeons. New types of surgery with ultraviolet excimer lasers and high powered pulsed-Nd laser energy transmitted through an optical fibre is now used in the treatment of liver cancer.

Recently there has been considerable interest in the use of lasers for medical diagnosis and non-surgical therapeutic applications. Laser induced fluorescence (LIF) from human tissues has recently been used for diagnosis of cancer.

The current and emerging medical applications are given in Table 19.3.

TABLE 19.3: Applications of various lasers in medicines

	Laser Source	Treatment
1.	CO_2 laser	Skin resurfacing, office based surgery
2.	Erbium laser	Dentistry, skin resurfacing, ophthalmology
3.	Ruby laser lesions	Hair removal, tattoo removal, pigmented
4.	Excimer laser	Glaucoma, Pacemaker lead removal
5.	Dye laser	Dermatology
6.	Diode laser	Surgery and micro surgery, hair removal, dermatology, ophthalmology, dentistry, urology, tissue welding
7.	Diode pumped	Photo coagulation, dermatology Nd-YAG laser

3. **Isotope Separation:** Lasers have been successfully used in separating the isotopes present in an isotopic-mixture. The technique is of immense importance for nuclear power engineering. Natural uranium ore (used for nuclear power plants) contains mainly ^{238}U and only about 0.7% of ^{235}U, which is in fact required to fire the plants and hence needs to be separated from the ore. The technique is based on the fact that different isotopes absorb light at different frequencies.

19.31 APPLICATIONS OF LASERS IN INDUSTRY

The precision properties of laser light have been of immense help in industry.

The beam which is coming out of a laser is usually a few millimetres in diameter, hence for most material processing applications, we need focusing elements, e.g. lenses, to increase the intensity of the beam. The beam from a laser has a well defined wavefront, which is either plane or spherical. When this beam passes through a lens, then (according to geometrical optics) it should get focused to a point. In actual practice, however, diffraction effects need to be considered and it can be shown that if λ is the wavelength (of laser light), a, the radius of the beam and f, the focal length of the lens, then, the incoming beam will get focused into a region of radius,

$$b = \lambda f / a$$

i.e. the smaller the λ, the smaller is the size of the focused spot and smaller the a, the larger is b.

Let P represent the power of the laser beam, then the intensity I obtained at the focused region is given by

$$I = \frac{P}{\pi b^2} = \frac{P a^2}{\pi \lambda^2 f^2}$$

For example, if we focus a 1 W laser beam of $\lambda = 1.06$ μm and radius of about 1 cm, by a lens of focal length 2 cm, then

$$I = \frac{1}{3.14 \times (1.06 \times 10^{-4} \times 2)^2} \ \text{W/cm}^2$$

$$\simeq 7 \times 10^6 \ \text{Wcm}^{-2}$$

The electric field strengths produced at the focus of a 3 MW peak power pulsed ruby laser are of the order of 10^7 V/cm. Such large intensities are produced in an extremely small regions of radii ~ 2×10^{-4} cm. Further, the larger the value of a, the greater will be the intensity. Then, we may use a beam expander to increase the diameter of the beam (Fig. 19.49).

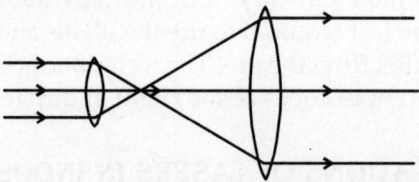

Fig. 19.51: A beam expander consisting of two convex lenses

It is to be noted that while obtaining small focused laser spectra, the beam has a large divergence, hence, near the focused region, the beam expands again within a very short distance which defines the depth of focus. This is also an important parameter which needs to be considered in choosing the parameters in a laser processing application.

Applications

(1) **Laser Welding:** High power lasers have found an important application in welding, e.g. a weld of 0.25 inch thick stainless steel was carried out by a CO_2 laser having an output power of 3.5 kW. The material was moved at a speed of 2 cm/sec in the focal plane of a lens of focal length 25 cm.

Pulsed ruby lasers have also been used in welding.

Laser welding has found important applications in the field of electronics and microelectronics which require precise welding of very thin wires or welding of two thin films together.

(2) **Hole-Drilling:** Drilling of holes in various substances is an interesting application of lasers. For instance, a laser pulse having a pulse width of ~10^{-3} of a second and an energy of ~ 0.05 J can burn through a 1 mm thick steel plate leaving behind a hole of radius 0.1 mm.

The Swiss watch industry in Europe has been using flash-pumped Nd-YAG lasers to drill ruby stones used in timepieces.

(3) **Laser-Cutting:** Lasers also find application in cutting materials. The most common laser used for this purpose is CO_2 laser due to its high output power.

In the cutting process, one essentially removes the materials along the cut. When cuts are obtained using pulsed lasers, then the repetition frequency of the pulse and the motion of the laser across the material is adjusted so that a series of partially overlapping holes are produced. The efficiency of laser cutting can be increased by making use of a gas jet coaxial with the laser. (The gas jet assists in expelling the molten material from the cut).

19.32 USE OF LASERS IN SCIENTIFIC EXPLORATION

Because of highly directional property of laser beams, it is possible to measure distances very accurately with the help of laser beam, for example, measurement of the distance from the earth to the moon.

In 1965, RADAR was used for the first time to measure the earth-moon distance and the accuracy was 0.7 mile. Later in 1967, LASER light was reflected from the moon and the distance was measured to within an accuracy of 600 feet. Then, the major breakthrough in measuring earth-moon distance came in 1969 when the astronauts placed a lunar laser reflector on the surface of the moon which enabled to measure the earth-moon distance with an accuracy of 6 inches.

It has been thought that the gravitational constant may not really be constant but may be decreasing with time due to the expansion of the universe. This probability can be checked with the use of lunar laser reflector.

To obtain energy from a controlled nuclear fusion process, a temperature of $\sim 10^7$ K is required. Because of the ability to obtain extremely high energy densities with lasers, they are being attempted to yield these high temperatures. At present the necessary temperatures are not being obtained because, some of the energy goes to expand the plasma. Magnetic fields that could confine the plasma and reduce the energy wastage in expansion are not yet available. When this would be feasible the lasers may lead to a solution to our energy crisis.

19.33 NEODYMIUM-GLASS LASER

One of the most useful laser systems is that which results when the Nd^{3+} ion is present as an impurity atom in glass [E. Snitzer and C.G. Young, Lasers, Vol 2, A.K. Levine (ed). Marcel Dekker, New York, p. 191 (1968)]. The energy level involved in the laser transition in a typical glass are shown in Fig. 19.52.

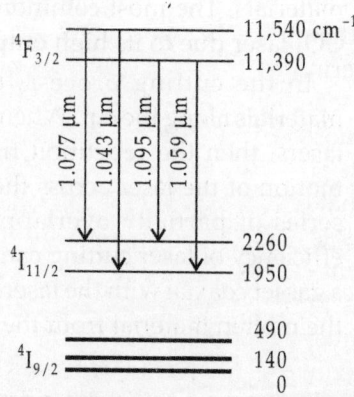

Fig. 19.52: Energy level diagram for the ground state and the states involved in laser emission at 1.059 μm for Nd^{3+} in a rubidium potassium barium silicate glass. The unit 1 cm^{-1} corresponds to $\nu = 30$ GHz or $h\nu = 1.24 \times 10^{-4}$ eV.

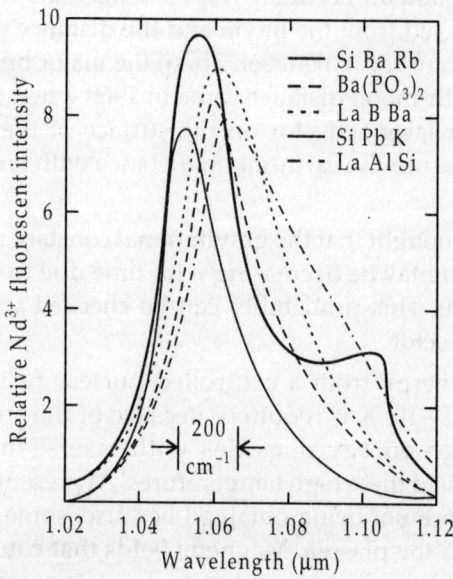

Fig. 19.53: Fluorescent emission of the 1.06 μm line of Nd^{3+} at 300 K in various glass bases. [After: E. Snitzer and C.G. Young, Lasers Vol 2, A.K. Levine (ed.), Marcel Dekker, New York, p. 191 (1968)].

The laser emission wavelength is at $\lambda = 1.059$ μm and the lower level is approximately 1950 cm^{-1} (0.24 eV) above the ground state. As in the case of Nd^{3+} : YAG, we have here a four-level laser, since the thermal population of the lower laser level is negligible. The fluorescent emission near $\lambda_0 = 1.06$ μm is shown in Fig. 19.53. The fluorescent line width ranges (for the glasses shown, around 300 cm^{-1} (9 THz). This width is approximately a factor of 50 larger than that of Nd^{3+} in YAG.

This is due to amorphous structure of glass, which causes different Nd^{3+} ions to "see" slightly different surrounding. This causes their energy splitting to vary slightly. Different ions consequently radiate at slightly different frequencies, causing a broadening of the spontaneous emission spectrum. (This is an example of homogeneous broadening). The lifetime of the upper laser level depends on the host glass and on the Nd^{3+} concentration.

Questions

19.1. What are the different modes of vibration for a CO_2 molecule? With the help of an energy level diagram, describe the working of a CO_2, laser.

19.2. Obtain the threshold condition for laser oscillations. Describe the different methods of pumping in lasers.

19.3. What are Einstein coefficient? Derive a relation among them.

19.4. Explain how population inversion is obtained in a diode laser.

19.5. Explain the working of a resonator cavity in a laser. What different types of cavities are used.

19.6. Explain briefly the formation of axial and transverse modes in a laser cavity.

19.7. Distinguish between three level and four level lasers. Describe the construction and working of a He–Ne laser.

19.8. What is a dye laser? What are its advantage?

19.9. What is meant by Q-factor? Explain Q-switching and mode-locking in lasers.

19.10. Explain the working of a Nd-YAG laser.

19.11. Why do non-linear phenomena occur in optics? Discuss with necessary theory second harmonic generation.

19.12. What is an Excimer laser? Explain the working of a XeF excimer laser.

20

X-ray Spectra

X-rays were discovered in late 1895, by Wilhelm C. Röntgen, almost accidently, while investigating the properties of cathode rays.

X-rays are produced whenever high energy electrons impinge on a solid element and it was found by Röntgen that heavy elements (like platinum) are more efficient than light elements (like Aluminium). Further, the faster the original electrons, the more penetrating the resulting X-rays, and the greater the number of electrons, the greater is the intensity of the beam of X-rays.

It is to be noted here that X-ray production is the result of inverse photoelectric effect. In photoelectric effect, photons of light transfer energy to the emitted (photo) electrons, whereas in the X-ray production, the kinetic energy of impinging electrons is converted into (X-ray) photons.

20.1 PRODUCTION OF X-RAYS

After his initial experiments, Röntgen designed a newer type of discharge tube (than Cathode Ray Tube) for the generation of X-rays which was later improved and modified by subsequent investigators. The pressure within the tube had to be maintained ~ 10^{-3} mm of mercury so that cathode rays arise in a cold discharge in the residual gas inside the tube under the influence of a high potential' difference (30 to 50 kV), some air molecules get ionized and heavier positive ions are attracted toward the cathode whereupon after their impact, electrons are emitted. These electrons are attracted by the anode and hit it with a high velocity, consequently, X-rays are emitted from the anode. This type of tube was called a soft X-ray tube because it was capable of producing only soft (i.e. less penetrating) X-rays. Owing to difficulties in regulating inside pressure and in controlling current in cathode rays, it has become almost absolete. The X-ray tube used nowadays is a development from the hard (or high vacuum) X-ray tube devised by W.D. Coolidge in 1913.

20.2 PROPERTIES OF X-RAYS

   (*i*) X-rays travel in straight lines like ordinary light.
  (*ii*) X-rays are not deflected by electric or magnetic fields. (They are not charged particles).
 (*iii*) X-rays ionize the gas through which they travel (depending on the intensity of the beam).
  (*iv*) X-ray can be reflected and refracted, they can also be polarized.
   (*v*) X-rays can be diffracted with the help of crystalline substances.
  (*vi*) X-rays can produce fluorescence in different substances (e.g. barium platino cyanide).
 (*vii*) X-rays can blacken photographic plates.
(*viii*) X-rays can penetrate through most of the substances. (They penetrate more in low density substances and less in high density substance).

20.3 THE CONTINUOUS AND CHARACTERISTIC X-RADIATION

It is known that X-rays are generated when electrons travelling at sufficiently high speed collide with the atoms of the target material (called anticathode) in an X-ray tube. Now, depending on the target potential and the nature of the target-element, two types of interaction occur between the incident electron beam and the target which result in two types of X-ray spectra—

   (*i*) continuous spectrum and
  (*ii*) characteristic spectrum

When an electron is abruptly slowed down at the target, it suffers a loss of energy as it enters the electric field of the nucleus of a target atom. This results in emission of a quantum of X-radiation of wavelength λ given by

$$\Delta E = \frac{hc}{\lambda}$$

where ΔE is the energy-loss of electron. Since, an electron accelerated by V volts has an energy eV electron volts, if the electron is instantaneously stopped, the quantum of energy radiated is equal to the electron's energy, i.e.

$$eV = \frac{hc}{\lambda}$$

Therefore, wavelength of emission,

$$\lambda = \frac{hc}{eV} = \frac{6.6 \times 10^{-34}\text{J} - \text{s} \times 2.99 \times 10^{8}\text{ms}^{-1}}{(1.6 \times 10^{-19}\text{C})V}$$

$$= \frac{1.24 \times 10^{-6}}{V}\text{ m}$$

or $$\lambda_{min} = \frac{12400}{V}\text{ Å} \qquad ...(1) \qquad (\because 1\text{Å} = 10^{-10}\text{ m})$$

Normally, an electron's energy is lost in a stepwise fashion, through a series of encounters, giving rise to a number of quanta whose wavelengths are longer than if the energy were instantaneously lost. As a result, from a stream of electrons, we get a continuous spectrum which exhibits a continuous range of wavelengths – there is a short wavelength limit which represents the maximum electron-energy transformed into X-radiation in a single encounter, given by expression (1).

Figure 20.1 shows the intensity of the X-rays plotted against the wavelength for the radiation from an X-ray tube with tungsten target for various constant values of potential difference V between the target and the cathode. The intensity vs. wavelength plot begins abruptly at a short wavelength limit λ_{min} (for maximum quantum energy), rises to a maximum (average quantum energy) and gradually decreases with increasing λ (minimum quantum energy). It has been found that

(i) The cut off wavelength λ_{min} is independent of the nature of the target, but is inversely proportional to V.

(ii) For a given target material, the total intensity of the continuous X-radiation (given by area under the curve) is proportional to V^2.

(*iii*) With different target-materials, the total intensity of the continuous X-radiation at a given value of V is proportional to the atomic number of element forming the target. Therefore, the target for X-ray emission is always manufactured from a heavy element.

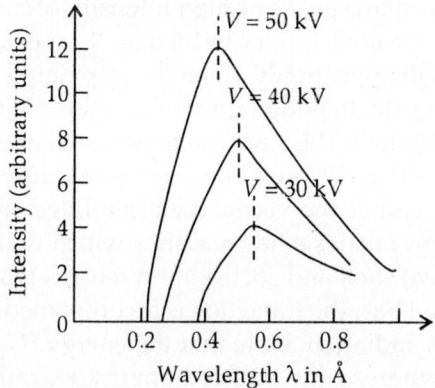

Fig. 20.1: Continuous spectrum of X-rays from a tungsten target

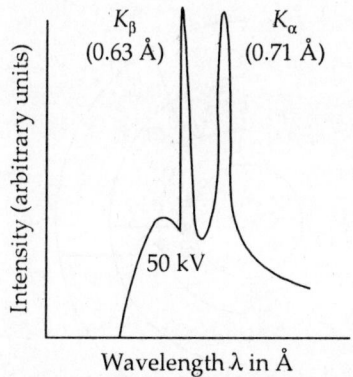

Fig. 20.2: Spectrum of X-rays from a Mo target showing the characteristic *K*-lines

Characteristic spectra are produced when bombarding electrons have sufficient energy to penetrate to interior of the target atoms thereby causing electronic transitions in the target atoms. Though, the continuous spectrum is produced only at fairly low voltages, when the voltage is raised to a critical level (depending on the element of the target), spectra in the form of sharp peaks (characteristic of the target-element) appear, superimposed on the smooth curve of continuous spectrum. The peaks for a given target element always occur at the same λ and usually appear in groups known as *series* which are designated as K, L, M and N (in the order of increasing wavelength).

Figure 20.2 shows X-ray radiation curve for a Molybdenum target. In this case, two sharp peaks of high intensity occur at 0.71 Å and 0.63 Å. These are called K-lines termed as K_α and K_β respectively. These K-lines will appear only when the impinging electrons have sufficient energy to dislodge and release an electron from the innermost (i.e. K) shell. The atom can remain in the resulting excited state for not more than 10^{-8} s and for expelled electron to be replaced, the probability is that the vacancy will be filled from next outer (e.g. L) shell. This creates a new vacancy which will be filled from next outer (i.e. M) shell and so, the atom returns to normal state in a series of steps. Thus, the transition is accompanied by emission of characteristic, X-radiation. Note that the energy required is the K-shell energy, whereas the emitted energy (X-radiation) is that appropriate to L to K (or M to K) transition, the wavelength being fixed by the energy difference between the shells.

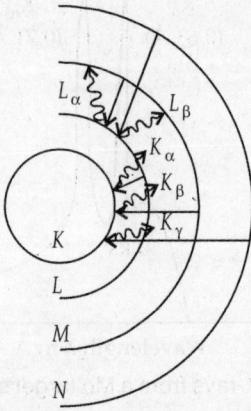

Fig. 20.3: Kossel's diagram to illustrate the production of K and L characteristic peaks

20.4 X-RAY ENERGY LEVELS AND CHARACTERISTIC X-RAY SPECTRA

We know that the X-ray spectra arise from the activity of the inner electrons, and the condition for the excitation is almost the complete removal of an electron from one of the inner (closed) shells.

If $K(L)$ electrons are replaced while falling to the $K(L)$ shell from various other outer shells, $K(L)$ series of characteristics peaks are generated (Fig. 20.3). If the kinetic energy of the incident electron is eV and the binding energy of the electron in the K-shell of the atom is E_K, then the electron emitted from K-shell will have kinetic energy

$$\frac{1}{2} mV^2 = eV - E_K \qquad \qquad ...(2)$$

Obviously, the characteristic X-radiation appears only when the (incident) electron energy is somewhat greater than the energy of the characteristic K-radiation.

Now, if in a X-ray tube, an electron emitted from the cathode strikes the target with a tremendous velocity, it penetrates well inside an atom of the target. If it ejects an electron from the K-shell of the atom, immediately an electron from the outer L-shell jumps to the K-shell, emitting an X-ray photon of energy equal to the energy difference between the two shells. Similarly, if an electron from the M shell jumps to the K-shell, X-ray photon of higher energy is emitted. The X-ray photons emitted due to the jump of electron from the $L, M, N, ...$shells to the K-shell give $K_\alpha, K_\beta, K_\gamma ...$lines of the K-series of the spectrum.

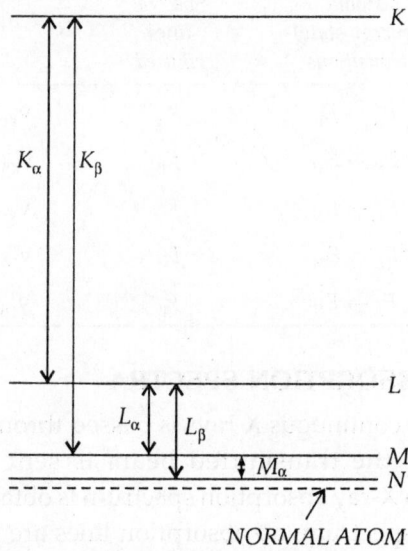

Fig. 20.4: X-ray energy levels and some transitions responsible for production of characteristic X-rays.

If the electron striking the target ejects an electron from the *L*-shell, an electron from the *M, N* ..shell jumps to the *L*-shell. In this process, X-ray photons of lesser energy are emitted. These form the *L*-series of the spectrum. Obviously, the frequencies of the *L*-series are smaller than those of the *K*-series. In a similar way, the formation of *M*-series, *N*-series etc. may be explained. The emission of various X-ray series can be represented on an energy level diagram as in Fig. 20.4.

When all the electrons are present in the atom in their respective shell, the energy of the atom is taken to be zero. The K-level is shown highest because it requires largest amount of energy to remove a K-electron. Since $E_K > E_L > E_M$ ... The levels *L, M,* ...etc. come successively below the former. The spectral lines accompanied by energy transitions of the atom from K-state to the *L, M, N,* ..states are designated as K_α, K_β etc. (Table 20.1). A selection rule limits such transitions which may be stated as

$$\Delta l = \pm 1, \quad \Delta j = 0 \quad \text{or} \quad \pm 1 \qquad \qquad ...(3)$$

TABLE 20.1: Spectral lines accompanied by energy transitions from *K* state to *L, M, N* states

Electronic transition	Atomic (energy state) transitions	Spectral lines emitted	Frequency of lines
$L \to K$	$E_K \to E_L$	K_α	$v_{K\alpha} = (E_K - E_L)/h$
$M \to K$	$E_K \to E_M$	K_β	$v_{K\beta} = (E_K - E_M)/h$
$N \to K$	$E_K \to E_N$	K_γ	$v_{K\gamma} = (E_K - E_N)/h$
$M \to L$	$E_L \to E_M$	L_α	$v_{L\alpha} = (E_L - E_M)/h$
$N \to L$	$E_L \to E_N$	L_β	$v_{L\beta} = (E_L - E_N)/h$

20.5 X-RAY ABSORPTION SPECTRA

When a beam of continuous X-rays is passed through the film of a substance, and the transmitted beam is sent into an X-ray spectrograph, an X-ray absorption spectrum is obtained and unlike X-ray emission spectrum, no absorption lines are seen. Instead, a continuous region of absorption bounded by a sharp edge in the position of the limit of the emission K-series is observed. (Fig. 20.5).

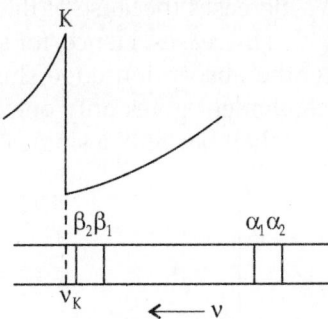

Fig. 20.5: X-ray absorption spectrum in K-region

When successive elements in the periodic table are used as absorbers, the edge shifts regularly toward the higher frequency side. These edges are characteristic of the absorber element and are called "absorption edges". In each case the frequency of the absorption edge v_K is called the K-absorption limit. Frequencies higher than v_K are strongly absorbed and lower frequencies are only slightly absorbed.

The absence of discrete lines in X-ray absorption spectrum is obvious. We know that in emission, the K_α line, for example occurs when an electron drops from the L-shell to the K-shell. In absorption, this line would occur only if an electron from the K-shell, after absorbing an X-ray photon of the required energy from the continuous beam, could go to the L-shell. But this does not happen because the L-shell has already its full quota of 8 electrons and so it cannot receive any further electron. Hence the K_α-line cannot be observed as absorption line.

The continuous absorption bounded by a sharp edge is also easy to explain. An X-ray photon cannot be absorbed by an atom when it has enough energy to remove an inner electron to infinity where the shells no longer exist. If W_K is the energy required to remove a K-electron, then a photon of frequency v can eject a K-electron provided $v \geq v_K$ where

$$h v_K = W_K \qquad \qquad ...(4)$$

Photons of frequency lower than v_K would be transmitted by the absorbing element. At the frequency v_K, absorption in the K-shell suddenly starts and thereafter continues for all frequencies higher than v_K. Thus, an absorption continuum is observed.

As Z increases, W_K increases (because of the increasing nuclear attraction) and so ν_K also increases. Hence, for successive elements in the periodic table, the absorption edge shifts towards higher frequency. Since each element gives only one K-absorption edge, we conclude that the K-shell has only a single energy level.

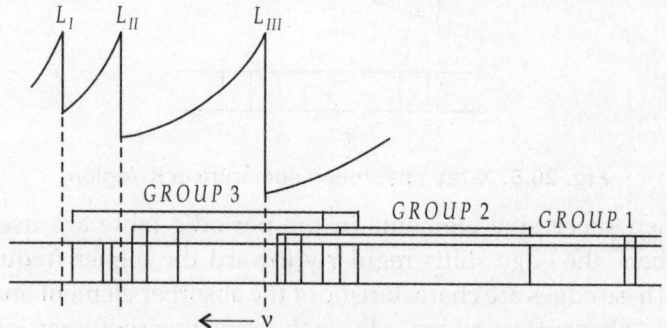

Fig. 20.6: X-ray absorption spectrum in L-region

When the X-ray absorption spectrum is taken in the L-region, again a continuum is observed but now with three absorption edges L_I, L_{II} and L_{III}. (Fig. 20.6). We know that the L emission lines form three groups according to their excitation voltages, indicating that the L-shell has three energy levels. The edges L_I, L_{II}, L_{III} directly give these energy levels.

Similarly, the M-absorption spectra of heavier elements show five absorption edges which give the five energy levels of M-shell, N-spectra give seven edges and so on. Thus, the absorption edges provide a direct means of determining the X-ray energy levels of elements.

20.6 MOSELEYS LAW

After the advent of Bohr's theory of atomic structure in 1913, English physicist H.J.G. Moseley began experiments on atomic transitions. At that time, the Z and A values for elements in the periodic table were not known; today we understand these are respectively, the number of protons and the number of neutrons plus protons in the nucleus. The A-value could be measured by experiments on atomic masses, but then, no one knew how to determine the Z-values other than by a very uncertain inference based on the chemistry of the element. The Bohr's model could not settle this issue.

Moseley found a way to measure the Z-values systematically, using what are known as secondary X-rays. When a beam of X-rays passes through matter, atomic electrons can absorb the photons in the beam and be ejected from the atom. Electrons from higher shells, then, can make transitions to the newly created vacancies. The frequencies of the radiation emitted (the secondary X-radiation), when these transitions take place covers a range that corresponds to the difference in energy between the levels in question; Moseley measured the K_α-series X-rays, which, in retrospect, correspond to the ejected atomic electrons having come from the $n = 1$ (*i.e.* 1s) shell and the electrons replacing them having come from the $n = 2$ shell. (For these transitions, the radiation is in the X-ray range).

Moseley found a simple empirical result for the K-series X-ray frequencies. Let the Z-value that marked the place of an element in the periodic series be Z_{inf} where the subscript "inf" abbreviates "inferred". Then Moseley found empirically that the frequency of the K_α series radiation was proportional to $(Z_{inf} - 1)^2$. Since Moseley's measurement had nothing to do with chemistry, which is the origin of the periodic table, he argued (on the basis of the Bohr model) that one could explain at least part of this result if Z_{inf} was actually the charge on the nucleus-Z itself. In other words, he confirmed that the assignment of Z by chemical properties of the periodic table was correct*. Further, there were gaps in his plot at Z = 43, 61, 72 and 75. These elements were all later discovered. At that time, it was not understood why Z – 1 appeared rather than Z.

How do the idea of the Bohr model "explain" Moseley's data?

In a hydrogen-like atom of charge Z, the difference in energy between the $n = 2$ level and the $n = 1$ level is given by

$$(13.6 \ eV)\left[\left(-\frac{Z^2}{2^2}\right)-\left(-\frac{Z^2}{1^2}\right)\right] = \frac{3}{4}(13.6 \ eV) \ Z^2 \qquad ...(5)$$

This is the K_α photon's energy, the frequency of this is, in this picture.

$$f_{K\alpha} = \frac{3}{4}(13.6 \ eV)\frac{Z^2}{h} \qquad ...(6)$$

---

* For instance, it was inferred by Moseley that Z = 27 for cobalt and Z = 28 for nickel, but their respective atomic masses are 58.93 and 58.71. Thus Moseley's work established the correct sequence of elements in the periodic table.

$$= \frac{3 \times (13.6 \; eV) \times Z^2}{4 \times 4.14 \times 10^{-15} eV.s} = (2.46 \times 10^{15} \text{ Hz}) Z^2$$

Here, in the present case, according to a naive argument, the second electron in the 1s level screens the nuclear charge by one, so that Z in the expression for the photon frequency is changed to $Z - 1$ and this is Moseley's empirical effect.

It was also observed by Moseley that the K-radiation consists of two distinct lines K_α and K_β. Later investigators showed that for the heavier elements, each of the K_α and K_β lines is in itself a close doublet (Fig. 20.7). These four lines are designated as K_{α_1}, K_{α_2}, K_{β_1} and K_{β_2}. Moseley observed that the K-lines of elements shift toward higher frequency in a regular way with increasing atomic number.

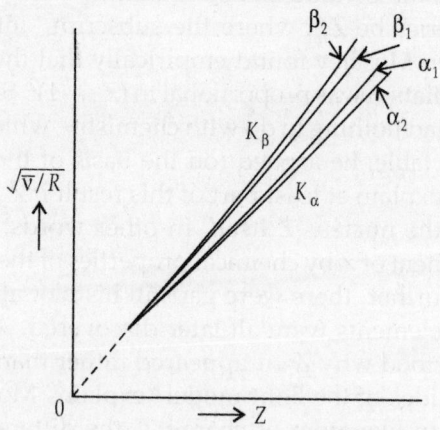

Fig. 20.7: Illustrating Moseley's law.

This is expressed in Mosley's law, which states that

"the frequency of each corresponding K-line is approximately proportional to the square of the atomic number of the emitting element."

Figure 20.7 shows a plot of $\sqrt{\bar{\nu}/R}$ (where $\bar{\nu}$ is wave number and R is Rydberg's constant) against the atomic number Z. These are very nearly straight lines for K_{α_1} and K_{α_2} and K_{β_1} and K_{β_2} and approach each other for lighter elements. For each of these curves, we may write

$$\sqrt{\frac{\bar{\nu}}{R}} = A \, (Z - \sigma)$$

or $$\bar{\nu} = RA^2 \, (Z - \sigma)^2$$

where A and σ are constants for a given transition of the K-series. For the K_α line $\sigma = 1$ and $A^2 = 3/4$. For this, Moseley wrote:

$$\bar{\nu} = R(Z - 1)^2 \left(\frac{1}{1^2} - \frac{1}{2^2} \right)$$

(If however graph were plotted using atomic weight instead of the atomic number, there would have been an appreciable departure from the straight line, which led Moseley to the conclusion that it is the atomic number and not the atomic weight, which is more fundamental to the atom).

Like K-radiations, the L-radiation is also composed of a number of lines and this number increases in going to heavier elements Moseley's law also holds for L-series lines. For example, Moseley found that for L_{β_1} line, $\sigma = 7.4$ and $A^2 = \frac{5}{36}$ and thus

$$\bar{\nu} = R (Z - 7.4)^2 \left(\frac{1}{2^2} - \frac{1}{3^2} \right)$$

Other series may also be similarly represented using different values of σ and A^2.

***Example* 20.1:** Estimate the value of the wavelength of K_α-line of silver ($Z = 47$). Given $R = 1.9737$ cm^{-1}.

Solution: From Moseley's law, for K_α-line,

$$\frac{1}{\lambda} = R(Z - 1)^2 \left(\frac{1}{1^2} - \frac{1}{2^2} \right)$$

$$= \frac{3}{4} R(Z - 1)^2$$

$$= \frac{3}{4} \times 109737 \text{ cm}^{-1} \times (46)^2$$

$$= 1.74 \times 10^8 \text{ cm}^{-1}$$

$$\therefore \qquad \lambda = \frac{1}{1.74} \times 10^{-8} \text{ cm} = 0.57 \text{ Å}$$

***Example* 20.2:** If the K_2 radiation of Mo ($Z = 42$) has a wavelength of 0.71 Å, calculate the wavelength of the corresponding radiation of Cu ($Z = 29$).

Solution: From Moseley's law, for K_α line, we have

$$\frac{1}{\lambda} \propto (Z - 1)^2$$

$$\therefore \quad \frac{\lambda_{Cu}}{\lambda_{Mo}} = \frac{(Z_{Mo} - 1)^2}{(Z_{Cu} - 1)^2} = \frac{(41)^2}{(28)^2}$$

$$\therefore \quad \lambda_{Cu} = (0.71 \text{ Å})^2 \times \frac{(41)^2}{(28)^2} = 1.52 \text{ Å}$$

20.7 COMPARISON BETWEEN X-RAY SPECTRA AND OPTICAL SPECTRA

We now summarise the striking differences between characteristic X-ray spectra and optical spectra.

(i) The frequencies in X-ray spectra are about a thousand times higher than optical frequencies. This indicates that very firmly bound electrons of the inner complete shells are responsible for the emission of X-ray spectra. The optical spectra on the other hand, arise from transitions of the outermost electrons of the atom.

(ii) The X-ray spectra arise as a result of the complete removal of an electron from an inner shell of the atom. For the emission of optical spectra, an outer electron is to be simply raised to a higher level.

(iii) The most striking feature of X-ray line spectra is that the frequencies of the lines increase steadily not periodically, from element to element. The reason is that the characteristics of X-ray spectra depend on the binding energies of the electrons in the innermost shells. With increasing Z, these energies simply increase uniformly owing to the increasing nuclear charge, and are not affected by the periodic changes in the number of outer electrons of the atom. This is why the X-ray spectra of elements of nearby atomic numbers are qualitatively very similar. The optical spectra, on the other hand, show abrupt, periodic changes from one element to the next. This is because the electronic structure at the surface of the atom changes from element to element periodically.

(iv) A further contrast to optical spectra is found in the relation of X-ray emission spectra to absorption spectra. The two are not identical. The X-ray emission spectra consist of continuous regions bounded by sharp edges. The simple explanation is that X-ray radiation of energy just sufficient to remove an electron from an inner shell will be absorbed, as will the radiation of higher energies, but that of lower energies will not be absorbed. Line absorption does not occur because the upper energy levels to which the electron would have to be raised are already occupied.

In case of optical spectra, emission and absorption spectra are identical. We get absorption lines at the same frequencies as the corresponding emission lines.

Questions

20.1. Explain the characteristic X-ray spectra in emission and absorption. Discuss how it differs from optical spectra.

20.2. Give the origin of X-ray spectra and explain why the absorption spectra are continuous with edges.

20.3. What is the importance of absorption-edges in X-ray spectra?

20.4. Explain why are emission and absorption X-ray spectra different. Give importance of absorption edges.

20.5. Account for:

(i) The X-ray spectra of elements of nearby atomic numbers are qualitatively very similar, while the optical spectra of these elements may differ considerably.

(ii) In optical atomic absorption, generally discrete lines are observed whereas X-ray absorptions show continua with edges.

20.6. If the K-absorption limit of uranium is 0.107 Å, find the minimum potential difference required across an X-ray tube to excite the K-series.

$$\left[\mathbf{Hint}: eV_K = h\nu_K = \frac{hC}{\lambda_K}\right]$$ [**Ans.** 116 kV]

Selection Rules
in Spectroscopy

21.1 INTRODUCTION

In preceding chapters we have given selection rules at some places. Here we will discuss a summary of selection rules for various spectroscopies. By now, it may be stated that a system will absorb or emit light if the energy of the light equals the energy difference between the two states involved in a spectroscopic transition. But there is no guarantee that the transition will occur if this condition is met. That is, the Bohr frequency condition is a necessary but not a sufficient condition for transition to occur. Criteria in addition to this is whether an absorption or emission satisfy the *selection rules*. Selection rules are of two types: quantum mechanical (or descriptive) which can be explained using wavefunctions, operators and quantum numbers, and others, which can be explained by describing what the atom or molecule is doing.

If a transition is favoured by a selection rule, we say that the transition is *allowed*. If a transition does not follow a selection rule, we say that the transition is *forbidden*.

In this chapter, we will first consider the more formal mathematical perspective (for quantum mechanical origin of selection rule) and then, we will give a review of a simpler approach based on change in quantum numbers.

21.2 DIPOLE MOMENT BASED SELECTION RULES

Spectroscopists use selection rules to keep track of what transitions are allowed and what are forbidden. Although most selection rule are derived by assuming quantum-mechanically ideal systems, in reality most selection rules are strictly not followed. But selection rules are still helpful in understanding the spectra.

Some selection rules are given in terms of allowed changes in quantum numbers. These are fairly specific: $\Delta v = \pm 1$ for vibrations or $\Delta J = \pm 1$ for rotations. But some selection rules are a bit more general. Two of them, in particular, are based on the dipole moment of a molecule:

1. In order to have a pure rotational spectrum, a molecule must have a permanent dipole moment. (e.g. HCl does have but H_2 and Cl_2 do not).

2. In order for a vibration to absorb light and appear in a spectrum, there must be a change in the dipole moment of the molecule associated with that vibration. (e.g. All three vibrations of H_2O appear in water's vibrational spectrum, but not all vibrations of CO_2 do).

These general selection rules have their origin in the following reason:

Quantum-mechanically, a spectroscopic transition is electric-dipole allowed if the following integral is non-zero

$$M = \int \psi_1^* \times \hat{\mu}\, \psi_2 \, d^3r \qquad \qquad ..(1)$$

where ψ_1 and ψ_2 are the upper and lower wavefunctions of the (electronic, vibrational or rotational) state, and $\hat{\mu}$ is the dipole moment operator given by

$$\hat{\mu} = e\hat{r} \qquad \qquad ..(2)$$

e is the charge separation and $\hat{r}$, the position operator. The quantity M is called the *transition-moment*. M may be exactly zero, in which case the transition between ψ_1 and ψ_2 is forbidden. If it is not zero, the transition is allowed.

The integral in eqn. (1) may be zero or non-zero depending on the properties of the product $\psi_1 \cdot r \cdot \psi_2$. The evaluation of this integral and the quantum-mechanical conditions under which it is zero or non-zero, ultimately lead to specific selection rules in terms of changes in quantum numbers. However, presently, (in the next section), we are looking for selection rules in terms of the dipole moment, and not the quantum numbers.

21.3 ROTATIONAL TRANSITIONS

Consider a molecule rotating in three-dimensional space (x, y, z). The dipole moment vector can be written as

$$\vec{\mu} = \mu_x \, \vec{r} + \mu_y \, \vec{y} + \mu_z \, \vec{k} \qquad ..(3)$$

where $\vec{r}, \vec{j}$ and $\vec{k}$ are unit vectors along the x, y, and z axes respectively and μ_x, μ_y and μ_z are components of $\vec{\mu}$ along the three axes. In spherical polar coordinates, if μ_0 is used to denote the scalar magnitude of the dipole moment.

$$\vec{\mu} = \mu_0 \, [\sin \theta \cos \phi \, \vec{r} + \sin \theta \sin \phi \, \vec{j} + \cos \theta \, \vec{k}] \qquad ... (4)$$

and the expression for the transition moment becomes

$$M = \int \psi_1^* \, \mu_0 \left(\sin \theta \cos \phi \, \vec{i} + \sin \theta \sin \phi \, \vec{j} + \cos \theta \, \vec{k} \right) \psi_2 \, d^3r$$

and since μ_0 is a constant, it can be written outside the integral and

$$M = \mu_0 \int \psi_1^* \left(\sin \theta \cos \phi \, \vec{i} + \sin \theta \sin \phi \, \vec{j} + \cos \theta \, \vec{k} \right) \psi_2 d^3r \qquad ..(5)$$

According to this equation, if a molecule does not have a permanent dipole, then μ_0 is exactly zero and $\therefore$ M exactly zero. Therefore, a molecule must have a non zero dipole moment in order to have a pure rotational spectrum.

21.4 VIBRATIONAL TRANSITIONS

In a vibration, the positions of the atoms are oscillating about some equilibrium position. Since the dipole-moment of a molecule depends on the positions of the atoms, the exact value of μ is a function of the atomic positions, i.e.

$$\mu = f\left(\vec{r} \right). \qquad ..(6)$$

where $f\left(\vec{r} \right)$ is some as-yet an unknown function of $\vec{r}$. Since $\vec{r}$ is changing in a vibration, therefore μ will also vary with $\vec{r}$ and it can be expressed in terms of a Taylor series as

$$\mu = \mu\left(\vec{r_e} \right) + \frac{\partial \mu}{\partial r}\bigg|_{r_e} \Delta \vec{r} + \frac{1}{2} \frac{\partial^2 \mu}{\partial r^2}\bigg|_{r_e} \left(\Delta \vec{r} \right)^2 + \dots\dots \qquad ..(7)$$

where each derivative is evaluated at $\vec{r} = \vec{r_e}$, the equilibrium position of vibration. Retaining the first two terms, we have the transition moment for the vibrational transition.

$$M = \int \psi_1^* \left[\mu\left(\vec{r_e}\right) + \frac{\partial \mu}{\partial r}\bigg|_{r_e} \Delta r \right] \psi_2 d^3r \qquad ..(8)$$

and since $\mu(\vec{r_e})$ and $\dfrac{\partial \mu}{\partial r}\bigg|_{r_e}$ are constants, we have

$$M = \mu\left(\vec{r_e}\right) \int \psi_1^* \psi_2 \, d^3r + \frac{\partial \mu}{\partial r}\bigg|_{r_e} \int \psi_1^* \Delta r \, \psi_2 \, d^3r \qquad ..(9)$$

Now, since different vibrational wavefunctions are orthogonal, the first integral in the above eqn. is zero. Thus,

$$M = \frac{\partial \mu}{\partial r}\bigg|_{r_e} \int \psi_1^* \Delta r \, \psi_2 \, d^3r \qquad ..(10)$$

Now, consider this equation like that for rotations. The partial derivative $\dfrac{\partial \mu}{\partial r}\bigg|_{r_e}$ is evaluated at the equilibrium position $\vec{r_e}$. It has a particular value, say, some number. However, if the dipole moment μ does not change, the value of this derivative at the equilibrium distance will be exactly zero and the transition moment (M) is exactly zero. Therefore, in order for a vibrational transition to be allowed, there must be a changing dipole-moment associated with that vibration.

21.5 SELECTION RULES FOR ELECTRONIC SPECTROSCOPY

For hydrogen atom, the wavefunctions can be described by four quantum numbers: the principal quantum number n, the angular momentum quantum number l, the z-component of angular momentum quantum number m_l and the z-component of spin angular momentum quantum number m_s. (There is also spin angular momentum quantum number $s = \pm 1/2$). These four quantum numbers arise from the solution of the Schrödinger equation for the hydrogen atom. When a hydrogen atom changes state by emitting or absorbing a photon, the wavefunction changes and thus, goes from one set of quantum numbers to another, i.e. there is a change in one or more quantum numbers. For allowed transitions, the following are the allowed changes in the various quantum numbers:

$$\left.\begin{array}{l} \Delta n = \text{anything} \\ \Delta l = +\,1 \text{ or} - 1 \\ \Delta m_l = 0 \text{ or} + 1 \text{ or} - 1 \\ \Delta m_s = 0 \end{array}\right\} \begin{array}{c} \text{Selection rules} \\ \text{for H atom} \end{array} \qquad ..(11)$$

There are similar selection rules for multielectron atoms, but the quantum numbers are defined in a little different way. For atoms having $Z < 20$, there is LS coupling and the resulting total electronic angular momentum is represented by the quantum number J, which has a z-component represented by M_J. For these atoms, the selection rules are

$$\left.\begin{array}{l} \Delta L = 0 \text{ or} + 1 \text{ or} - 1 \\ \Delta S = 0 \\ \Delta J = 0 \text{ or} + 1 \text{ or} - 1 \\ \Delta M_J = 0 \text{ or} + 1 \text{ or} - 1 \end{array}\right\} \begin{array}{c} \text{Selection rules} \\ \text{for multielectron} \\ \text{atoms } (Z < 20) \end{array} \qquad ..(12)$$

except that $\Delta L = 0$ is forbidden if the initial value of L is 0 and $\Delta J = 0$ is forbidden if the initial value of J is zero.

For larger atoms $(Z > 20)$, we get a better understanding of electronic behavior if we assume jj coupling.

Still however, even for large atoms, it is common for the individual electronic states to be labelled according to the L and S quantum numbers, so, the above selection rules are still applicable.

Electronic spectra of diatomic molecules follow similar selection rules except that the relevant quantum numbers are labelled Λ, Σ and Ω. These represent the magnitude of the orbital electronic momentum along the molecular axis, the projection of the spin angular momentum along the molecular axis, and the vector combination of these two, respectively. For diatomic molecules, the selection rules are

$$\left.\begin{array}{l} \Delta \Sigma = 0 \\ \Delta \Lambda = 0 \text{ or} + 1 \text{ or} - 1 \\ \Delta \Omega = 0 \text{ or} + 1 \text{ or} - 1 \end{array}\right\} \qquad ..(13)$$

Selection rules for electronic transitions in molecules are based on group theory and depend on the symmetry of the molecule.

21.6 SELECTION RULES FOR PURE ROTATIONAL AND VIBRATIONAL SPECTROSCOPY

We have seen that the rotation of molecules can be treated quantum mechanically by considering the molecule as a rigid rotator. In doing so, we arrive at rotational states that depend on a rotational quantum number J. There is also the z-component of the rotation, indexed by the quantum number M_J.

A molecule can have three unique axes of rotation depending on the symmetry of the molecule. CH_4 (for example) is tetrahedral and has all equivalent axes of rotation, so it can be described by a single rotational quantum number. Such molecules are referred to as *spherical tops*. The H_2O molecule has a bent shape and three different rotational axes have different rotational behavior, molecules of this type are called *asymmetric tops*. The pyramidal shaped NH_3 molecule needs only two unique axes to be defined and are called *symmetric tops*. There are two categories of symmetric tops: Prolate symmetric tops have their two larger rotational moments of inertia equal, while *oblate* symmetric tops have their two smaller rotational moments of inertia equal. Symmetric tops also have quantized angular momentum about a molecular axis specified by the quantum number K.

Now, we know that molecules must have a permanent dipole moment to show a pure rotational spectrum. Because of their symmetry, spherical tops do not have a dipole moment and therefore do not show a pure rotational spectrum. Prolate and oblate symmetric tops have a permanent dipole moment and so, may show a pure rotational spectrum. Then, we can describe them in terms of the following selection rules:

$$\left. \begin{array}{l} \Delta J = 0 \text{ or } + 1 \text{ or } - 1 \\ \Delta M_J = 0 \text{ or } + 1 \text{ or } - 1 \\ \Delta K = 0 \end{array} \right\} \qquad ..(14)$$

However, there is the exception that $\Delta J = 0$ I not allowed if the initial value of J is 0. If the rotational spectrum is measured using Raman scattering, we have the rule that $\Delta J = + 2$ or $- 2$.

For asymmetric tops, J and M_J are useful quantum numbers, so the first two equations (eqn. 14) are good selection rules. However K is no longer a useful quantum number and selection rules get more dependent on group theory. (We do not consider such cases here).

In case of vibration, the vibrations of a molecule with N atoms can be broken down into $3N - 6$ normal modes of vibration. These normal modes collectively, describe all possible vibrations of the molecule. Each vibration can be treated independently as if it were an ideal harmonic oscillator. Then, quantum mechanics gives us analytic solutions for the ideal harmonic oscillator system, and in doing so introduces a vibrational quantum number v for each vibration. For vibrations, the selection rule is

$$\Delta v = + 1 \text{ or } - 1. \qquad ..(15)$$

We have seen that this selection rule derives from the Taylor series expansion of $\mu = \mu\left(\vec{r}\right)$. We had truncated our expansion after only two terms. The above selection rule is based on the second terms. The above selection rule is based on the second term in eqn. (7) (*i.e.* $\psi^*_1 (\Delta r) \psi_2$). However, if we take into account the next term, we would get a transition moment integral that would have $\psi_1 (\Delta r)^2 \psi_2$ in it. This term would ultimately yield some additional allowed transitions that follow the selection rule

$$\Delta v = + 2 \text{ or } - 2. \qquad ..(16)$$

These transitions are called *overtone transitions* and are typically less intense than those that obey $\Delta v = + 1$ or $- 1$.

Appendix A

THE AUGER PROCESS (SPONTANEOUS RADIATIONLESS DECOMPOSITION OF ATOMS)

Two energy levels of an atomic system which belong to different term series but happen to lie close together, due to perturbation effects, appear to influence each other when we consider higher approximations: there is a shift of the two levels in the sense of a repulsion and there is a mixing of the eigen functions of the two states, each of the actual levels being a hybrid of the two original nearby coinciding levels.

If one term in a discrete series has the same energy as a term of a continuous term spectrum, all the higher terms of the series have the same energy as correspondingly higher terms of the continuous range, and so, all the higher terms of this series may be perturbed. In such a perturbation by a continuous term, the shift of the original discrete level can assume a continuous series of values, i.e. the level becomes *diffuse* (Fig. A.1)

The three uppermost levels of the series *A* are overlapped by the continuum of the series *B*. The radiationless transitions from the discrete to the continuous state are indicated by horizontal arrows. A spectral line corresponding to (or from) such a diffuse level is not sharp but more or less strongly broadened (diffuse).

The true state is a hybrid. Part of the time the system is in the discrete state and part of the time in the continuous state. So, when, as a result of mutual perturbation, the system has once gone from the discrete into the continuous state, it cannot return to the discrete state. Consequently, if an atomic system is transferred to such a diffuse state, e.g. by light absorption—it undergoes a radiationless decomposition after a certain lifetime. This process was first observed by Auger in the X-ray region and is referred to as the *Auger process*.

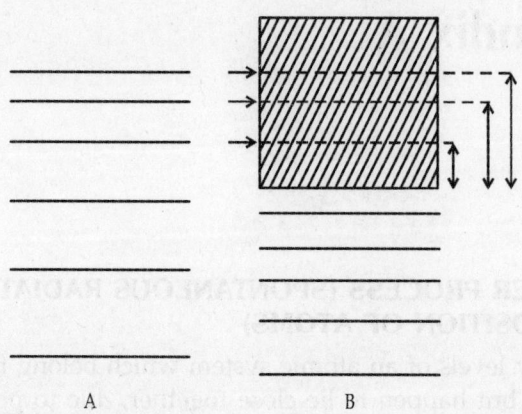

A   B

Fig. A.1: Energy levels diagram for the Auger process

Auger Transitions

Auger processes are due to occurrence of Auger transitions. When a vacancy is created in an inner shell of an atom, it may be filled by an electron from a level with higher energy with emission of radiation. These radiative transitions compete with radiationless transitions in which one electron from a level with higher energy occupies the initial vacancy (created by photoionization or by particle impact) with the simultaneous ejection of a second electron (Fig. A.2).

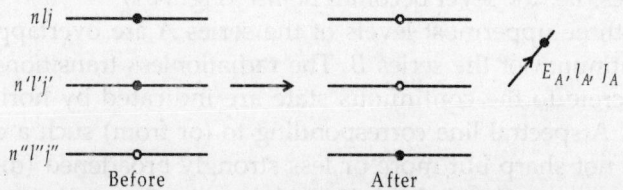

Fig. A.2: Schematic representation of an Auger transition

Before the transition a vacancy with quantum numbers $n'' \, l'' \, j''$ is created. The upper level (nlj) and $(n'l'j')$ are initially occupied.

After the transition one electron occupies the original vacancy and the other is ejected (with energy E_A and quantum numbers l_A and j_A) leaving two vacancies in the upper energy levels.

The transition can be denoted by

$$(n'' l'' j'') \rightarrow (nlj) \, (n' l' j')$$

Alternatively, the notation employed for X-ray spectra can be used, e.g. the $(1s) \rightarrow (2s) \, (2p_{1/2})$ and $(1s) \rightarrow (2s) \, (2p_{3/2})$ transitions are written as

$$K \rightarrow L_I L_{II} \text{ and } K \rightarrow L_I L_{III}$$

respectively.

If the ejected electron has a kinetic energy E_A, then

$$E_A = E_{n'' l'' j''} - E_{nlj, \, n' l' j'}$$

where $E_{n'' l'' j''}$ is the energy of the atom with the vacancy $(n'' l'' j'')$ relative to the energy of the neutral atom. $E_{nlj, \, n' l' j'}$ is the energy of the atom with two vacancies relative to the energy of the neutral atom. These radiationless transitions are known as Auger transitions after their discovery by P. Auger in 1925.

When vacancies are created in a sample, either by collision process or by radiation, the kinetic energies and angular distributions of the resulting Auger electrons can be measured using electron spectrometers and much information can be obtained in this way about the structure of inner shells of atoms.

Appendix B

THE AMMONIA MASER

In what follows, we will discuss the application of quantum mechanics to a practical device which is called the ammonia maser. MASER is an acronym for Microwave Amplification by Stimulated Emission of Radiation. The ammonia maser generates electromagnetic waves and the operation is based on the properties of ammonia molecule.

When the NH_3 molecule is in any specific state of rotation (or transition) it may be considered as a two state system, as shown in Fig. B.1 (i.e., the molecule flips over from one configuration to another).

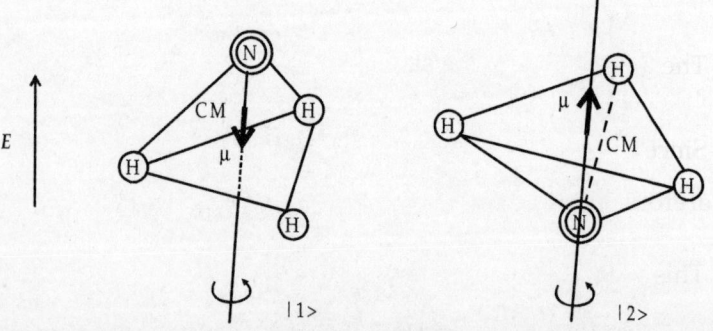

Fig. B.1: Two base states of ammonia molecule. CM is centre of mass and μ is the electric dipole moment

The two base states are represented by $|1\rangle$ and $|2\rangle$. Any state $|\psi\rangle$ of the system can be described as

$$|\psi\rangle = |1\rangle C_1 + |2\rangle C_2$$

where $\qquad C_1 = \langle 1|\psi\rangle$ and $C_2 = \langle 2|\psi\rangle$.

The two amplitudes C_1 and C_2 vary with time according to the Hamiltonian equations.

$$i\hbar \frac{dC_1}{dt} = H_{11}C_1 + H_{12}C_2$$

$$i\hbar \frac{dC_2}{dt} = H_{21}C_1 + H_{22}C_2 \qquad ..(B.1)$$

By symmetry, we have $H_{11} = H_{22} = E_0$

and $\qquad H_{12} = H_{21} = -A$. So,

$$C_1 = \frac{a}{2}e^{-(i/\hbar)(E_0 - A)t} + \frac{b}{2}e^{-(i/\hbar)(E_0 + A)t} \qquad ..(B.2)$$

$$C_2 = \frac{a}{2}e^{-(i/\hbar)(E_0 - A)t} - \frac{b}{2}e^{-(i/\hbar)(E_0 + A)t} \qquad ..(B.3)$$

We can construct an amplitude C_{II} as

$$C_{II} = C_1 + C_2 = \langle 1|\Phi\rangle + \langle 2|\Phi\rangle \qquad ..(B.4)$$

(Here we have defined a state $|\psi_{II}\rangle$ for which coefficient $b = 0$. Then at $t = 0$ the amplitudes to be in states $|1\rangle$ and $|2\rangle$ are identical for all times. Similarly we may define C_I). C_{II} is just the amplitude to find the state $|\Phi\rangle$ in a new state $|II\rangle$ in which the amplitudes of the original base state are equal.

$$C_{II} = \langle II|\Phi\rangle$$

and $\qquad \langle II| = \langle 1| + \langle 2|$

or $\qquad |II\rangle = |1\rangle + |2\rangle \qquad\qquad\qquad ..(B.5)$

The amplitude for the state $|II\rangle$ to be in state $|I\rangle$ is

$$\langle I|II\rangle = \langle I|1\rangle + \langle I|2\rangle \qquad ..(B.6)$$

Since $\langle X|\Phi\rangle = \sum_i \langle x|i\rangle\langle i|\Phi\rangle$

therefore

$$\langle II|II\rangle = \langle II|1\rangle \langle 1|II\rangle + \langle II|2\rangle \langle 2|II\rangle \qquad ..(B.7)$$

This will be equal to one only if

$$C_{II} = \frac{1}{\sqrt{2}}(C_1 + C_2) \qquad ..(B.8)$$

Similarly $\quad C_I = \frac{I}{\sqrt{2}}(C_1 - C_2) \qquad ..(B.9)$

and $\qquad < I \,| = \dfrac{1}{\sqrt{2}}[< 1\,| - < 2\,|]$

$$< II\,| = \dfrac{1}{\sqrt{2}}[< 1\,| + < 2\,|] \qquad \qquad ..(B.10)$$

(It can be seen that $<I\,|\,I> = <II\,|\,II> = 1$ and $<I\,|\,II> = <II\,|\,I> = 0$).

If now, the molecule is in static electric field ε, then

$$H_{11} = E_0 + \mu\varepsilon, \; H_{22} = E_0 - \mu\varepsilon. \qquad ..(B.11)$$

Assuming that the field does not affect appreciably the geometry of the molecule, we have

$$H_{12} = H_{21} = -A. \qquad\qquad ..(B.12)$$

We are interested in general solution of the pair of equations (B.2) and (B.3), we use the trial functions

$$C_1 = a_1 e^{-iwt} \equiv a_1 e^{-(i/\hbar)Et}$$

$$C_2 = a_2 e^{-iwt} \equiv a_2 e^{-(i/\hbar)Et} \qquad\qquad ..(B.13)$$

Substituting these values in eqn (B.1), we get

$$Ea_1 = H_{11}a_1 + H_{12}a_2$$
$$Ea_2 = H_{21}a_1 + H_{22}a_2$$

or $\qquad (E - H_{11})a_1 - H_{12}a_2 = 0$

$$-H_{21}a_1 + (E - H_{22})a_2 = 0 \qquad\qquad ..(B.14)$$

or $\qquad \begin{vmatrix} (E - H_{11}) & -H_{12} \\ -H_{21} & (E - H_{22}) \end{vmatrix} = 0 \qquad\qquad ..(B.15)$

$\Rightarrow \qquad (E - H_{11})(E - H_{22}) - H_{12}H_{21} = 0$

$$\Rightarrow \qquad E = \dfrac{H_{11} + H_{22}}{2} \pm \sqrt{\dfrac{(H_{11} - H_{22})^2}{4} + H_{12}H_{21}} \qquad ..(B.16)$$

The upper energy we denote by E_I and the lower energy by E_{II} so that

$$E_I = \dfrac{H_{11} + H_{22}}{2} + \sqrt{\dfrac{(H_{11} - H_{22})^2}{4} + H_{12}H_{21}} \qquad ..(B.17)$$

$$E_{II} = \dfrac{H_{11} + H_{22}}{2} - \sqrt{\dfrac{(H_{11} - H_{22})^2}{4} + H_{12}H_{21}} \qquad ..(B.18)$$

Hence

$$| \psi_I > = | I > e^{-(i/\hbar)E_I t}$$

and $| \psi_{II} > = | II > e^{-(i/\hbar)E_{II} t}$ ..(B.19)

For the system to be in one of the stationary states, we must have

$$|C_1|^2 + |C_2^2| = 1 \text{ or } |a_1|^2 + |a_2^2| = 1 \qquad ..(B.20)$$

Using the values of H_{11}, H_{22}, H_{12} and H_{21} and from eqns (B.11) and (B.12) in eqns (B.17) and (B.18), we get

$$E_I = E_0 + \sqrt{A^2 + \mu^2 \varepsilon^2}$$

$$E_{II} = E_0 - \sqrt{A^2 + \mu^2 \varepsilon^2} \qquad ..(B.21)$$

Now, the *principle of operation* of NH_3–MASER is as follows:

First we find a way of separating molecules in the state $| I >$ from those in state $| II >$. Then the molecules in the higher energy state $| I >$ are passed through a cavity which has a resonant frequency ~24000 MHz. The molecules can deliver energy to the cavity, and then leave the cavity in the state $| II >$. Each molecule making such a transition will deliver the energy $E = E_I - E_{II}$ to the cavity. Thus the energy from the molecules will appear as electrical energy in the cavity.

How to Separate two Molecular States

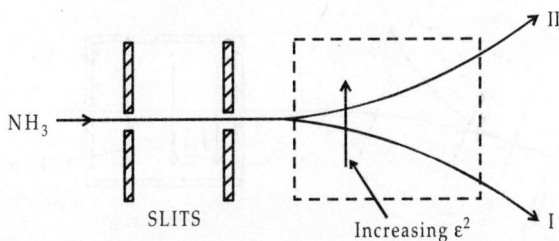

Fig. B.2: The ammonia beam separated by an electric field in which ε^2 has a gradient perpendicular to the beam

The ammonia gas (in form of a little jet) after passing through a pair of slits is set through a region where there is a large transverse electric field. The electrodes are so shaped that $\varepsilon \cdot \varepsilon$ has a large gradient perpendicular to the beam. Now with fields that can be generated in the lab, $\mu\varepsilon$ is always much smaller than A. Then, for all practical purposes, (from eqn (B.21)), we have

$$E_I \simeq E_0 + A + \frac{\mu^2 \varepsilon^2}{2A}$$

$$E_{II} \simeq E_0 - A - \frac{\mu^2 \varepsilon^2}{2A} \qquad\qquad ..(B.22)$$

i.e., energies vary (approximately) linearly with ε^2. The force on the molecules is

$$F = \frac{\mu^2}{2A} \nabla \varepsilon^2$$

$\left(\dfrac{\mu^2}{2A} \right.$ is the polarizability of the molecule and ammonia has an unusually high polarizability because A is small).

Thus due to gradient of ε^2 perpendicular to the beam, a molecule in state $|I\rangle$ has an energy which increases with ε^2 and therefore (because of higher energy) will be deflected toward the region of lower ε^2. A molecule in state $|II\rangle$ will, on the other hand, be deflected toward the region of large ε^2 (because its energy decreases as ε^2 increases).

Passage through a Time-dependent Field

In the ammonia maser, after separation, the beam with molecules in the state $|I\rangle$ (and energy E_I) is then sent through a resonant cavity (as shown in Fig. B.3).

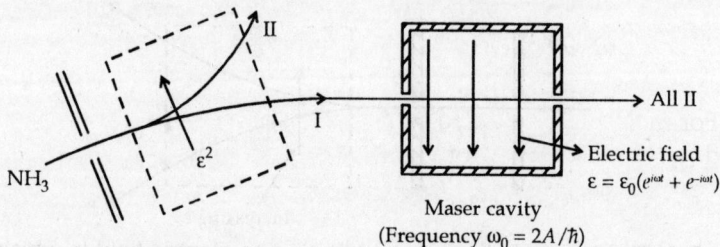

Maser cavity
(Frequency $\omega_0 = 2A/\hbar$)

Fig. B.3: Schematic representation of the ammonia maser

The other beam is discarded. Inside the cavity there is a time varying field represented by

$$\varepsilon = 2\varepsilon_0 \cos \omega t = \varepsilon_0 (e^{i\omega t} + e^{-i\omega t})$$

We may represent

$$C_I = r_I e^{-i(E_0 + A)t/\hbar} = r_I e^{-iE_I t/\hbar}$$

$$C_{II} = r_{II} e^{-i(E_0 - A)t/\hbar} = r_{II} e^{-iE_{II}t/\hbar}$$

i.e., the probability of being in state $|I\rangle$ or in $|II\rangle$ is just $|r_I|^2$ or $|r_{II}|^2$ respectively. If $\mu\varepsilon < A$, then r_I and r_{II} become slowly varying functions of time (in comparison with exponential functions). It can be shown that

$$i\hbar \frac{dr_I}{dt} = \mu\varepsilon_0 e^{-i(\omega - \omega_0)t} r_{II}$$

$$i\hbar \frac{dr_{II}}{dt} = \mu\varepsilon_0 e^{i(\omega - \omega_0)t} r_I$$

If $\omega = \omega_0$, we have

$$\frac{dr_I}{dt} = -\frac{i\mu\varepsilon_0}{\hbar} r_{II}$$

and

$$\frac{dr_{II}}{dt} = -\frac{i\mu\varepsilon_0}{\hbar} r_I$$

Thus each of the above equations satisfies the differential equation of simple harmonic motion

$$\frac{d^2 r}{dt^2} = -\left(\frac{\mu\varepsilon_0}{\hbar}\right)^2 r$$

The following equations are a solution

$$r_I = a\cos\left(\frac{\mu\varepsilon_0}{\hbar}\right)t + b\sin\left(\frac{\mu\varepsilon_0}{\hbar}\right)t$$

$$r_{II} = ib\cos\left(\frac{\mu\varepsilon_0}{\hbar}\right)t - ia\sin\left(\frac{\mu\varepsilon_0}{\hbar}\right)t$$

For molecular system to be in state $|I\rangle$ at $t = 0$, we require $r_I = 1$ and $r_{II} = 0$. This holds for $a = 1$ and $b = 0$. Then probability

$$P_I \equiv |r_I|^2 = \cos^2\left(\frac{\mu\varepsilon_0}{\hbar}\right)t$$

Similarly

$$P_{II} \equiv |r_{II}|^2 = \sin^2\left(\frac{\mu\varepsilon_0}{\hbar}\right)t$$

If it takes the molecule time T to go through the cavity and if we make the cavity just long enough so that $\mu\varepsilon_0 T/\hbar = \pi/2$, then a molecule which enters in state $|I\rangle$ will certainly leave the cavity in state $|II\rangle$, i.e. if it enters the cavity in the upper state, it will

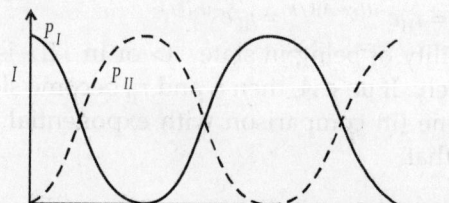

Fig. B.4: Probability for two states of the ammonia molecule in a field $2\varepsilon_0 \cos \omega t$

leave the cavity in the lower state, and the loss of energy can go nowhere but into the cavity which generates the field and sustained oscillations could be produced due to resonance (Molecular energy would be converted into the energy of external electromagnetic field).

Appendix C

MODE LOCKING IN LASERS

Mode locking is a technique for producing periodic, high power, short duration laser pulses. A typical inhomogeneously broadened laser cavity may support oscillations in many modes simultaneously. The output of such a laser depends on the relative phases, frequencies and amplitudes of the modes. The total electric field can be written as

$$\varepsilon(t) = \sum_{n=0}^{N-1} (\varepsilon_0)_n \exp\left[i(\omega_n t + \delta_n)\right] \qquad ..(C.1)$$

where $(\varepsilon_0)_n$, ω_n and δ_n are respectively the amplitude, angular frequency and phase of the nth mode. Usually these parameters are all time varying, so, the modes are incoherent and the total irradiance is simply the sum of the irradiances of the individual modes, i.e.

$$I = N\varepsilon_0^2$$

where, for simplicity, we have assumed that all N modes have the same amplitude ε_0.

Now, suppose that we force the various modes to maintain the same relative phase δ to one another, i.e. we mode lock the laser such that $\delta_n = \delta$. The total irradiance must now be found by adding the individual electric fields rather than the irradiances. Using eqn (C.1), the resultant electric field is

$$\varepsilon(t) = \varepsilon_0 \exp(i\delta) \sum_{n=0}^{N-1} \exp(i\omega_n t) \qquad ..(C.2)$$

For convenience, we write

$$\omega_n = \omega - n\delta\omega \qquad ..(C.3)$$

where ω is the angular frequency of the highest frequency mode and $\delta\omega$ is the angular frequency separation between two nearby modes, which is given by

$$\delta\nu = \frac{c}{2L} \quad \text{or} \quad \delta\omega = \pi\frac{c}{L} \qquad ..(C.4)$$

(L = optical path length between the two mirrors of the cavity). Eqn (C.2) can be written as

$$\varepsilon(t) = \varepsilon_0 \exp(i\delta) \sum_{n=0}^{N-1} \exp\left[i(\omega - n\delta\omega)t\right]$$

$$= \varepsilon_0 \exp\{i(\omega t + \delta)\} \sum_{n=0}^{N-1} \exp[-\pi inct/L]$$

or $\qquad \varepsilon(t) = \varepsilon_0 \exp\{i(\omega t + \delta)\}\left[1 + \exp(-i\phi) + \exp(-2i\phi) + \ldots\right.$

$$\left. + \exp\{-(N-1)i\phi\}\right] \qquad ..(C.5)$$

where $\qquad \phi = \dfrac{\pi ct}{L} \qquad\qquad\qquad\qquad ..(C.6)$

The term within bracket in eqn (C.5) is a geometric progression and so, we can write

$$\varepsilon(t) = \varepsilon_0 \exp\{i(\omega t + \delta)\}\frac{\sin(N\phi/2)}{\sin(\phi/2)}$$

and the irradiance

$$I(t) = \varepsilon(t) \cdot \varepsilon^*(t)$$

or $\qquad I(t) = \varepsilon_0^2 \dfrac{\sin^2(N\phi/2)}{\sin^2(\phi/2)} \qquad\qquad ..(C.7)$

Figure C.1 illustrates this for $N = 3$. Thus $I(t)$ is periodic ($\Delta\phi = 2\pi$) in the time internal $t = 2L/c$ (i.e., the round trip transit time for light within the cavity). The maximum irradiance occurs for values of $\phi = 0$ or $p\pi$, (p being an interger). Then $\dfrac{\sin^2(N\phi/2)}{\sin^2(\phi/2)} = N^2$, i.e. the maximum value of the irradiance is $N^2 \varepsilon_0^2$.

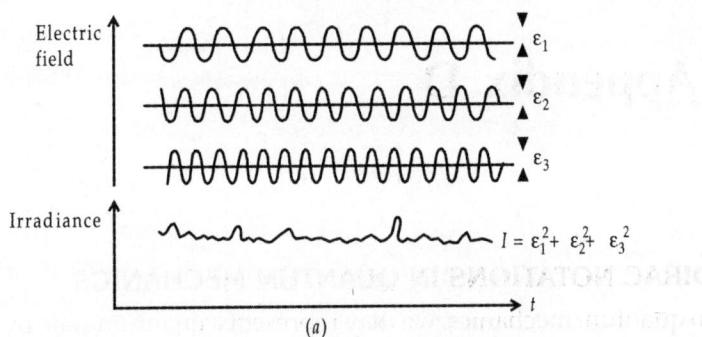

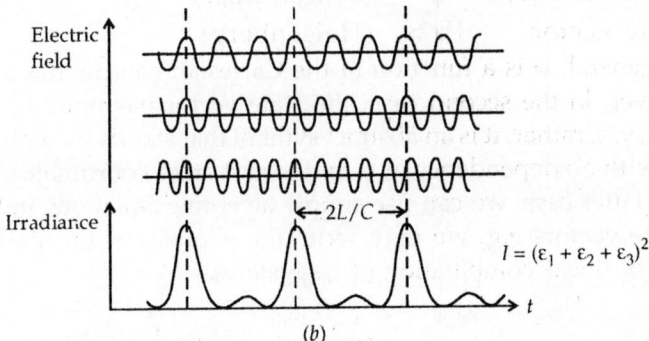

Fig. C.1: Output irradiance I of (a) non-mode locked and (b) mode locked lasers. In (a), there are random fluctuations whereas in (b) the phases of the adjacent modes differ by a constant and output is a series of narrow intense pulses of time spacing $2L/C$ and duration $2L/(Nc)$

Similarly, the irradiance has a minimum (or zero) value when $N\phi/2 = p\pi$, p being a non-zero integer, i.e. when $\phi = 2p\pi/N$ or $t = (1/N)(2L/c)p$. Thus, the time duration of the maxima, which is the time taken for the irradiance to fall from its maximum value to an adjacent zero ($p = 1$) is $(1/N)(2L/c) = 2L/(Nc)$. *To obtain high power short duration pulses therefore, there should be a large number of modes, and this requires a broad laser transition and a long laser cavity.*

Appendix D

DIRAC NOTATIONS IN QUANTUM MECHANICS

In quantum mechanics, we may represent a quantum state by either of these two notations:

Wave functions ψ (Schrödinger)

State vector $|\psi>$ (Heisenberg)

In general, ψ is a function of the Cartesian coordinates x, y, z. However, in the second case, the state vector has nothing to do with x, y, z, rather, it is an abstract symbol that stands for a physical state with no dependence of ψ on the system of coordinates. Thus in the latter case, we can use simple algebraic equations in terms of state vectors e.g. we may write the expression for a state in terms of linear combination of base states

$$|\psi> = \sum_i C_i |\psi_i> \qquad ..(1)$$

where coefficients C_i form a set of ordinary complex numbers and $|\psi_i>$ are base states. The base state vectors form what is so called a Hilbert space. The base states are such that their complex conjugates satisfy

$$C_i = <\psi_i|\psi>$$

The base states are like components of a vector in ordinary vector space with amplitudes C_i as a measure of the magnitudes of the components (base states).

Equation (1) shows that any state can be described in terms of the base states by giving the amplitudes to be in each of the base states.

494

The notation $|\psi>$ corresponds to the Dirac notation and is known as a ket vector. Its complex conjugate $<\psi|$ represents a bra vector.

Thus, according to Dirac, any state $\psi(x)$ of a quantum system may be represented as

$$\psi(x) = <x|\psi> \qquad\qquad ..(2)$$

The base states are orthogonal, i.e.

$$<\psi_i| = |\psi_i>^*$$

or

$$<\psi_i|\psi_j> = \delta_{ij} \qquad\qquad ..(3)$$

where δ_{ij} is Kronecker delta. (= 1 for $i = j$ and = 0 for $i \neq j$).

Now,

$$<x|\psi_i> = (<\psi_i|x>)^* \qquad\qquad ..(4)$$

This is like the relation $\vec{A} \cdot \vec{B} = \vec{B} \cdot \vec{A}$ of vector algebra. But, there is a difference, however; in the dot product of $\vec{A}$ and $\vec{B}$, the order of $\vec{A}$ and $\vec{B}$ does not matter, whereas in eqn. (4), we must keep straight the order of the terms.

Further, we may write a vector $\vec{A}$ as

$$\vec{A} = \sum_i \vec{n}_i (\vec{n}_i \cdot \vec{A}), \; (n_i \text{ is an integer}).$$

In an analogous way, any state vector $|\psi>$ can be expanded in terms of the complete set of orthonormal vectors $|F_m>$ corresponding to the operator F as

$$|\psi> = \sum_m |F_m><F_m|\psi>$$

The amplitude to go from any state to another can, in general, be written as a sum of products, each product being the amplitude to go into one of the base states times the amplitude to go from that base state to the final condition, with the sum including a term for each base state:

$$<x|\psi> = \sum_i <x|\psi_i><\psi_i|\psi>$$

Further, we may expand any operator A in term of the complete set of operators $|F_m><F_n|$ as

$$A = \sum_{m,n} A_{mn}|F_m><F_n|$$

If an operator A operates on the state function $|\psi>$, then, the matrix elements A_{mn} are given by

$$A_{mn} = <\psi_m|A|\psi_n>$$

Bibliography

1. *Introduction to Atomic Spectra*, H.E. White McGraw Hill, (1934).
2. *Atomic Spectra and Atomic Structure*, G. Herzberg, Dover Publications (1944).
3. *Spectra of Diatomic Molecules*, G. Herzberg, Van Nostrand Reinhold Company (1939).
4. *Solid State Spectroscopy*, H. Kuzmany, Springer (1998).
5. *Instrumental Methods of Analysis*, Willard, Merritt, Dean and Settle, CBS Delhi (1986).
6. *Principles of Instrumental Analysis*, D.A. Skoog and J.J. Leary, Harcourt Brace College Publishers (1999).
7. *Fundamentals of Molecular Spectroscopy*, C.N. Banwell, Tata McGraw Hill, New Delhi (1990).
8. *Instrumental Methods of Chemical Analysis*, G.W. Ewing, McGraw Hill, Singapore (1985).
9. *Laser Fundamentals*, W.T. Silfvast, Cambridge University Press (1998).
10. *Lasers (Theory and Applications)*, K. Thyagarajan and A.K. Ghatak, Macmillan India Ltd. (1991).
11. *Lasers and Non linear Optics*, B.B. Laud, Wiley Eastern Ltd. (1985).
12. *Atomic and Molecular Spectra*, Raj Kumar, Kedar Nath Ram Nath, Meerut, (India) (2003).
13. *Spectroscopy (Atomic and Molecular)*, G. Chatwal and S. Anand, Himalaya Publishing House, Mumbai (2001).
14. *Molecular Spectroscopy*, P.S. Sindhu, Tata McGraw Hill, New Delhi (1990).
15. *Introduction to Modern Physics*, F.K. Richtmyer, E.H. Kennard and J.N. Cooper, Tata McGraw Hill, New Delhi (1997).

16. *Optical Physics*, S.G. Lipson, H. Lipson, D.S. Tannhauser, Cambridge University Press (1995).

17. *Fundamental of Optics*, F.A. Jenkins and H.E. White, McGraw Hill Book Company (1981).

18. *The Feynman Lectures on Physics*, R.P. Feynman, R.B. Leighton and M. Sands, Narosa Publishing House, New Delhi (1996).

19. *Physics of Atoms and Molecules*, B.H. Bransden and C.J. Joachain, Pearson Education Ltd. (2004).

20. *Optoelectronics, An Introduction*, J. Wilson and J.F.B. Hawkes, Prentice Hall of India (2001).

Index